The Changing Social Geography
of Canadian Cities

CANADIAN ASSOCIATION OF GEOGRAPHERS
SERIES IN CANADIAN GEOGRAPHY
General Editor: Cole Harris

Canada's Cold Environments
Hugh M. French and Olav Slaymaker, editors

The Changing Social Geography of Canadian Cities
Larry S. Bourne and David F. Ley, editors

The Changing Social Geography of Canadian Cities

EDITED BY
LARRY S. BOURNE
AND
DAVID F. LEY

McGill-Queen's University Press
Montreal & Kingston • London • Buffalo

© McGill-Queen's University Press 1993
ISBN 0-7735-0926-7 (cloth)
     0-7735-0972-0 (paper)

Legal deposit second quarter 1993
Bibliothèque nationale du Québec

Printed in Canada on acid-free paper

McGill-Queen's University Press acknowledges the finan-
cial support of the Government of Canada through the
Canadian Studies and Special Projects Directorate of the
Department of the Secretary of State of Canada.

Canadian Cataloguing in Publication Data

Main entry under title:
    The Changing social geography of Canadian cities
    (Canadian Association of Geographers Series in Canadian
    Geography)
    Includes bibliographical references and index.
    ISBN 0-7735-0926-7 (bound) – ISBN 0-7735-0972-0 (pbk.)
    1. Cities and towns – Canada. 2. Sociology, Urban – Can-
    ada. I. Bourne, L. S. (Larry Stuart), 1939–   . II. Ley,
    David
    GF511.C43 1993   307.76'0971   C92-090718-0

This book was typeset by Typo Litho composition inc.
in 10/12 Times.

Contents

Contributors

Larry S. Bourne	University of Toronto
Trudi Bunting	University of Waterloo
Wayne K.D. Davies	University of Calgary
Michael J. Dear	University of Southern California
Len J. Evenden	Simon Fraser University
John C. Everitt	Brandon University
Alison M. Gill	Simon Fraser University
Richard Harris	McMaster University
Daniel Hiebert	University of British Columbia
Deryck W. Holdsworth	Pennsylvania State University
Donald G. Janelle	University of Western Ontario
Audrey L. Kobayashi	McGill University
James T. Lemon	University of Toronto
David F. Ley	University of British Columbia
Suzanne Mackenzie	Carleton University
John R. Miron	Scarborough College
Eric G. Moore	Queen's University
Peter W. Moore	City of Scarborough
Robert A. Murdie	York University
Sherri H. Olson	McGill University
Geraldine J. Pratt	University of British Columbia
Damaris Rose	INRS-Urbanisation, Université du Québec à Montréal
Mark W. Rosenberg	Queen's University
Peter J. Smith	University of Alberta
S. Martin Taylor	McMaster University
Marie Truelove	Ryerson Polytechnical Institute
Paul Villeneuve	Université Laval
Gerald E. Walker	York University
Jennifer Wolch	University of Southern California

The Changing Social Geography
of Canadian Cities

CHAPTER ONE

Introduction: The Social Context and Diversity of Urban Canada

D.F. LEY AND L.S. BOURNE

This is a book about the places, the people, and the practices that together comprise the social geography of Canadian cities. Its purpose is both to describe and to interpret something of the increasingly complex social characteristics of these cities and the diversity of living environments and lived experiences that they provide. While the chapters that follow are a selection, albeit purposeful, of the work of social geographers, and their treatment both of places and of issues, our objective as editors is to give readers a deeper understanding of social patterns and recent trends in urban Canada and of social geography as a field of study.

THE GEOGRAPHIC CONTEXT: SOCIETY, TERRITORY, AND URBANIZATION

Cities everywhere are social and geographical prisms for the societies in which they have evolved. They serve as both mirrors and moulders of a nation's society, culture, and politics. Cities also enjoy disproportionate geographical concentrations of a nation's productive capacity, fixed assets, and mobile resources of capital and labour, all situated on small and unique pieces of territory. Not least, cities are centres of innovation and technological change and, in a modern economy, serve as the control points of the economic system and its political organization, as well as its window on the outside world.

For Canada this context is particularly important in that it helps us to identify some of the background elements necessary for an understanding of urban growth and urban living. This country, with its relatively sparse population, and its export-led and resource-based economy, both spread over a vast territory, has always exhibited considerable external economic dependence, substantial regional diversity, and strong attachments to local places

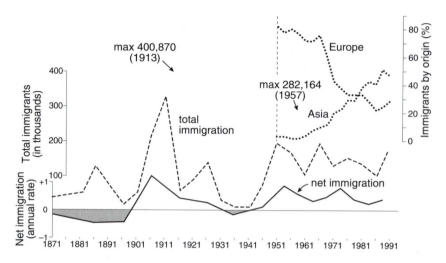

Figure 1.1
Immigration flows, Canada, 1871–1991

and spaces. Each of its regions and cities conveys not just distinctive land-scapes for its visitors and artists but different living and working environments for its citizens. The country was also colonized by Europeans from east to west, initially through waves of immigration, with each wave adding a new layer of diversity to the social fabric and the local economic base. Cities in each region differ from those in other regions in terms of their relative inheritance of cultural and social attributes, buildings, institutions, and landscapes from each period in this sequential history of nation-building. Immigration has continued to play a significant and frequently unpredictable role in Canada's population growth, contributing more than eight million persons over the last century, more than compensating for emigration, particularly to the United States, and more recently for a declining birthrate (Figure 1.1). Most of these immigrants have flowed into and through the cities. Europeans and other immigrants entered a territory already settled by long-established aboriginal cultures, which contributed sites and place names to a number of current urban centres. Now in excess of three-quarters of a million, native people have also been part of the urbanization of recent decades and are present in identifiable communities in many cities, particularly in western Canada, and especially in Winnipeg, Regina, and Edmonton.

A starting-point in the study of contemporary cities is to document the massive growth of the population of urban Canada in the twentieth century

Table 1.1
The urban transformation of Canada: rural and urban population distributions, 1901–91

Year	Total Pop. (000s)	Rural farm (RF) (000s)	(%)	Rural non-farm* (RNF) (000s)	(%)	Urban† (000s)	(%)	(Tot.)	RF	RNF	Urban
1901	5,371	3,357	62.5	n.a.‡	n.a.	2,014	37.5	n.a.	n.a.	n.a.	n.a.
1921	8,788	4,436	50.5	n.a.	n.a.	4,352	49.5	3.2	1.6	n.a.	5.8
1941	11,507	3,113	27.1	2,123	18.4	6,271	54.5	2.5	0.9	n.a.	2.2
1961	18,232	2,073	11.4	3,465	19.0	12,700	69.6	2.9	−1.7	3.2	5.1
1981	24,343	1,040	4.3	4,867	20.0	18,436	75.7	1.7	−2.5	2.0	2.3
1991	27,296	806	3.0	5,481	20.0	21,008	77.0	1.2	−2.3	1.3	1.4

(Annual growth rates (%) span the last four columns.)

Sources: Census of Canada, various years.
* Not identified separately until the census of 1931.
† Definitions of urban populations from 1901 to 1941 were not the same as those from the 1951 census to the present (see text for details).
‡ n.a. = not available.

(Table 1.1 and Figure 1.2). Between 1901 and 1941 Canada's population more than doubled, from 5.4 million to 11.5 million. Between 1941 and 1991 the population doubled again, to 27.3 million. More dramatic still was the growth of the urban population. In 1901 only 37 per cent of the population, or 2.0 million people, lived in urban areas.[1] In 1941 that proportion had increased to 54.5 per cent, or 6.3 million people, and by the census of 1991 over 77 per cent, or 21.0 million Canadians, lived in urban areas. Conversely, the rural population declined from 63 per cent in 1901 to 20 per cent in 1991, while the rural farm population fell to below 3 per cent. In less than five decades, indeed over barely two generations, the population of urban Canada has tripled. No other single fact conveys so succinctly the social and geographical transformation of the nation in the post-war period.

Equally significant, the particular kinds of urban environments in which Canadians have lived and worked have been continuously redefined and reconstituted. Prior to 1931 no Canadians lived in urban areas with populations of more than one million. In 1941 only Montreal surpassed that figure, and just 30 per cent of Canadians lived in all of the nation's large cities (of more than 100,000). By 1991 over 8.6 million Canadians, or 31 per cent of the total, resided in a metropolitan setting of more than one million people, and over 61 per cent lived in the twenty-five census metropolitan areas (CMAs).[2] This transformation in itself conveys another impression of the scale of potential reorganization of urban areas in Canada. As a corollary

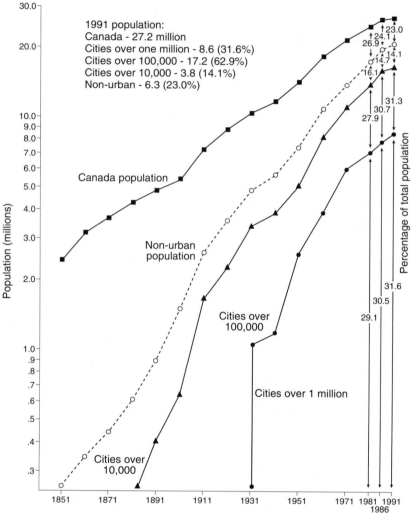

Figure 1.2
Historical growth of urban populations, by size category, Canada, 1851–1991

there are now also many more places that contribute to the social geography that we address in this volume (Table 1.2). Prior to the Second World War, Canada had only seven urban places with over 100,000 people, and only thirty-six other cities and towns that could be regarded in any sense as urban settlements. In the post-war period the concentration of population in larger urban places accelerated. By 1991 there were twenty-seven places with over

Table 1.2
Growth of the Canadian urban system: number of urban places and levels
of metropolitan concentration, 1921–91

	1921	1941	1961	1981	1991
Number of urban places by population size*					
1 million and over	–	1	2	3	3
100,000–999,999	6	7	16	23	24
30,000–99,999	12	19	25	52	57
10,000–29,999	25	36	60	74	63
Total	43	63	93	152	147[†]
Levels of metropolitan concentration[‡]					
% population in CMAs[‡]	35.4	40.2	48.3	55.3	61.1
% population in 3 largest CMAs[‡]	18.8	22.2	25.0	29.1	31.6
Total population (000s) in 3 largest CMAs	1,651	2,551	4,725	7,095	8,622

* Defined as functional urban places – e.g. census metropolitan areas (CMAs) and census agglomerations
(CAs) – not by political or administrative boundaries.
† An extensive reorganization of (and increase in the number of) CAs resulted in a decrease in the number
of census-defined urban places in the 1981–86 period (see Simmons and Bourne 1989).
‡ CMAs as defined at each census.

100,000 people and 147 with populations of over 10,000. Given this four-fold expansion of the urban system, any review of the social geography of its member cities must by definition be selective in terms of the range of examples provided. Most of the chapters to follow in fact focus on the changing social structure of individual metropolitan areas, though to offset a possible metropolitan myopia, several chapters allude to the social characteristics, viability, and changing roles of small towns, and one chapter is dedicated wholly to their study.

Two additional sets of data serve as part of the stage set on which the individual chapters in this volume are played out. Table 1.3, summarizing the growth of the country's twenty-five CMAs from 1961 through 1991, merits two brief observations that are relevant to our subsequent discussions. First, although Canadian urban areas are for the most part small by world standards, we have been creating large urban places at a relatively rapid rate. Second, urban growth rates have varied widely, from as high as 50 per cent (Calgary) in one decade to a low of −2 per cent (Sudbury), but as these places become larger their growth rates have been decreasing in both magnitude and variability. Differential growth performances are usually accompanied by a distinctive combination of social changes, particularly through selective in- and out-migration. Migrants tend to be better educated and

Table 1.3
Growth of census metropolitan areas (CMAs), Canada, 1961–91

Rank (1991)	CMA†	Population (000s)				Annual growth rate* (% change)		
		1961	1971	1981	1991	1961–71	1971–81	1981–91
1	Toronto	1,825	2,602	3,130	3,893	4.3	1.5	2.4
2	Montreal	2,110	2,729	2,862	3,127	2.9	0.4	0.9
3	Vancouver	790	1,082	1,268	1,603	3.7	1.7	2.6
4	Ottawa-Hull	430	620	744	921	4.4	1.6	2.4
5	Edmonton	337	496	741	840	4.7	3.3	1.4
6	Calgary	279	403	626	754	4.4	4.7	2.1
7	Winnipeg	476	550	592	652	1.6	0.6	1.0
8	Quebec	358	501	584	646	3.9	1.5	1.1
9	Hamilton	395	503	542	600	2.7	0.8	1.0
10	London	181	253	327	382	3.9	1.2	1.7
11	St Catharines	217	286	342	365	3.1	0.7	0.7
12	Kitchener	155	239	288	356	5.4	2.1	2.4
13	Halifax	184	251	278	320	3.6	1.1	1.6
14	Victoria	154	196	242	288	2.7	1.9	1.9
15	Windsor	193	249	251	262	2.9	−0.1	0.4
16	Oshawa	81	120	186	240	4.8	2.2	2.9
17	Saskatoon	96	127	175	210	3.2	2.8	2.0
18	Regina	112	141	173	192	2.5	1.7	1.1
19	St John's	91	132	155	172	4.5	1.7	1.1
20	Chicoutimi-J.	105	126	158	161	2.0	1.4	0.2
21	Sudbury	111	158	156	158	4.2	−0.2	0.1
22	Sherbrooke	84	98	125	139	1.7	2.0	1.1
23	Trois-Rivières	91	109	125	136	2.0	1.4	0.8
24	Saint John	96	107	121	125	1.2	0.7	0.3
25	Thunder Bay	92	115	122	124	2.5	0.6	0.2

Sources: Census of Canada, various years; Simmons and Bourne (1989).
* Based on CMA as defined at the end of each census period.
† CMA as defined in 1991.

younger than non-migrants (except of course for those moving to retirement areas), and migrants in turn become part of the local population pool that influences fertility levels, growth rates, and social change in subsequent periods.

Table 1.4 identifies a selection of socioeconomic, labour-force, and demographic characteristics of these metropolitan areas. Some indices differ quite considerably from place to place, while others do not. Average household incomes and indices of relative poverty, for instance, essentially differentiate larger from smaller urban areas (an effect of city size) and cities in the east of the country from those in Ontario and the west (a regional effect). Both patterns are outcomes of basic differences in economic structure, occupational profiles, and rates of labour-force participation. Rates of natural population increase and net migration also differ sharply, as noted above, reflecting the age structure of the residents and the effects of past migration decisions. Some cities stand out because of their unusual age distribution, either with large proportions over age 65 (in retirement centres such as Victoria and St Catharines) or under 14 (Calgary). Others stand out because of their relatively low incomes (Sherbrooke, Saint John, Trois-Rivières), rapid population decline (Sudbury), high unemployment (St John's, Chicoutimi), or proportionally large foreign-born populations (Toronto, Vancouver, Hamilton). In other instances, distinct regional groupings of cities with similar attributes, such as those in Quebec or the Atlantic region, are also evident. In these bald figures we can see the vague outlines of many of the social attributes and issues that are addressed with more finesse in the chapters to follow.

Despite this diversity, there are also numerous consistencies in the internal patterns and spatial organization of most Canadian cities. There are marked contrasts, for example, in population densities, the age of building structures, and types of shopping facilities among the central core, the old suburbs, and the new suburban fringe. Also, the poorest districts tend to be located in the east end, downwind from the major pollution sources, while the elite tends to occupy the higher ground. But how do these internal patterns actually vary within urban Canada? Do they differ systematically among cities of different size and by region, as suggested above?

Tables 1.5 and 1.6 illustrate some of these spatial regularities in the internal structure of Canadian cities taken as a whole. Table 1.5 compares selected socioeconomic indices for all census tracts in the twenty-seven largest urban areas grouped by city-size category; Table 1.6 compares the averages of inner-city tracts with those of older and newer suburban districts. Note in Table 1.5 that some indices of urban form vary systematically across size categories: population density, household income, mobility, average dwelling value, and percentage foreign born tend to be greater in larger cities, while household size and homeownership rates tend to be higher in smaller urban areas. Other indices, such as age of housing and rates of participation in labour markets, show little or no consistent variation by city size. In Table 1.6 the internal ecological gradations from inner city to outer suburb are clearly apparent in terms of density, household size, age structure, income, homeownership, and proportion of foreign born. It is these types of variations in social and physical structure, and of course debates about their interpretation, that provide grist for discussion in the chapters that follow.

Table 1.4
Selected socio-economic attributes of Canada's metropolitan areas, 1986

Rank* (1986)	CMA†	Number of households (000s)	Average household‡ income ($000s)	Index§ of low income (%)	Rate of natural increase#	Net internal migration 1981–86"	Pop. aged 65+ (%)	Foreign born (%)	Labour force part. rate female (%)	Unemp. rate male (%)
1	Toronto	1153.6	43.0	11.4	3.5	2.6	9.6	36.0	52.0	4.8
2	Montreal	1124.5	33.0	16.3	2.9	0.6	10.2	15.7	43.7	10.2
3	Vancouver	532.2	36.1	11.0	3.4	2.6	12.2	28.4	47.1	11.7
4	Ottawa-Hull	291.1	40.9	12.5	4.0	4.7	8.9	13.4	50.8	7.1
5	Edmonton	256.0	37.0	10.7	6.3	-2.1	7.3	18.2	51.9	11.0
6	Calgary	236.2	40.8	10.4	6.9	-1.0	6.8	20.6	50.7	10.1
7	Winnipeg	233.3	33.5	14.1	3.3	0.8	12.1	17.9	47.7	7.5
8	Quebec	216.3	32.9	15.0	3.1	0.4	9.4	2.3	42.7	10.9
9	Hamilton	201.3	36.7	12.3	3.2	0.9	11.5	24.1	44.5	6.0
10	St Catharines	110.9	33.8	12.8	1.1	-1.5	13.3	20.4	43.0	7.6
11	London	113.7	35.2	11.7	3.2	0.6	11.4	18.2	51.1	6.4
12	Kitchener	110.2	35.8	10.7	5.3	3.5	9.7	20.7	51.5	4.8
13	Halifax	103.8	36.5	12.8	4.6	2.5	8.8	6.9	46.4	8.0
14	Victoria	96.6	31.7	9.4	0.5	4.2	17.9	20.7	43.2	11.4
15	Windsor	89.9	36.2	14.9	2.5	-0.8	11.7	20.0	41.1	8.2
16	Oshawa	57.1	40.4	13.8	5.4	4.0	8.3	17.2	46.1	4.1
17	Saskatoon	66.4	33.7	9.2	3.9	4.4	9.4	9.3	49.1	8.7
18	Regina	63.1	36.6	11.4	5.9	0.8	9.8	9.4	50.9	8.0
19	St John's	48.7	35.8	15.9	6.5	0.1	8.9	3.1	41.1	15.0
20	Chicoutimi-J.	45.2	32.1	18.5	4.1	-3.8	7.3	0.7	33.8	14.9
21	Sudbury	49.7	33.0	13.7	3.0	-5.2	8.8	10.0	39.9	11.7
22	Sherbrooke	45.6	28.9	16.5	3.5	0.0	10.3	3.4	45.7	9.9

23	Trois-Rivières	42.8	28.7	21.5	2.5	−2.6	10.4	1.4	38.0	14.6
24	Thunder Bay	43.9	36.4	9.2	2.5	0.5	11.8	15.2	44.6	10.3
25	Saint John	39.4	30.4	14.6	3.3	−0.6	11.6	4.7	40.1	14.0

Sources: Census of Canada, various years.

‡ Private households only.

§ Based on Statistics Canada low-income cutoff (LIC), the minimum income necessary to sustain an average-size household in each locality.

From Statistics Canada, Bulletin 91–251.

" Estimated (in %) using 1986 census definitions; excludes foreign immigration and emigration.

* Ranked by total population.

† CMA (census metropolitan area) as defined in 1986.

Table 1.5
Descriptive statistics, comparative urban structure, by size category, 1981
(means for census tracts)

Variable	All urban areas	Urban size groups			
		> 1 million	500,000 to 1,000,000	200,000 to 500,000	< 200,000
1 Population density (per km²)	4,047.8	5,709.9	3,030.9	1,906.2	1,853.1
2 Mobility (% movers)	49.0	50.2	49.7	46.9	45.5
3 Household size (average)	2.76	2.73	2.76	2.77	2.91
4 Household income (median)	22,483	22,812	22,957	21,814	20,932
5 Average dwelling value ($)	83,299	100,105	70,298	72,002	55,646
6 Dwellings owned (%)	55.3	50.6	57.1	64.3	61.0
7 Foreign born (%)	20.7	26.0	18.9	19.3	9.0
8 Service employment (%)	32.0	31.8	31.5	32.3	33.6
9 Female labour-force part. rate (%)	54.1	54.6	55.5	53.1	50.2
10 Age of housing (% before 1946)	24.8	25.8	21.9	26.6	24.9

n = 2,754 census tracts (334 deleted) in 24 CMAs and 3 CAs.

CANADIAN CITIES: THEMES AND VARIATIONS

The very idea of the "Canadian city" is a recent one, at least as a scholarly construct.[3] Textbooks in urban geography and other social sciences have spoken generically of a North American city without acknowledging any distinctive character to urban areas in Canada. As Evenden and Walker, for example, note in chapter 12, the classic sociological study of Forest Hill ("Crestwood Heights") could find nothing to distinguish this exclusive inner suburb of Toronto from comparable elite districts in American cities.[4] The perceived homogeneity of urbanization in North America (if sometimes moderated by the observation that cities in Canada were "twenty years behind" trends south of the border) led all too easily to the uncritical transfer not only of urban models but also of urban policies from the United States. But of course such stated homogeneity masked important differences between cities on both sides of the border that are as real as the distinctions between the nations themselves.[5]

Table 1.6
Descriptive characteristics, comparative urban structure, by intraurban location, Canada,
1981 (means for census tracts)

		Intraurban location (group means)		
Variable*	All urban areas	Inner city (n = 795)	Older suburbs (n = 1076)	New suburbs (n = 1156)
1 Population density (per km²)	4,007.4	8,030.2	3,317.1	1,838.5
2 Mobility (% movers)	49.8	52.9	44.5	52.5
3 Household size (average)	2.80	2.34	2.75	3.17
4 Population over 65 (%)	10.1	14.6	11.3	5.8
5 Household income (median)	23,068	16,590	23,619	27,094
6 Average dwelling value ($)	84,255	79,142	84,530	87,584
7 Dwellings owned (%)	56.3	31.8	58.7	71.3
8 Foreign born (%)	21.0	23.8	21.7	18.5
9 Service employment (%)	31.8	37.0	31.2	28.6
10 Female labour-force part. rate (%)	54.8	51.9	54.1	57.6

* Age-of-housing variable deleted because it was used as part of the criteria for delimiting the three zones.
n = 3,003 census tracts (83 deleted) in 24 CMAs and 3 CAs.

If, as is often said, Canadian cities have a more European ambience, this is because Canadian society itself established a more complete institutional affinity with European, notably British, models of governance. So, for example, the regulatory role of the state in Canadian cities has imposed firmer planning and land use templates on the urban landscape and has led to stronger commitments to public services, including transit, housing, parks, libraries, and recreation. Furthermore, the introduction of regional (and metropolitan-wide) governments in many parts of Canada has facilitated the more equal distribution of these services throughout the metropolitan area than is generally the case in the more politically fragmented American metropolis. While the American state, particularly the federal government, has been a major force in US cities, its role has frequently been one of aiding private rather than public solutions to urban problems. The differential role of transit (a public mode of transportation) and freeway-building (a private mode) is a case in point. In comparable urban areas, public transport in Canadian cities recorded about 2.5 times the revenue miles per capita of its US counterpart, while American cities had four times the urban expressway capacity per capita of Canadian metropolitan areas.[6] The effects of these transportation differentials are considerable: Canadian urban areas tend to be more compact, exhibiting higher average densities, lower levels of decentralization (e.g. of employment), and fewer blighted inner-city areas.

It is this regulatory role of government that brings, as Holdsworth suggests in chapter 2, one of the distinctive senses of place to Canadian urbanization. But, as he observes, corporatism extends beyond the public sphere to the private realm also. Canadian institutional culture and public policy have long favoured development of large, private corporations with a national range in transportation, resource development, distribution, and, perhaps most significant, banking. As a result, the Canadian central-city landscape is occupied by a familiar set of icons across the country, including the château roofs of the railway hotels, with the white or gold towers of the Royal Bank frequently facing the black towers of the Toronto-Dominion Bank, as both loom above a covered mall anchored by the Eaton's department store. In the United States, in contrast, a much larger economy has permitted entry of a more diverse set of corporate actors and a competitive system policed by various forms of anti-trust legislation and state legislatures, tendencies that establish regional but not national domination for many corporations.

An additional characteristic that again positions Canada closer to northern Europe is its commitment to the welfare state, developed later than in Europe, but nonetheless moderating extremes of wealth and poverty that are deemed acceptable in the United States. The consequences show up powerfully in social patterns within metropolitan areas. Concentrations of poverty exist in American inner cities and are associated with a range of social problems, crime, racial tensions, and environmental deterioration, which have prompted middle-class withdrawal (aided by freeway-led dispersal) to the suburbs. In the Canadian metropolis, in contrast, inner cities are more diversified and to date at least exhibit a higher quality of life; every large city contains an enduring old elite district near downtown which has provided an anchor for waves of inner-city reinvestment, including gentrification, which in turn encourages further private investment downtown. It seems likely also that ethnic and racial homogeneity made Canadian central cities more predictable and therefore, for some residents, more liveable. The restrictive immigration policy in force until the late 1960s privileged the well-established immigrant channels from northwest Europe; as late as 1971, two-thirds or more of metropolitan residents identified themselves with British or French ancestry. In the United States a far more diverse immigrant pool, together with the structural marginality of large racial minorities and the continued deterioration of many public services (e.g. schools), made the American central city an altogether less familiar and indeed less safe and financially secure option for the middle class.

But in establishing the existence of a distinctive urban sense of place in Canada, one should be cautious not to press that homogeneity too far. We have already noted from Table 1.4 the local and regional variations that occur around the national theme of the "Canadian city." These differences are revealed in a more sophisticated multivariate analysis of census-tract data for metropolitan areas reported by Davies and Murdie in chapter 3. To anticipate one of their findings here, one in seven metropolitan census tracts across the country experienced severe impoverishment. However, there was marked regional scatter around this benchmark; in several cities in Quebec and Atlantic Canada the incidence of chronic impoverishment was twice this level or more, while in the major cities of Toronto, Vancouver, and Calgary the rate was much lower. Indeed, Calgary's rate of one tract in forty suffering chronic impoverishment was one-tenth that of some cities in eastern Canada.

Moreover, there is evidence that such gaps in well-being between metropolitan areas have intensified over the past twenty years. If we consider, for example, the innermost census tracts around the central business district, quite different processes prevail between Winnipeg or Saint John and Toronto or Saskatoon. In Winnipeg growing poverty is the dominant trend in the inner city, with deterioration of income and employment and problems of housing affordability both in absolute terms and relative to the entire metropolitan area. Local commentators observe a disturbing downward spiral: "Inequities between the inner and non-inner city are increasing. Moreover, those groups which are already the most disadvantaged – Natives and single parents – are experiencing the greatest level of worsening conditions."[7] But if deepening poverty is a dominant element of much of Winnipeg's inner city, in other metropolitan areas the trend is in the opposite direction. In Toronto, for example, a planning commissioner made the claim, albeit somewhat exaggerated, that the inner city is becoming "a ghetto for the rich."[8] Some support for these local observations is offered by Table 1.7, which examines the changing income of families living in the districts immediately adjacent to downtown between 1970 and 1985. Differences between cities at the extremes widened appreciably in terms of the incidence of low- and high-income families living around the urban core. In Winnipeg, more than half fell into the low-income category in 1985, a slight increase over 1970, with hardly any families in the upper-income group. In Toronto, in contrast, the low-income proportion fell to one-third by 1985, with a large increase of high-income families to nearly one-quarter of the total. Despite these shifts, however, inner cities in general continue to be poorer than their suburban fringes, although the contrast is much less stark than in the United States.

As we shall discuss shortly, the 1980s showed an intensification of market forces and a steady withdrawal of the tempering arm of the state. This is one

Table 1.7
Percentage of inner-city families in selected income groups, 1970 and 1985
(constant 1985 dollars)

Inner city	Less than $20,000		$20,000 – $60,000		Over $60,000	
	1970	1985	1970	1985	1970	1985
Toronto	43	33	48	45	9	22
Montreal	54	42	39	42	7	16
Vancouver	46	43	50	48	4	9
Ottawa-Hull	39	28	55	53	6	19
Edmonton	42	32	53	55	5	13
Calgary	47	41	50	49	3	10
Winnipeg	54	56	44	41	2	3
Quebec	50	45	45	45	5	10
Halifax	56	42	43	50	1*	8
Saskatoon	47	37	52	50	1*	13
Regina	51	39	48	49	1*	12
Saint John	51	55	48	51	1*	1*
All inner cities	47	39	47	46	6	15
Rest of CMAs (non-inner city)	24	21	67	60	9	19

Source: Ram, Norris, and Skof (1989).
* Approximate figure (data suppressed because of small number of households involved).

of the important factors in the growing polarization between cities experienced during the decade. It also tends to homogenize the institutional context of Canadian and American urbanization. This convergence may have been hastened by the Free Trade Agreement between the two countries signed in 1988, with its rhetoric of creating a "level playing field" in terms of economic and public policy. Movement in Canada toward American institutional culture cannot avoid influencing the dominant processes shaping Canadian cities. Ironically, no sooner has the distinctiveness of Canadian cities been recognized than they face compromise and likely evolution toward a more American model.

CONTINUITY AND CHANGE: THE POST-WAR EXPERIENCE

An elderly visitor arriving by boat in the old harbour of Montreal in the summer of 1992, after an absence of almost a half-century, would immediately notice immense changes in the skyline and landscape of that historic city. Yet there would also be no question in the mind of the visitor that this place was indeed Montreal rather than somewhere else. Similar reactions would

follow from a return visit to Halifax, or Vancouver, or perhaps even Toronto. Most Canadian cities, especially those with an imposing physical site,
a distinct culture and language, and a historically defined core of older structures, maintain a unique sense of place. Moreover, each poses a fascinating
intellectual dilemma, juxtaposing a sense of continuity, stability, and persistence, if not inertia, with a feeling and appearance of rapid social and economic change. Part of the challenge in this book is to document the simultaneous presence of both continuity and change.

Canadian cities have indeed witnessed a series of interrelated structural
and life-style transitions since the end of the Second World War. Not only
have most cities, as noted above, become substantially larger in population
and even more so in geographical area and social diversity, they are now
much more functionally complex and less easily categorized. They all, however, have been subjected to similar processes of change, though in varying
degrees and with uneven effects.

Perhaps the most fundamental social transition has been in the composition of the population – the demography – and in the ways in which urban
Canadians choose (and are obliged) to organize their living arrangements,
their working environments, and their social and family relations. Immediately after the war, Canada entered a period of unusually high fertility, increasing marriage rates, and high levels of household formation and foreign
immigration. Similar trends were evident in most other Western industrial
countries, but the Canadian experience was exceptionally marked. The resulting "baby boom," especially from 1948 to 1963, produced a very large
generation whose size, composition, and behaviour are still sending ripples
through the social order, the labour force, and the economy as it ages. The
relative size of this population cohort was made even more apparent by the
scale of the subsequent marriage and fertility declines (the "baby bust") of
the 1970s and 1980s. Although fertility levels have now stabilized, the
effects of the ageing baby-boom population will be felt well into the next
century. The three million Canadians aged 65 and over in 1991 will more
than double to 6.4 million by 2020, adding substantially to the need for social services for the elderly.

Parallel shifts in living arrangements have redefined the kinds of households that Canadians live in, and this in turn has altered the social services
and housing that they need and demand. Average household size declined
from over 4.1 in the 1940s to 2.7 in 1991, and more than half of all households are now of non-traditional form. Among members of the household,
new configurations of roles and relationships have also emerged, most emphatically in terms of the proportion of married women in the paid labour
force. In nuclear families, the two-income household has become the norm,
further reordering family incomes and decisions about residential location.[9]
In 1986, 65 per cent of all families in Canada had two or more income earn-

ers, compared with 33 per cent in 1951. Rose and Villeneuve report in chapter 8 how in nuclear families with pre–school-aged children the share of mothers in waged employment rose from 35 per cent in 1976 to 63 per cent in 1988. This feminization of the work-force has introduced a profound shift in the home and in neighbourhood rhythms of daily life. It has created increased demand for public services, notably daycare,[10] provided opportunities for expansion of private services such as home cleaning or fast food, and often produced a staffing dilemma for community agencies dependent on volunteers.

Urban development on any significant scale is conditional on an enhanced role for the local state. The public sector in urban Canada has been forced to adjust to increasing demands for new and improved public goods and services from a population that is larger, more diverse, considerably wealthier, and more discriminating about its needs. Obvious examples include the level and quality of infrastructure and community services required in a new suburban neighbourhood of the 1990s in comparison with those typically provided (or not provided) in the 1950s. Other examples include provision of educational and recreational facilities and services for those with special needs, such as homes for the elderly and handicapped and group homes for transients, as well as daycare for the children of working parents. Although few such services would at present be deemed sufficient or entirely satisfactory, their presence constitutes part of the myriad of functions that have created a "public city."

Post-war construction of the welfare state has provided an important context to the social geography of the Canadian city.[11] From a slow beginning, the federal state's reach into the urban environment deepened and broadened, attaining its most explicit form in the 1970s with the creation of the short-lived Ministry of State for Urban Affairs. The ministry's mandate was to co-ordinate those activities of federal government departments and agencies that had (or were seen to have) significant effects on urban development and the quality of life. Although it was not successful in that task, for reasons too numerous and complex to discuss here, it raised political awareness of the mounting problems of the cities and the need for greater policy co-ordination and innovation. Following the ministry's demise in 1979 the locus of responsibility for formulation of urban policy shifted almost exclusively to the provincial governments, some of which have effectively exercised that authority. The direct role of the state in planning and regulating urban developments is now played out largely in the arena of municipal-provincial relations.

The federal government's intervention has been more sustained in the realm of shelter. For example, it constructed on average 830 units of public housing per year between 1949 and 1963, but from 1964 to 1978 production levels expanded to 11,680 units per year.[12] In addition, in the early 1970s a new range of federal and provincial non-profit social housing programs contributed on average a further ten thousand units each year to the mid-1980s. However, through the 1980s the supply of new federally assisted housing fell ever further behind growing demand, contributing to the acute problems of housing affordability discussed in later chapters.[13] One indicator among many of an intensifying problem of affordability is provided by the declining access of renters of prime home-buying age to the average-priced home; whereas 50 per cent of such tenant households could afford homeownership in 1971, this figure fell to only 7 per cent in the housing-price boom of 1984-88. But in this context of growing need, government programs have been cut back severely. The public housing program was largely phased out in the late 1970s, and allocations for new social housing have been continuously pruned. In British Columbia, for example, with the nation's most chronic affordability problems in the early 1980s, allocations for new social housing were cut by half between 1980 and 1984.[14] Meanwhile, the costs associated with maintaining the existing social housing stock, and of subsidizing rents for an increasingly impoverished resident population, have continued to mount.

The combination of a growing public debt and the increasing prominence of neo-conservative economic and tax policy served to discipline and constrict the welfare state during the 1980s. Restraint budgets, service cutbacks, user fees, and privatization have redefined the delivery of social services.[15] The coincidence of limited services and the severe recession of the early 1980s gave a public face to poverty that had scarcely been observed for fifty years. Many of those trapped by poverty during the recession of the early 1980s were still there at the onset of the next recession in the early 1990s, especially in Ontario, Quebec, and the Atlantic region. In chapter 16, Dear and Wolch note estimates of homelessness which range upward of 100,000 across Canada, with figures for Toronto, the nation's wealthiest city, of up to 20,000. The shortfall of public provision in social welfare has led to the reappearance of a necessary but sometimes erratic voluntary sector to meet essential needs. The first food bank was opened as the earlier recession struck home in 1981; by the end of 1984, seventy-five were active across Canada, and, despite the economic recovery of the late 1980s, by 1989 an even larger number of food distribution centres were reported in Metropolitan Toronto alone.[16] Since 1989 and the severe economic downturn, demands on these facilities (supporting 120,000 people in 1992) have intensified.

Poverty, hunger, and homelessness have therefore become more visible components of urban life. Their consequences, in combination with escalating street crime and the drug culture, are tarnishing the long-valued liveability of large cities, particularly Montreal, Toronto, Vancouver, and Winnipeg. A new and very disturbing element is the emergence of a pattern of violent teenage gangs. In Vancouver, police estimate there are some 500 members of such gangs, most of whom are from non–English-speaking immigrant homes. Public services fell far behind the needs of a large influx of Third World immigrants and refugees during the 1980s. Although half the students in Vancouver's schools are not native English speakers, in 1991 the school board employed only one Chinese-speaking counsellor, one who speaks Punjabi, and one fluent in Spanish. Despite the traumas of war and refugee-camp internment, the city's Vietnamese refugee community of over 17,000 is served by only one native-speaking psychiatrist; in the entire metropolitan area there are four Chinese-speaking psychiatrists to serve a potential ethnic community of 200,000.[17]

The market-oriented social policy of the 1980s saw increased employment as the most appropriate response to urban poverty, and the different levels of the state have been heavily implicated in priming local economic development. Now almost every city in Canada has established an economic development office seeking out new investment in response to the increasingly competitive economic climate. As is also implied by the giddy array of hallmark events (Olympic and Commonwealth games, world's fairs, and so on) which have been sought and secured by Canadian cities, tourism, recreation, and leisure are seen as a significant mechanism of economic growth, both materially and in image-building.[18] This new culture of consumption is making large demands on public budgets and central city landscapes, while also raising fundamental political and moral questions about the allocation of scarce resources. In 1989 Toronto opened its $560-million covered sports stadium, while by 1990 fund-raising was well along for a new ballet and opera house ($230 million). In the same year, the city made bids for the 1996 Olympics (estimated cost $2.52 billion) and the World's Fair in 2000 ($1.3 billion). The opportunity cost of this scale of investment is quite breathtaking, not least in a city with serious environmental problems, high living costs, and a crisis of affordable housing. An anti-poverty coalition, Bread Not Circuses, assailed the media, the provincial government, and city and Metropolitan Toronto councils about the justice and wisdom of such conspicuous consumption; their efforts were said to be one reason for the failure of the bids for the Olympics and World's Fair and subsequent postponement of the ballet-opera house project. These special events require much closer scrutiny to determine, among other things, the relative distribution of costs and benefits across space and through society. One important question is the extent to which such megaprojects are shaping not only new landscapes but also growing social polarization within the city, as well as between cities.

The polarization argument rests on two pillars. First, it is claimed that the economic restructuring of the 1980s and early 1990s has significantly eroded middle-class and blue-collar jobs, particularly unionized ones. The pattern of job creation has favoured a substantial minority of highly skilled, white-collar positions, in management and the professions, and a majority of less-skilled service jobs, often part-time and with limited career prospects, held disproportionately by women. Second, in addition to the working poor, a virtual underclass is set apart by a range of circumstances, including family disintegration, age, physical and mental ill-health, structural unemployment, and cultural and racial marginality, and penalized by reductions in social services. Its numbers include teenage runaways, other victims of family violence, and groups of single parents, the elderly, Natives and certain new immigrants, the mentally ill and physically handicapped, and those suffering from drug and alcohol dependence. Representatives of each of these groups are to be found in the poorest areas of public and private housing.[19] Once concentrated only in the inner city, they are being displaced outward by the pressures of gentrification and redevelopment of the urban core and by the availability of cheaper land in the suburbs for the construction of social housing. In 1985 the Social Planning Council of Metropolitan Toronto identified sixteen areas of poverty and serious social need, half of them in the city of Toronto and half in the surrounding municipalities. In this sense at least, the suburbs and the city are becoming more alike.

Patterns of household income, such as those in Table 1.7, offer some clear evidence of increasing social polarization in the districts surrounding downtown. Extending the analysis to an entire metropolitan area reveals that this pattern is sustained. A study of disparity in household incomes showed that social polarization has become more accentuated in both the city of Toronto and the Toronto census metropolitan area (CMA) between 1970 and 1985.[20] While inequality of income is twice as marked in the city as in the metropolitan area, in both places it has risen by some 40–50 per cent. Over a longer period, the full extent of deepening inequality is apparent: in 1950 the average income of the wealthiest census tract in the city of Toronto was three to four times that of the poorest; by 1985 the gap had broadened to nearly 14 to 1. The recession of the early 1990s will probably accentuate these contrasts.

The economic restructuring of Canadian cities is heavily implicated in these changing social distributions, and it is the landscape of the new urban economy that would most powerfully disorient our elderly returnee to Montreal as she viewed it from either the St Lawrence River or, more likely in the 1990s, along a flight path to one of the metropolitan airports. The waterfront, formerly a busy site of shipping, wholesaling, and manufacturing, has grown strangely quiet. Indeed on some waterfront areas of other cities, including Toronto's Harbourfront and Vancouver's False Creek, the former port and its related functions are unrecognizable. The old manufacturing and wholesaling base has been stripped from the central city landscape – with part relocated to cheaper and more modern sites in the suburbs and part shut down, a victim of obsolescence and global competition. Replacing blue-collar jobs, which dominated the old urban core, are an army of office workers – managers, professionals, technical workers, and clerical staff – and an entourage of shop assistants, waiters, taxi drivers, hotel workers, hospital and university staff, and other service employees.[21] But the office building is their dominant place of work, and it is the office building that has been the major element in the remaking of the downtown landscape. In Toronto's financial district, for example, a work-force of 100,000 people is crammed into an area of less than one square kilometre, 93 per cent of them employed in 124 office buildings covering 2.6 million square metres of office space.[22] Even in Montreal, with its slower post-war growth, the amount of downtown office space more than tripled between 1960 and 1987. In other major centres, such as Calgary and Ottawa, growth was even faster.

The new geography of employment has been transmitted to a new geography of housing and neighbourhoods. A portion of the professionals and managers working downtown has chosen an inner-city residence, leading to the embourgeoisement of many older districts like Centretown and Sandy Hill in Ottawa, Don Vale and the Annex in Toronto, Plateau Mont Royal in Montreal, and Kitsilano and Fairview in Vancouver. Between 1971 to 1986, there was a net gain of 116,000 employees in the advanced services sector living in the inner cities of six large Canadian metropolitan areas. Following the pattern of deindustrialization they replaced 215,000 workers in manufacturing and other sectors.[23] The combined processes of replacement and displacement have seriously eroded the stock of affordable housing in major cities. Poorer households are increasingly found in residual areas, particularly in the protected enclaves of assisted housing; the private market has largely left them behind. These social changes affect more than housing; they extend to the whole fabric of community. In Vancouver's Kitsilano neighbourhood, for example, some 1,500 condominium units were built between 1971 and 1976; by the latter date, 50 per cent of the population in the redeveloped district was between 25 and 34 years of age. The main commercial thoroughfare of Fourth Avenue underwent commercial gentrification, as fewer than a quarter of the shops and services existing in 1966 survived ten years later. Other features of community fabric have been eroded; school enrolment in Kitsilano fell by 55 per cent between 1965 and 1985, while attendance at local Protestant churches declined by 38 per cent from 1971 to 1985. The site of a Pentecostal church has been transformed into a neighbourhood pub, a workers's hall has been converted into an experimental theatre, and a union office has been replaced by a medical-dental building.[24]

Overlaid on these changes to the social geography is a radical reworking of the ethnic, racial and cultural composition of Canadian cities. Immigration is one facet of a broader internationalization which brings business and property investment as well as tourism from Europe, Asia, and the rest of the Americas. This process has been encouraged by overseas advertising, trade missions, and the special attractions of a cultural event like the Olympic or Commonwealth games, planned to show off the advantages of an aspiring "world city." Canada's cities have of course always been international, but since 1945 the dominant source region of immigrants in northwest Europe (notably Britain) has given way to a more diversified set of origins. In 1957, 95 per cent of immigrants were from Europe or the United States, with the largest single group from the United Kingdom. This immigrant share fell in consecutive decades until by 1990 it accounted for only 29 per cent of new arrivals, compared to over 49 per cent from Asia. Indeed in 1990 the leading countries of origin of immigrants were Hong Kong, Poland, the Philippines, Lebanon, India, and Vietnam. Destinations within Canada continue to be overwhelmingly urban; in 1990 the metropolitan Toronto region was the destination of one-third of all new immigrants, while Vancouver was the recipient of another 15 per cent. As a result, for the first time in Canadian history, a high proportion of immigrants and immigrant children in the major cities are neither Caucasian nor English-speaking. By 1991 the "visible-minority" population made up 24 per cent of the 3.9 million people in the Toronto CMA. Their arrival has enriched Canadian cities but has coincided with other pressures on jobs and the provision of public services, leading inevitably to some of the stresses and social problems mentioned earlier.

Adjustment problems are also shared by some native-born residents who have grown up with a standard of a bicultural nation springing from the British and French "charter groups." Multiculturalism, ethnic rights, and charges of institutional racism have sometimes caused both confusion and resentment.[25] Defensive reactions have taken a number of forms. In Quebec thorough-going language legislation has sought to prevent any challenge to the French cultural and linguistic realm. Elsewhere different kinds of issues have come to the fore. In Toronto and particularly Vancouver, for example, disagreement has surfaced over "monster houses," large and ostentatious dwellings built in mature, inner suburbs. The Anglo-Canadian landscape taste of these upper–middle-class districts has favoured Tudor and other European styles, moderate-sized homes on leafy green streets, and richly landscaped lots. The newly arrived households who are the principal owners of these large houses are mainly business people with an alternate set of landscape values which emphasize large, new, visible homes, bright light, and minimal vegetation. Demolition of existing houses and the felling of mature trees have brought local disagreements to a head and are indicative of the nuanced frictions that so easily arise in a society extolling a new-found cultural pluralism.

INTRODUCTION TO THE CHAPTERS

The changing emphases of social geographers represent an attempt to keep pace with the dynamic nature of urbanization at the end of the twentieth century. Those who use the various transition labels, – post-industrial, post-

Fordist, and post-modern – whatever else they may disagree on, concur that some fundamental shifts in urban society are afoot. Of course, the inertia in the built environment and institutional forms inherited from past decades tempers the pace of absolute change in the spatial structure of cities,[26] though the trends we have discussed certainly indicate that the social geography of Canadian cities is undergoing some critical transitions. Economic and occupational restructuring, a new demography combined with changes in living arrangements, shifts in household composition and family relations, increasing ethnic diversity, and revised political priorities are but the most obvious of these shifts.

A change has also occurred in social geography over the past twenty years, from pattern to process-oriented research, from descriptions of map distributions to explanatory and interpretive accounts of the processes and meanings that underlie those patterns. This shift has inevitably extended the theoretical range of research from the implicit micro-economic theory of human ecology to consideration also of social and, more recently, cultural and political theory.[27] The same period has seen a blurring of disciplinary boundaries and a regrouping of traditional empirical work – a binding together of formerly fragmented literatures. One of the long-standing divorces in human geography has been the separation of home from work, isolating the realm of consumption and reproduction from that of production. On a number of fronts this separation is under attack, as is evident from several chapters that follow. Feminists have correctly pressed for a reconceptualization of domestic work, and others have pointed out the intersecting relations of housing and labour markets, relations that are reciprocal, not one-directional; indeed there is abundant evidence that for mothers, home-based considerations strongly constrain both the type and the location of employment they seek.[28] Other theoretical integration is attempting to bridge the separations between the social and the political and between the social and the cultural. Recent work on the role of the state and the politics of externalities has underscored the political nature of the map of social well-being across the city; consider, for example, some of the negative aspects of increased community control, the so-called NIMBY ("not in my back yard") syndrome, in restricting the location of necessary social services such as group homes. Integration of social and cultural themes is also evident in several chapters below, notably those concerned with the meaning of urban places and housing.

Part I of the book begins with an evocative interpretation of the sense of place in Canadian cities. Deryck Holdsworth (chapter 2) offers a series of persuasive vignettes portraying the variations and continuities of Canadian urban landscapes, from metropolis to small town and from the corporate office district to the residential areas which comprise the territorial bases for urban residents. In a wide-ranging chapter, he identifies a mix of regional diversity as well as corporate standardization and the distinctive contributions of both the public and the private sectors as elements moulding the meaning of Canadian urban places. In a more analytical vein, but with an equally broad sweep across the nation, Wayne Davies and Robert Murdie (chapter 3) break new ground with a factorial ecology combining almost 3,000 census tracts in twenty-four metropolitan areas in 1981, an innovative attempt to identify common factors so that the social areas of cities may be directly compared against each other. Several original findings are presented, including the comparative distribution of urban wealth and poverty at the intra-metropolitan scale. But the authors also point to apparent changes in the relations of people and place, a growing complexity of family types in the city, and the separate emergence of an impoverishment factor. The latter, a troubling dimension seemingly emergent in the late twentieth century, incorporates low income and unemployed- and female-headed households as the locus of a deepening social problem.

The relationship between a changing society and its spatial patterning is explored further by John Miron (chapter 4), who provides important new longitudinal data on the population characteristics of Canadian households since the 1940s. The changing demography of households is expressed directly by adjustments in the housing stock, which has improved in quality and has presented more varied tenure options for a majority of residents. This historical evolution is conceptualized in a micro-economic perspective which emphasizes demand-side factors, while directing our attention to the specific choices from which households must select. The insights derived from this theory are then supplemented by discussion of supply factors and government policy to account for the dramatic transitions that have occurred in living arrangements. It appears that households, the basic spending units in society, are elastic; they create demand for housing and services and in turn are created by the available supply. Donald Janelle then explores, in chapter 5, the temporal rhythm as well as the spatial patterning of human activity. His chapter examines the dynamic nature of daily and weekly activity patterns in Halifax which are of course concealed by the static picture of homo dormiens as presented in the census. The description of the city around the clock highlights the ebb and flow of urban life and illustrates some of the time-space regularities of urban movement and activity patterns for different social groups, regularities that reveal constraints as well as opportunities for urban residents, particularly women.

The patterns of everyday urban life are framed by some of the contexts discussed in part II. The pervasive mobility of urban Canadians is analysed by Eric Moore and Mark Rosenberg (chapter 6). Following a review of different perspectives on mobility and migration, they point to several of the complexities and weaknesses of a data base (the census) that is used widely but often uncritically. The strikingly variable rates of mobility among several subgroups of the population have much to do with the differential character of social areas in the city. So too the destination specificity of immigrants toward large cities, Toronto and Vancouver in particular, during the period 1981–91 is rapidly changing the face of the largest cities and redefining their social problems. The shifting ethno-cultural mosaic is explored further by Sherri Olson and Audrey Kobayashi (chapter 7), who similarly relate it to the destination specificities of successive immigrant waves which have broken over the Canadian nation, initially from east to west, though also in the 1980s increasingly from west to east, and in a hierarchical pattern diffusing downward through the urban system from larger to smaller centres. From these migration flows have emerged distinctive social clusters and neighbourhoods within cities, a geographic base that has permitted not only staged acculturation but also in some instances economic and political empowerment. Drawing on some focused and colourful examples, the authors demonstrate that in Montreal both the landscapes and the sentiments around ethnicity assume a particularly sharp profile.

Montreal is also the principal setting for the expansive chapter by Damaris Rose and Paul Villeneuve (chapter 8) which examines the complex, shifting, and interrelated contexts of work, labour markets, and family life. The first major trend is what they call the "tertiarization" of the economy – the growing shift toward the diverse service sector in metropolitan, particularly central-city, employment, a development that has its own geography, favouring the major cities, especially Toronto. Closely related is the "feminization" of the work-force, a trend also with multiple origins and an integral part of which is occupational polarization – the tendency toward a dual labour market contrasting managerial and professional decision-makers with office staff and service workers performing routinized tasks. All of these trends, it is shown, affect households' choice of residence and the changing social character of both inner city and suburbs. Turning specifically to the changes in housing stock and local communities, Larry Bourne and Trudi Bunting (chapter 9) note several important developments: the emergence of an integrated residential development industry, renovation as a significant alternative to new building, introduction of the condominium, and the variable role of government housing policy, all of which vary widely across the country. Among those agencies and institutions creating residential landscapes, the home-building and real estate industries and the state act as gatekeepers, seeking to match households and housing units. Together their actions influence the appearance, composition, and evolution of urban neighbourhoods. Although it is suggested that the majority of neighbourhoods have experienced relative stability rather than rapid change for much of the period since 1945, major structural shifts are nonetheless being overlaid upon this stable pattern. These include a growing mismatch of supply and demand, the difficulty of access from far-flung suburbs, rising maintenance costs in the older stock, income polarization within cities, and continued problems of affordability, and these will probably limit further increases in housing standards in the future.

In part III the chapters turn from a systematic review of contexts to a more synthetic interpretation of selected urban places. Daniel Hiebert also brings a theoretical project to his study (chapter 10) of Toronto's Spadina district from 1901 to 1931 – the need to bring together the realms of production and consumption not only theoretically but also in detailed empirical study. To accomplish this, however, requires that the character of social groups themselves be animated, so that class, ethnicity, and gender, rather than appearing as more or less independent, given factors, be treated as relational and also as constitutive. That is, one should think of these statuses as active, evolving, and open-ended rather than as passive and completed. This is a tall order for a short chapter, but Hiebert is able to bring together the evolution of the Spadina garment industry and its Jewish residential community during their formative years, and thus to illustrate some of the crossovers among ethnic status, work, and class composition. Very different from the immigrant industrial districts are both new and traditional higher-status districts in the inner city which are an abiding feature of Canadian cities from coast to coast. In chapter 11, David Ley is concerned primarily with traditional elite neighbourhoods and secondarily with newer districts of middle-class and upper-middle-class settlement, the latter often associated with the processes of gentrification and redevelopment. Traditional elite districts have distinctive landscape signatures extolling patrician values and owe their longevity to effective, if often discreet political mobilization, itself a product of tightly bonded social networks. They have frequently provided an anchor for gentrification, which has diffused in wave-like fashion from their outer margins into adjacent lower-income neighbourhoods. In their meticulous renovations and new designer buildings, gentrified districts also express no less of a mutual reinforcement between the physical landscape and a privileged social identity.

In terms of population size, the suburbs are now quantitatively more prominent than the central city. While researchers have been slow in recognizing this trend, it is pointedly identified in the title of chapter 12, by Len Evenden and Gerald Walker. Their review of a broad and scattered literature integrates research in an important field where much study remains to be done. In opposition to the conventional stereotype of suburbia as an unplanned landscape of single-family dwellings designed for middle-class nuclear families, most of whom work in the central city, reality shows considerably more complexity: a degree of public and private planning from the 1950s at least, growing social and ethnic mix with multi-family structures, childless, and two-wage-earner households, and new centres of employment, shopping, and leisure. These trends are combined with new concerns regarding the level of public services and a quest for functional and symbolic centrality, for a sense of place that expresses more than their traditional status as dormitory suburbs.

The small town also must be retrieved from the stereotypes of outsiders, even if these images are often shrouded in a mood of romantic nostalgia. Part of the difficulty in the analysis of small towns is definitional, note John Everitt and Alison Gill in chapter 13. Aside from a common attribute of size (but what size should be a threshold?), small towns show great diversity in their history, function, and viability. Some are declining; others are benefiting from a population turnaround which is bringing new migrants, including the elderly, and new jobs. Diversity extends also to appearance and street grids, with earlier, unplanned patterns contrasting significantly with the carefully planned site design of most current resource and resort towns, even if the clear spatial imprint of a single dominant employer (and property owner) adds its own constraints. It is part of the peculiar histories of many small towns that economic cycles and instability impede the smooth flow of property relations and the provision of public and private services. Thus, through regional development and social programs, the state can help dampen often extreme economic oscillations and sustain populations that remain loyal to the small town's quality of life even under unfavourable economic conditions.

In part IV, the principal foci are the role of the state in shaping patterns of social well-being in Canadian cities and the role of public policy in guiding market processes and seeking to ensure at least minimal service levels and living standards for all Canadians. Integrating a wide literature, James Lemon (chapter 14) demarcates five historical stages in the evolution of the welfare state in Canada. Social policy, at all periods highly politicized, has since the mid-1970s become a principal target of restraint-minded right-wing political platforms. We have also seen pressures to shift the locus of responsibility for social welfare from senior levels of government to the local level, with the inevitable inequalities resulting from that shift. The result is a safety net that is uncomfortably strained and, as is implied by the growth of food banks and homelessness, may no longer be working for many of those at the bottom of the income ladder.

In exploring the meanings of the home and homeownership, Richard Harris and Geraldine Pratt observe in chapter 15 that the home is for some people the locus of paid employment, for others a locale that nurtures values in contradistinction to the world of work, but for all residents, whether consciously or unconsciously, the home conveys messages of status and identity. The powerful sentiments evoked by the home evince a strong desire among most households for ownership, and housing policy in Canada, as in Australia, Britain, and the United States, has sought to advance homeownership. It has also supplied an investment with substantial returns to most purchasers, although inflated prices have decreased affordability for those

wishing to enter the market. The powerful meanings of home have frequently triggered political reactions which governments at all levels cannot
ignore and have sought to direct through varied policy guidelines and subsidy programs. In contrast, homelessness is a sharp and visible expression of
the new urban reality of the 1980s and early 1990s. Michael Dear and Jennifer Wolch report in chapter 16 on the changing profile (and growing numbers) of the homeless in Canada, who now include women and children,
with perhaps one-third of the total suffering some form of mental disability.
While the path to homelessness is a diverse one, the authors identify as
causes a set of social policies that have led to a contraction of welfare services and social housing, together with deinstitutionalization of the mentally
ill and the lack of alternative community-based services. At the same time
economic restructuring has increased unemployment and inflated the number of the working poor, while eroding the affordable inner-city housing
stock through redevelopment and gentrification.

The geography of health offers a powerful testimony to the presence of environmental problems and social inequalities in cities. Martin Taylor, in
chapter 17, reviews several studies which show that health inequalities are
persisting and may even be widening. A distinctive threat to urban health
comes from environmental contamination, including air and land-based pollution, although a series of vexing measurement and inferential problems
make definitive results in this important area difficult to achieve. A final
focus of the chapter concerns the geography of community mental health,
where the use of both quantitative and qualitative methods provides clear evidence of the effects of the community environment on the well-being of the
mentally disabled. A significant contribution to urban well-being is the differential access of social groups to public services, including health care. In
this volume we include an example of this approach in chapter 18 by Suzanne Mackenzie and Marie Truelove on changes in access to non-family
childcare services. The need for this service is a result of women's growing
participation in the work-force discussed in several earlier chapters. Childcare services have been chronically under-provided, in part because of un-
challenged assumptions about the roles of individual family members.
Childcare spaces did not increase substantially until the 1970s, and despite
an eight-fold growth between 1973 and 1990 severe shortages remain: in
Metropolitan Toronto, for example, the authors estimate that less than one-
quarter of the need is served by formal daycare, leaving a substantial shortfall to be absorbed by the informal sector.

How has post-war urban planning affected the social geography of Canadian cities? Peter Smith and Peter Moore begin their assessment (chapter 19)
with two preliminary considerations: the political nature of the physical
planning process, with the competing pressures this raises, and the pervasive
fragmentation of metropolitan jurisdictions, meaning that planning too is

usually spatially fragmented. Suburban growth has been addressed by planners, but successful management at a larger spatial scale – the urban region – has been rare. For the planning of individual subdivisions, in contrast, the record is stronger, and there have been important achievements in the rationalization of traffic and land use and the provision of community services and infrastructure in new suburban neighbourhoods. In the central city, redevelopment planning over the past twenty years has been fraught with political activism and political solutions. The clearest results of planning have come with publicly initiated transit and redevelopment schemes – the urban renewal projects of the 1950s and 1960s and the rehabilitation and neighbourhood enhancement programs of the 1970s and early 1980s. Most recently, other approaches more suited to the political mood of the past decade have been introduced: partnerships of public agencies and private interests have co-operated in a number of revitalization projects. Whether these varied strategies can respond to the future challenges of an evolving urban society remains to be determined.

What does not appear in the chapters in this volume is a single perspective or all-purpose model of the social geography of Canadian cities. Instead, and far more important, what emerges is a realistic sense of the diversity of the social fabric, as well as a measure of the dynamic processes that underlie changes to it. Canadian cities are continually being made and remade, and, as a result, both their landscapes and our interpretations remain unfinished and partial. To understand urban social geography is therefore a continuing challenge, in terms of setting priorities for research, information, and public discourse. It is our hope that this collection will contribute to such a process of learning, and perhaps to the quality of ensuing public debate and its attendant consequences.

PART ONE

Patterns: People and Place in Urban Canada

CHAPTER TWO

Evolving Urban Landscapes

D.W. HOLDSWORTH

Canadian urban landscapes defy easy categorization. Most places have distinctive "signatures" – their own melange of site attributes and settings – but at the same time there are also many similarities. These unique and generic elements are the products of both the inertia of regional historical development and more recent standardizing forces associated with the land market, government policies, and nationally organized enterprises. This chapter offers three vignettes of recent changes in Canadian urban places which illustrate the Canadian mix of public and private initiatives that have made (and persist in remaking) distinct yet similar urban landscapes.[1] My goal is to move back and forth between the contemporary and past "sense of place" of particular urban realms, to emphasize elements within the palimpsest of built environments that are imbued with significant social meaning. I shall then attempt to suggest the more widespread, pan-Canadian incidences of such elements without forcing the examples into generic models. It is in the textures of distinctive places, the lived worlds of streets, houses, shops, and offices, that memories accumulate, attitudes are shaped, and cultural and political preferences are defined.[2]

Three distinctive milieus are emphasized: metropolitan downtowns, small-town "Main Streets," and metropolitan residential neighbourhoods. The changing built environments in these three settings reflect the inextricably linked political and economic forces that have been at work; in Canada the economic is very often political. Federal intervention to ameliorate or compensate for provincial marginality vis-à-vis Ontario wealth (and American wealth at the continental scale) has fostered agencies whose work finds expression in the built environment. And a Canadian cultural predilection for more rather than less government has made a growing set of local, provincial, and federal interventions seem normal and expected. The urban landscapes thus become distinctively Canadian, even if the Maple Leaf is not visible on a rooftop.

DOWNTOWN

Forces of corporate agglomeration have dramatically reshaped the cores of Canadian cities since 1945, especially through the addition of millions of square metres of office space packaged in International-Style office towers. Although the scale of this recent change is remarkable, non-locally owned structures have long dominated the urban skyline (for instance, the "château-esque" hotels built by central-Canadian railway companies mostly before the Depression). Even so, public initiatives – including planning districts, transit systems, parking authorities, and heritage protection – have shaped the pace and appearance of these transformations. Similarly, while tentacles of metropolitan dominance emanating from Montreal and Toronto have redeveloped regional downtowns such as Halifax and Vancouver, the cores of these and other centres also reflect public input that retains some local distinctiveness.[3] In this vignette, Toronto-centred activities are highlighted, for although Montreal was the pre-eminent city in the late nineteenth century and co-primate in the first half of the twentieth, Toronto has pulled away as the dominant financial and management node for the Canadian economy since the Second World War.[4]

King and Bay and Canada

Fifteen minutes before noon, on a cold February day, at the corner of King and Bay in the heart of Toronto's downtown. The wind sweeps down from the office towers and in from Lake Ontario. Some office workers brave the cold on their way to early lunch. Many more prefer the warmth and convenience of the underground malls that link several dozen office buildings in the city core. Messengers and taxis scurry along the streets, taking people and special delivery packages between buildings. Streetcars trundle by, linking the blue-collar, ethnic, and newly gentrified neighbourhoods that fan out on both sides of the densely packed office towers of downtown.

Here, in a 360-degree vista, is much of the management centre of Canada: all five of the big banks are in view – the Toronto-Dominion, the Bank of Montreal, the Canadian Imperial Bank of Commerce, the Bank of Nova Scotia, and the Royal Bank.[5] All are associated with strikingly monumental building schemes of the last twenty-five years that have dramatically transformed Toronto's downtown. They are the crown jewels in a rebuilding of the financial core that contains over 3.0 million square metres of office space.[6] The Canadian National (CN) Tower, visible along a corridor to the south, is another symbol of Toronto's role of nexus in Canada's urban system, while the Eaton Centre, a giant four-block retail and office complex north on Yonge Street, is symptomatic of the vitality of the core. Old landmark buildings, such as Old City Hall, the St Lawrence Market, and Union Station, hint at some modest attempt to retain the best of the past.

For many non-Torontonians, however, King and Bay is the evil heart of the financial system, the home of the faceless decision-makers who shape *their* regional and local economies; the coldness of the weather echoes their perception of a cold-hearted place, and they shake their heads at the social values that must go with such an environment.[7] Politicians seeking to evoke the dreams of ordinary people juxtapose Main Street with Bay Street.[8] Novelist Heather Robertson confesses that when she moved east from Winnipeg, "as a westerner I used to curse the Toronto bank towers. I saw them as malevolent tombs built out of the bled corpses of western Canadians stacked up, layer on layer, 100 storeys high."[9]

It is not simply the glass and the steel and the cold concrete sidewalks that trigger this response. This small section of King and Bay has long been a part of the Canadian identity. The Toronto Stock Exchange, relocated to Bay Street in 1914, and rebuilt in 1937 in the depths of the Depression, seemingly underlined the separateness of Bay Street, and Toronto, from circumstances elsewhere.[10] The Stock Exchange moved from its Bay Street location in 1985, but by then Bay Street would be a disappointment for those who came to see the Canadian "Wall Street." It was never a canyon of key office towers (although its visual terminus at Old City Hall gives the street an imposing focus). The true axis of financial power is, and always has been, King Street, and the fulcrum has shifted steadily westward along King to, and then beyond, Bay over the last century.[11]

The shift of office development that created the King Street landscape of the 1990s can be summarized as the ongoing search for larger and more efficient towers for increasingly complex corporations. The tall seven-storey bank towers of the 1890s that rose above a Victorian retail strip contained opulent banking halls, space for the bank's offices, and speculative rental areas for other tenants. By the 1910s, a dozen or so banks were vying for national markets through an extensive series of bank branches. The head-office function of these concerns required more space for departmental operations as well as bigger banking halls in the competitive Toronto market. King Street was rebuilt in the Edwardian era to include several towers over ten storeys. Four separate buildings near the intersection of King and Yonge streets, three named after banks, claimed in succession the title "tallest building in the British Empire" between 1906 and 1914. Most of the tenants were not bank-related but were the land development, mining, forest, and transportation companies that harnessed the wealth of the Canadian Shield, the Prairies, the Rockies, and the Maritimes, as well as Ontario.

Between the late 1920s and the 1950s the city's office frontier had shifted to King and Bay.[12] Several buildings rose over twenty stories. Though it was the financial sector that "named" the new landmarks, other tenants reflected the resource and industrial sector. Mixed in were the accountants, stockbrokers, consultants, and insurance firms that fed off the primary and secondary sectors. Buildings still occupied nearly all their ground-floor land, typically with their ornamental front doors right at the sidewalk. From the 1960s on, a new phase stabilized much of the growth around King and Bay. First, the Toronto-Dominion Centre, with its black towers and low banking hall all designed by Mies van der Rohe dramatically shifted the morphological frame for buildings. Some forty-eight separate property parcels and lanes were assembled with help from city council, and the "tall-towers-set-in-open-plazas" became a part of the downtown landscape. Across Bay Street, the new Canadian Imperial Bank of Commerce replaced the old Imperial bank head office and several other properties with a 57-storey tower, to create Commerce Court (the old 1930s building being retained as Commerce Court North). This southeast corner of King and Bay also became a mini-plaza, but its ornamental trees became twisted bonsai-sculptures in the new wind-tunnel intersection. Later in the 1970s, the 17-storey Bank of Montreal building on the northwest corner was replaced by the 72-storey First Bank Tower of First Canadian Place, with the Bank of Montreal a visible tenant.[13] And ScotiaBank, having outgrown its 1951 skyscraper (but retaining it under complicated heritage and bonusing zoning), built a 65-storey tower on the northeast corner. In this round of redevelopment, with banks competing like Florentine merchants through their urban towers, the Royal Bank won kudos for its lower-height, twin-towers-linked-by-spacious-atrium solution; its gold-leafed windows still signal its privileged status as one of the decision-making points for the Canadian financial and corporate sector. The new Canada Trust Tower on Bay adds another dimension, post-modern in design and of lower height, but also a departure from the box-like structures around it.

Beyond aesthetics, what do these towers means? Two Montreal-based banks, the Bank of Montreal and the Royal Bank, have moved much of their executive function to Toronto, reflecting a further inexorable shift within the Canadian metropolitan system. Three blocks to the west, at King and University, the move of the huge Sun Life Insurance Co. from Montreal to a new complex in Toronto underlined the shift. And, as in earlier decades, many other firms, including conglomerates for the mining and resource sector, the real estate/industrial/retail magnates, and the Canadian offices of enterprises based in the United States, Japan, or Europe, are tenants in this new forest of glass skyscrapers.

The banks have certainly concentrated their executive functions in these towers, as well as a significant portion of their head-office operation, but some of their routine clerical and record-keeping activities are now located in suburban back-offices. The decentralization of office work had started in the 1920s in the case of insurance companies: Canada Life and Manufacturers Life moved their head offices to just beyond the fringe of downtown, where large parcels of land would allow for later expansions; other insurance companies followed in the 1950s. In the 1970s and 1980s, a new complexity of workplaces emerged with the development of a new generation of computers and telecommunications technologies that pushed large segments of the office population to distinctly suburban offices in and beyond the metropolitan region. Some of this decentralization has been encouraged by planning policies of the city of Toronto in an attempt to lessen the strain on road and transit infrastructure.

Other forces have reshaped downtown, beyond the free-market mechanisms of real estate and corporate decision-making. It is a popular interpretation to credit a major role to the reform movement on city council that came to power in 1973. Certainly city hall imposed limits on density, setback, and building configuration. When the proposed Spadina freeway was stopped in 1970, a critique of the rapid increase in the size and extent of the downtown office district spawned a planning package to stimulate mixed land-use developments and decentralized growth. The Central Area Plan and the city's Official Plan were outcomes of a decade of debate between private development interests and a pro-neighbourhood city council, and they shaped the context of subsequent growth. Some megaschemes were cancelled, while others such as the Eaton Centre were reduced in size but still built; many other building schemes were restated in less monumental form. The design for the new Sun Life complex on both sides of University Avenue at King Street exploits the plan by buying the density air-rights owned by the nearby historic St Andrew's Church – thus a heritage structure was saved and the texture of that end of King Street is more varied.

Preceding that phase, however, a longer concern by politicians and planners in the linkage between downtown growth and the wider city, and in the vitality of downtown and contiguous neighbourhoods, influenced other policies from city hall.[14] For example, immediate post-war concerns about the continued viability and accessibility of the downtown in an era of suburban growth produced a commitment to a mass-transit system. The north-south Yonge Street subway line opened in 1954; then the Bloor-Danforth east-west axis, beginning in 1963; and later the Spadina line, a revitalized streetcar fleet, and provincially supported GO Transit commuter trains. Parallel efforts by planners and downtown business interests to ensure adequate long- and short-term public parking led to the first parking authority in Canada in 1952.[15] Surface areas made available through the cut-and-cover operations of subway construction led to a bead of public lots that serviced neighbourhood merchants, and four major public garages in the core injected competition and quality into the operations of the largely private-sector parking industry.

Torontonians enjoy access to important public "fringes" of the city's vibrant office centre: to the north, the new City Hall of 1965; to the west, Roy Thomson Hall for symphonies and other concerts, a new Metrohall as the seat of metropolitan government (1992), the CBC Broadcast Centre, the

Metro Convention Centre, and the Skydome; to the south, the federally managed Harbourfront complex of cultural facilities and condominiums built beyond the Gardiner Expressway and bringing people back to the formerly industrial waterfront; and to the east, a renovated St Lawrence Market, the kernel of a revitalized mass of nearby housing and retail developments. In these latter projects, civic involvement in enhancing the attractions of locality are evident, all made easier by Toronto's dominant role as national finance centre and capital of the country's largest province.

Outposts of King and Bay

In 1969, demolition of buildings at the corners of Granville and Georgia streets began the development of Vancouver's Pacific Centre. The "black tower" of the Toronto-Dominion Bank and an almost windowless Eaton Centre were locally regarded as indicative of Toronto's insensitive impact on the core of the city. At the other end of the country, Stan Rogers's lament for what had happened to Halifax's Barrington Street, as it was transformed by "Upper Canadian concrete and glass, down to the water line,"[16] and the blunt intrusion of bank towers onto the St John's waterfront are other instances of the external forces that have reshaped local downtowns across Canada. Certainly since the 1960s and 1970s, mega projects involving office towers, shopping centres, and municipal forgiveness on taxes or parking fees imparted a new corporate quality to many Canadian cities. Interlocking directorates between banks and development companies (Royal Bank/Trizec in Vancouver, Calgary, and Montreal; Scotiabank/Trizec in Calgary, Winnipeg, and Saint John) created remarkably similar urban forms in once distinctly different cities.[17] It could be argued that the testing ground for such drastic renewal schemes was Montreal, with the giantism of Place Ville Marie in the 1960s, but invariably Toronto is seen as the culprit.

The enormous scale of some of these projects triggered local reform responses, often rallying around emotional concern for local landmarks. In Halifax, one group of citizens rallied to protect the views from the Citadel and Signal Hill that were fast disappearing with the new Barrington Street developments.[18] In Vancouver, the landmark Birks Building at Georgia and Granville was lost to the new Scotiabank development, but Christ Church Cathedral was saved. So too were Chinatown and the historic Gastown area; these were formerly slated for demolition associated with a waterfront freeway and an office complex called Project 200 that had been promoted from Montreal by the Canadian Pacific Railway.[19] The scheme was stopped, protests gave rise to Vancouver's 1970s reform council, and the areas were subsequently reinvigorated by a mixture of federal and local improvement schemes.

In these and other instances, revision of local by-laws to include bonuses for preserving designated buildings, as well as a stronger civic sense, protected and nurtured heritage structures in subsequent development schemes, but nonetheless a distinctly different scale of new urban landscapes has emerged. The projects of national development capital have been brokered into locally tolerable agreements, and the surrounding blocks have benefited from the significant input of public goods. Consequently, similar to the public elements of Toronto's core, stellar urban images have developed at the edges of these downtowns. In Vancouver, these ingredients include Canada Harbour Place and the hotel/convention centre complex built by the federal government during Expo 86, on the Burrard Inlet waterfront; Gastown and the CBC building/Queen Elizabeth Theatre (both federal), on the eastern edge; the provincial BC Place Stadium overlooking False Creek, to the south; and the provincial Law Courts/Robson Square, on the western fringe. In Halifax, a federal initiative that revitalized old merchants' wharfs as the Historic Properties provides a counterpoint to the concrete and glass of Barrington Street and the World Trade Centre north of the Parade. In Ottawa, the Rideau Centre (a hotel/retailing complex) followed rehabilitation of the historic Bytown Market, and development of conference and arts facilities along the Rideau Canal similarly records the mixture of forces at work in transforming the downtown cores of Canadian cities.

Space does not permit a more exhaustive survey of Canadian cities, but not all the changes can be seen as instances of metropolitan outreach, either from Toronto/Montreal or American/European/Asian sources. Regionally based capital has been important in the restructuring of downtowns. Prairie cities have received investments from the resource sector – local capitalists in Alberta and state-controlled crown corporations in Saskatchewan – and thus the head offices of credit unions, grain pools, provincial telephone and power utilities, and farm machinery suppliers became elements of this revitalization.[20]

These examples of recent transformations in Canadian downtowns reflect the consequences of a more planned and regulated development process. There has been thus far less freeway blight than that found in American cities, and a broader consensus on the social benefits of a publicly subsidized transit system has kept downtowns accessible. Rapid-transit systems in Toronto and Montreal, and then in Calgary, Edmonton, and Vancouver, were built on a base of extensive existing public transit. Parking authorities also attempted to keep the core viable for car-driving consumers and workers. In many centres, downtown retailing and cultural activities are part of the quilt of a varied landscape in a predominantly office region. As a place for business, shopping, leisure, and culture, the cores of Canadian cities have gained from the brokering of public and private interests. One testimony to that marriage has been that when Canadian venture capital moved south to develop real estate in Los Angeles, Minneapolis, and New York, for example, the "package" skilfully included a range of facilities in multi-use complexes.

MAIN STREET

For all the influence of metropolitan centres, much of the architectural, and even social, distinctiveness of Canadian regions can be found in its small towns, especially on their Main Streets.[21] The symbiotic relationship between people and place, architecture and social ritual that elevates modest centres to the status of "metropolis" could be found on Main Street. The locale is a backdrop to significant pieces of Canadian literature such as Stephen Leacock's Mariposa, Alice Munro's Jubilee, and Margaret Laurence's Manawaka (which include, it is argued, elements of Orillia, Wingham, and Neepawa, respectively).[22]

In smaller Canadian communities such as Chatham, New Brunswick, and Perth, Ontario, the distinctiveness of Main Street has been recently eroded by the development of fringe shopping strips and malls as well as by the marketing strategies of national bank branches and retail chains. All across the country, these communities have attempted, with government assistance, to revitalize their retail competitiveness. Such revitalization programs for Main Street provide another veneer of standardization, albeit one that preserves more regional nuances than the national and international styles of corporate or franchise businesses.

Chatham on a Saturday Afternoon

Anne Murray's "Broken Hearted Me" fills the fall air outside the Northumberland County Farmers' Market, held inside an old garage on Water Street in Chatham, New Brunswick. Running parallel to the Miramichi River, Water Street is Chatham's "Main Street." Three blocks to the west, the Ambassador Beverage Room awaits its evening customers. In between, where Cunard Street runs down the hill into Water Street, stands an impressive stone pile that used to be the Post Office. On its steps, three oldtimers sit and pass the time of day. All three wear Montreal Expos baseball caps, occasionally touched in response to the toot of a passing pick-up truck or to someone coming out of Joe's Corner Store across the road.

On this sunny Saturday afternoon in October, downtown Chatham is alive. The two flower shops are busy, with customers buying stems to decorate the town's churches and graves in the local cemetery. The skate-sharpening special at the hardware store is a reminder that winter is not too far away. The Downtown Beauty Salon, across from the Ambassador, does a steady business. Around the corner on Cunard Street, the fish and chips special at Ruby's competes with the "Chinese and Canadian cuisine" of the Cunard Restaurant. Several blocks up Cunard Street, past the town's movie house and more shops, the imposing silhouette of the old Catholic cathedral blocks the vista. Some two hundred cars are parked around the church, attending a monster yard-sale.

Chatham's "Main Street" seems a friendly place. The familiarity and human scale of the architecture are similar to that found in most Maritime downtowns. Many shops occupy the ground floor of detached, wooden, two-or-three-storey buildings, whose gable ends face the street; they look like houses and suggest a time when retailing shared the space with residential, wholesaling, and manufacturing uses. Those on Water Street seem to have a dual orientation – perhaps river frontage for a long-gone era of boat traffic – like similar buildings between road and river at Buctouche, Moncton, New Glasgow, and the dozens of other Water Streets in Atlantic Canada.[23]

If most of Chatham's downtown is wooden and dates from the lumber-trade era of the 1870s and 1880s, its few brick and stone buildings are the late Victorian jewels on the street. The Post Office also used to be the Customs Inland Revenue Building, a fact announced, along with its date – "V.R. 1900" – in fancy pediments at the roofline. Its importance is stated in the structure's stone façade, in the elaborate Romanesque arches around the doors, and in the ornate roofline. The structure is a reminder of the wider world, of the Dominion presence that regulated the timber trade and provided a variety of services for settlements up-river and along the coast. To the east of the old Post Office lies the Elkin Block, also built in 1902; it is only seven bays wide and three storeys high and would be commonplace on many an Ontario Main Street, but here it seems immense alongside the gable-end stores. The third noticeable departure from the wooden stores is the Ambassador Hotel. It is surprisingly ornate for a tavern, but long-time residents could tell you that this was the Bank of Montreal until 1963 and that there used to be a law firm upstairs. The brick cornice and other fine detailing on the upper floors do not match the small and curtained windows in the almost bricked-in ground floor. Prohibition may be long gone, but even in the early 1960s, when the tavern took over the bank building, its sentiments still dictated appropriately discreet public images for drinking.

The street has a strong unity. If one looks back down Water Street from the tavern, the brick, wood, gables, and cornices all blend together along a narrow, slightly curving street. The warp and weft of the street are its signs, jutting out like commercial laundry in whatever 1930s, 1950s, or 1970s fashions were current when the enterprise changed hands or the owner decided to invest in the neon or plastic of the times. Without consciously trying to do so, Chatham presents a strongly historical streetscape. And yet it is also an old, and somewhat tired, streetscape. Creaghan's Ladies Clothing, long a fixture in the town and "a dependable place to shop," closed in the late 1980s. The Post Office stands half-empty, abandoned when its new replacement was built on the edge of downtown. Part of the building now serves as a regional office for New Brunswick Agriculture, a branch of government that seems a little marginal in this part of the province. The only post-war buildings in the downtown house the two banks. The Bank of

Montreal is directly across the street from its old home at the Ambassador, but now in a steel and glass box, as is its competitor, Scotiabank. On the Elkin Block, the long plate-glass storefront is bare, and has been since 1981 when Schrier's Beauty Centre closed. Only the ghostly imprint of the name hints at what went before. Three other storefronts in a four-block stretch are vacant.

A glance down Water Street gives a hint why the stores are empty. High above the river, Centennial Bridge, a part of New Brunswick Highway 11, quickly channels modern road traffic north and south past Chatham. Adjacent to the highway, about 1.5 km south of the river, is a new shopping mall where Sobey's supermarket and Zellers department store are the anchor tenants. In their parking lot squats a 6-metre-high inflatable "Munchie," the promotional toy of Hostess Potato Chips, which would hardly fit between the buildings on Water Street! Further along the river, the stacks of the paper mills are quiet and smokeless, a victim of recession and New Brunswick's notoriously vulnerable forest industry. When the Canadian Forces Base (CFB) Chatham, about 6 km to the south, was cut back, a few more shops closed as well.

Will things get better? Will the good days come back? Will anyone invest in the downtown? For most Maritimers, these are not new questions.[24] There are grounds, in Chatham at least, for an optimistic answer. Many of the attractions are visual, especially the sheer delight of Victorian trim highlighted in different colours against the main façade. Other attractions are human ones, as storekeepers and customers "visit" while they finish their transactions or exchange community news in the line-up at the cash-register. There is a sense of continuity in the dealings on Water Street. And an optimism that downtown has a future. Plans are afoot to redevelop the land where the Department of Pay Services had been slated, including as well the former Creaghan's department store building, for a retail/office/library complex. Land behind Water Street along the river has been cleared and renovated for parking and "people spaces"; Chatham Business Development Corp. is trying to develop a shipbuilding museum for downtown; and a downtown logo incorporating stylized sails signals a desire to rekindle the connection with the town's river heritage.[25] Also in 1988, the Miramichi Pulp Mill at Newcastle announced a $480-million expansion program. Perhaps world markets for the region's mining and other primary products will increase. And perhaps the riding would be a critical seat for provincial or federal majorities, and more initiatives come to the region! Cunard Street and Water Street are waiting for them.

Revitalization of Main Street

All across Canada, there are towns such as Chatham. Their Main Streets have some delightful old buildings, and their stores present a wide range of goods and services. Some of them, through initiatives by merchant organizations and civic improvements, signal that they know who they are and how to compete in today's business climate. Far more are in real trouble. They have a declining economic base and an ageing set of merchants, and they face aggressive competition from new shopping strips and malls on the edge of town. They are in danger of becoming a marginal piece in the fabric of Canadian communities. If they did, it would be more than just an economic event. A considerable portion of Canadian identity is to be found in small-town society – the middle ground between metropolitan and rural ways.

Yet region by region, change has come. The decline of the prairie hamlet is perhaps the most noticeable. Historically these communities have been defined as two grain elevators near the train station, and then an agricultural implement outlet, a hardware store, a Chinese restaurant, and a rural post office, all in a two-block strip parallel to or perpendicular to the tracks.[26] But in a changing transportation system where the rail line is hardly used and it is cheaper to truck grain to the new inland terminal, the raison d'être of these low-order centres is disappearing. In the interior of British Columbia or the fishing/forestry/mining towns of Atlantic Canada and the Canadian Shield of Ontario and Quebec, similar economic underpinnings erode the community. By contrast, retail strips and the malls all across the country offer convenience, away from the weather, but rarely a sense of locality. Typically they present a bland, undecorated façade, seen distantly across a half-full parking lot set back from the road. Inside, the stores are predictably the same whether in Brandon, Manitoba; Cornwall, Ontario; or New Minas, Nova Scotia.

Is there still hope for Main Street, or has it become simply a device of the mall, to be invoked on special sidewalk sale days? Perhaps there is stasis developing in the retail market as urban places adjust to the era of the automobile. Suburban malls and strips have 40 per cent of all the retail trade in the country. The decline of Main Street has been inexorable. At the turn of the century, catalogue outlets began to bring urban goods to rural Canada. Then the metropolitan influences began to invade Main Street with incipient supermarkets, first in the form of Eaton's groceterias in the 1920s and then with wholesaler-organized chains. Banks have always been a symbol of outside interests, and their latest rounds of revitalization schemes – glass and concrete boxes designed from head office – underlined the gulf between the modern and the once-modern. Automobile traffic, especially since the 1950s, has encouraged merchants to transform their upper-storey façades into giant signboards for speedier passers-by, and then to knock down adjacent buildings for them to park on. Main Street was declining from within, as well as not meeting the threat from the fringe strip.

Programs to highlight and reinvigorate Main Street's variety, visually and economically, have recently attempted to stabilize the decline – another il-

lustration of the willingness of governments at all levels to intervene for social and welfare reasons. One example is certainly that of the Main Street
program of the Heritage Canada Foundation; parallel initiatives by both provincial governments and private foundations have tried to make a difference
in the look and livelihood of small-town communities. The foundation
funded demonstration projects in the 1980s in Nelson, British Columbia;
Fort Macleod, Alberta; Moose Jaw, Saskatchewan; Cambridge, Ontario;
Perth, Ontario; Windsor, Nova Scotia, and Bridgetown, Nova Scotia.[27] Co-
ordinators retained by the foundation worked as facilitators, helping business associations, individual merchants, and property owners address the
problems facing their downtowns. They also gave design advice, helped research economic profiles of consumers and hinterlands, and helped local
groups and individuals seek available provincial and federal funds for
downtown-related developments.[28]

Perth is a good example. Once a sleepy town in eastern Ontario, suffering
from mill-closure, a declining agricultural hinterland, and regional out-
migration, it is now almost within commuter reach of Ottawa. Tourism has
increased along the Rideau Canal corridor, and Parks Canada has recently
revitalized the turning basin on the canal in the town for tourist traffic. Heritage Canada began slowly. It managed to involve a local merchant in a sensitive repainting of his storefront. A window-dressing workshop made
merchants realize how better to market their goods. Bit by bit the local effort
snowballed, energized and encouraged by Heritage Canada's expertise. The
Bank of Montreal was persuaded to renovate its old stone bank, strip away
the modernization of the 1950s, so that it is now the bank that saves more
than money. On one block, an old department store had left a two-storey
signboard that obliterated the second- and third-floor windows of a magnifi-
cent stone structure built in the nineteenth century. The old signage was
taken down, and upper-floor use was revitalized for apartments that now had
sunlight. The old Carnegie Library was transformed into offices. The same
story goes for Nelson, Moose Jaw, and Windsor (Nova Scotia); all have harnessed an interest in the past, tourism, and a local desire to preserve an integral component of community.

Ironically much of this heritage intervention has made Main Street more
and more the same, everywhere in the country! Consultants bring the same
brickwork to the sidewalks, and the same planters, garbage receptacles, or
benches. Sometimes there is the same lettering on storefronts. Main Street
has come to look more like the mall, at the same time as the mall has come
to look like Main Street, with its "sidewalk days" promotions. These transformations parallel the forces that brought the same five bank towers to
every metropolitan centre. The forces of change are not just retail practices
in the era of the car. More standardized retailing practices, national fran-
chises, and mass media coverage have contributed to rural and small-town

Canada becoming more emphatically incorporated into the urban fields of metropolitan regions. There are, to be sure, landscape variants from region to region, but the Main Street/strip/mall of the 1980s is the latest version of a marketing system long at work.

RESIDENTIAL NEIGHBOURHOODS IN LARGE CITIES

Residential landscapes have also been recast by public as well as private forces. Federal funds for neighbourhood revitalization and private capital alike have refurbished many residential districts of Canadian cities. Other schemes, such as the St Lawrence neighbourhood in Toronto and False Creek in Vancouver, have created new residential areas in older industrial districts. Yet these are but two instances of the welfare state's imagination and creativity in the design of public spaces and in its policies of encouraging mixed social and market housing which have subsequently attracted private capital to formerly derelict areas. Municipal initiatives in social planning and commitment to public transit systems have also helped retain vitality. More broadly, the federal Canada Mortgage and Housing Corp. (CMHC) has, since its inception in 1947, made notable efforts to improve single-family neighbourhood settings and to humanize high-rise environments across the country. The parallel role of large corporate developers in creating the new "integrated" residential sphere is noteworthy. Planned developments were a component of elite suburbs earlier in the century, but Don Mills in Toronto can be seen as a role model for other post-war corporate residential developments. The latter were designed for a middle-class clientele, and the "package" included public amenity and variety, plus local jobs and services, rather than mere tract housing.

Vancouver: Life-style and Landscape in Kitsilano

Six o'clock in the early evening. A hot day in July. A young couple walk arm in arm back from Kitsilano Beach up Larch Street. They pass renovated houses, with diagonally applied cedar-cladding, cedar-chip/bamboo front yards, and perhaps a Japanese lantern in the corner that can be dated from certain years of *Sunset* or *Western Living* magazines. Then they walk past a Greek grandmother, dressed all in black, tending plants in the front garden of a white-stuccoed house (the back of which is constantly being renovated). Pride of place in the garden is given to a fountain where water trickles into the decorative pool from the cornucopia held by a cherub. Between the cedar- and the stucco-sided houses is an example of what all these houses looked like before the professionals and the Greeks modified them for their own images. It is a one-and-a-half-storey bungalow, with brick piers hold-ing up the front porch.[29] An old Anglo-Vancouverite sits on the porch, look-ing across his privet hedge. He struggles to comprehend the scene about him. The couple walking by are dressed in a way he would hardly see in his own bedroom; for the past six years, he has only been able to smile and nod, but never talk, to the vegetable gardener next door; and as for the cedar-clad house on the other side, what he has seen through the uncurtained side win-dows is greeted with disbelief by his buddies down at the Legion Hall. The notion of family as he knew it seems to have disappeared completely. Loom-ing in the background, a high-rise condominium, another intrusion; why couldn't they have stayed in the West End with all the other high-rises, and left his neighbourhood the way it used to be?

The couple walking up Larch Street are oblivious to all this, happy to have gotten sand in their shoes in the time after work. They couldn't have done that as easily in Toronto,[30] or Edmonton, where they both had worked for a few months. They are thinking more about where to go for supper: the converted Spanish Mission–style gas station, now a California-Mexican eat-ery, or the bar/fish restaurants on Granville Island, or up to a Greek restau-rant on Broadway? All are within walking distance. They choose Granville Island, so they can stay close to the water. Until a few years ago, someone told them in a bar the other night, False Creek had been bounded by saw-mills and ironworks, and the housing on the Fairview Slopes to the south was a mixture of rooming houses, communes, and tenements. Granville Is-land had been a set of corrugated warehouses and industrial plants. Now, following redevelopment by the city (and on Granville Island by CMHC), a seawall guided them past clusters of townhouses, past open park space, and let them fantasize about one day owning a yacht as nice as those they saw in the marinas on the Creek.

Kitsilano is old suburbia, now gentrified. It has been transformed by a few high-rises (before they were zoned out in 1974) and many three-storey apart-ments (most of them self-owned), and new townhomes, as well as by the upgrading of single-family houses. The new life-styles have also brought irrevocable changes in density and social composition. Some of these forces are clearly in the private sector, both individual and more corporate, but the shifts in Kitsilano, and in many other similar neighbourhoods, have been in-fluenced by planning guidelines and public money. By 1980, Kitsilano had received $8 million in government housing-rehabilitation funds.[31]

On the south side of False Creek (just east of Kitsilano), a mix of federal, provincial, and municipal money has helped to develop a distinctive, indeed remarkable, medium-density neighbourhood. In turn this injection of funds assisted in creating a positive climate for private-sector investment on the Fairview slopes to the south of the Creek.[32] On the north side of the Creek,

BC Place similarly was a transformation permitted by public agencies which cleared out old industrial uses, initially for Expo 86,[33] but to be developed later by the private sector as a distinctive residential enclave. Here too public initiatives have created enormous benefits for the community.[34] And in Kitsilano, revitalization of Broadway, the neighbourhood's local Main Street, with civic and merchant funds, together with cultivation of a Greek ethnic signature especially in restaurants, echoes the duality of revitalization efforts.

Clearly these changes include influences from state intervention: policies of multiculturalism that encourage retention of immigrants' identity rather than assimilation; policies of rehabilitation and property maintenance that arrest further decay; policies of zoning that have tried to balance density options; and massive public initiatives in the development of both sides of the False Creek basin.

Residential Density Options

Kitsilano's range of housing options is quite varied. It includes architect-designed large houses, contractor-built bungalows, and sweat-equity pioneer cottages that were the suburbia of their day early in the century; infill and replacement two storey apartments built in the 1940s and 1950s; three-storey-and-balcony apartments and tall high-rise towers put up in the 1960s and 1970s; and a series of co-ops, town-houses, and gentrified houses dating from the 1970s and 1980s. Such variety in close proximity is typical of many neighbourhoods in Canadian cities, and the juxtapositions reflect a diversity of builders, overlapping edges of planning zones, turfs of ratepayer groups, and transit options, as well as historical inertia.

Canadian neighbourhoods have changed in appearance as the costs of land and construction, the limits of zoning, and shifts in consumer demand have generated different forms and densities. For all the recent reinvestment in inner-city residential neighbourhoods, the dominant residential option since 1945 has been suburbia. Earlier, streetcar suburbs for both middle- and working-class communities had developed in piecemeal fashion and with a modest building scale. Production of the post-war suburb has been dramatically different, more corporate in organization and massive in scale. The prototype of the new scale can be seen in Toronto's Don Mills, built since the 1950s, where a single developer followed through from land acquisition to developing roads and then construction of a mix of housing types in cluster developments, with shopping centres, schools, and churches; separate districts were allocated for industry and offices.[35] This "packaged community, a backlands utopia away from the city,"[36] became a model for other instances of corporate land accumulation and suburb-making such as Erin

Mills and Bramalea in the Toronto region and for smaller mixed-housing schemes such as the CPR's Arbutus Village and Langara Gardens in Vancouver.

Most suburbs were not planned as well and with as much distinctiveness. Indeed, for the outer reaches of Metropolitan Toronto, the endless suburbia in Scarborough gave rise to the derisive label "Scarberia," with rows and rows of almost identical houses, with two-car garage snouts taking up most of the front yard, identical neo-traditional façade trim, and often no sidewalks.[37] The corporatization of urban landscapes has been as dramatic in suburbia as it has been in the downtown skyline, with the interchangeable set of contrived community names (Pheasant Run, Sandyford Place, Sherwood Forest, and so on) for almost identical tract houses and town-house clusters that back on to arterial roads.[38] Yet the presence of so many mixes of single-family homes, town-houses, and high-rise apartments in the suburbs – indeed a distinctive Canadian metropolitan landscape by the 1980s – suggests a strong element of planning.

Although the corporate sector is responsible for the look of many suburbs, an important sequence of public interventions has given a "subconscious" plan to these locales. As a consequence of federal incentives in the immediate post-war years to encourage creation of a house-building industry, the Central Mortgage and Housing Corp. (CMHC – later Canada Mortgage and Housing Corp.), with various amendments to the National Housing Act (NHA), developed guidelines for the approval of site plans and layout of subdivisions and even encouraged improved house design.[39] Although urban planning has been a provincial jurisdiction in Canada, the work of CMHC and the Canadian Design Council over the years has been an influential national force. Furthermore, many of the suburban high-rise nodes are a product of region-wide planning strategies and metropolitan-scale decentralization schemes.

Although Kitsilano had few instances of slum housing – certainly not enough to prompt large-scale renewal schemes – other neighbourhoods in Vancouver and other Canadian cities were renewed soon after the end of the war. In Toronto, Cabbagetown had long been the city's largest working-class slum[40] and had deteriorated during the Depression to the point of being targeted for wholesale clearance. In what was a pioneer initiative in municipal public housing, the city of Toronto created a Housing Authority in 1947 that oversaw the demolition of six blocks south and west of Parliament and Gerrard. Regent Park, the three-storey brick apartment buildings that replaced them, set amid ample open space, with cul-de-sacs and playgrounds, was hailed as innovative housing – indeed, a model for CMHC initiatives.[41] Subsequent phases of Regent Park (South) used 15-storey apartment buildings amid four-storey rows in a scheme less generous in architectural design and density, and the area has gradually deteriorated as repair funds shrink, occupants change, and housing lists get longer. Schemes such as Jeanne Mance Park in Montreal and the Strathcona renewal in Vancouver represented similar attempts to provide low-income housing with a modicum of design.

A parallel renewal effort was far larger and more emphatically in private hands, replacing houses with tall apartment blocks. Vancouver's West End, a former elite neighbourhood with large houses on large lots, was transformed into a forest of high-rise towers in the 1960s and 1970s.[42] Developers took advantage of by-law bonuses available through the provision of balconies and ground-level amenities to create one of the densest districts in Canada and helped produce a neighbourhood ambiance that triggered emphatic opposition to high-rises in most other neighbourhoods. In Toronto, to the north of Cabbagetown, the St James Town complex of some fifteen apartment towers was hailed in the mid-1960s as a city within a city, a private developer's option to urban renewal, built on about 13 hectares that had held decaying Victorian houses. Elsewhere in the city, such as the Quebec/Gothic avenues area, such high-rises intruded on more established middle-class neighbourhoods, triggered strong community opposition, and led anti-developer reform candidates to run for council. Similar responses can be chronicled for Halifax and Montreal.

The conflict over both public and private renewal schemes around the issue of density and type of occupants led to a phase of infill housing that was more sensitive to local scale. Complexes of co-op houses in town-house and row-house arrangements, filled with socially mixed (income-mixed) groups or targeted to the elderly, women, or ethnic groups, were built often on old industrial sites where there was minimal sense of (and disruption of) existing community. In the case of Vancouver's False Creek south side or Toronto's St Lawrence neighbourhood, these housing schemes have been part of a broader public contribution to the public city, through development of retailing and entertainment complexes that attract a city-wide clientele.[43]

If socially mixed infill housing was one response to the excesses of the bulldozer renewal projects, another response was careful renovation of remaining housing. In Toronto's Cabbagetown, the Victorian stock left after the Regent Park and St James Town schemes had been completed has undergone remarkable transformation. Some 80 per cent of the housing stock of Don Vale has been renovated in the last two decades, and with it has come social upgrading into a solidly middle-class enclave. In Toronto, gentrification was called "whitepainting," a label that started in the 1960s when houses were literally painted white; it now connotes a house sandblasted to its original Victorian polychromatic brick, with wooden details on porches, doors, and gables uncovered (or reproductions added) and the interiors significantly upgraded.[44] As with the Vancouver renovators, there is often sheer invention of the past, fabricating a neo-Victorian chic. The process is found in Toronto's Cabbagetown, Don Vale, the Annex, and parts of the west end and is a process largely orchestrated by real estate interests, although many individual new homeowners have also transformed their own houses. Here, too, although the agencies of change seem to be individual or corporate, the public umbrella is evident. Many Canadian cities now have heritage lists that protect or at least recognize older buildings and distinctive neighbourhoods. This embrace of heritage is usually associated with the reform-oriented city councils of the early 1970s, and there has often been a political alliance between councils and the new professional class that has occupied these areas.

It is not just the "new class" that has modernized the inner city. The case of the Greeks in Vancouver's Kitsilano can be seen repeated in Toronto by Italian, Portuguese, and Chinese immigrant groups which have transformed many streets to the west of downtown, most noticeably west of Kensington Market. Although once scheduled for clearance schemes, these houses now persist as renovated stock, modernized and refitted with bright exterior colours, yard art, cantinas, new façades, and new interiors that accord with their owners' cultural roots. The increasingly visible evidence of immigrant streams in Canadian neighbourhoods (once constrained to specialty retailing and places of worship) is testimony perhaps to significant initiatives in Canadian cultural policy; programs that encourage multiculturalism have since the 1970s given strong voice to heritage issues, some of which find manifestation in confident cultural signatures in domestic space, a confidence that increases with numbers.

CONCLUSIONS

If many Canadian tastes and activities are moulded or managed from Toronto, Toronto is itself a small player among world cities and indeed is a conduit and pale mirror of many things in New York and elsewhere. Toronto was plugged in by freeway to New York state before it was connected to Montreal, it participated in Great Lakes and manufacturing-belt circuits of management and design ideas, and it has been the recipient of the bulk of American branch-plant head-office functions.

Nonetheless it might be argued that there is a Toronto model of the public good, of brokering public and private interests in shaping the city.[45] Toronto has been, perhaps without it or the country realizing what was going on, a testing ground for projects elsewhere – Eaton Centre–type complexes now in cities the size of Guelph as well as Vancouver, design ideas for low-income public-housing schemes adopted by the federal government, and suburban landscapes that offer a social and land-use mix. The metropolitan dominance thesis should not, however, be taken too far. Ottawa is a very important place, followed by the respective provincial capitals. Federal and provincial agencies have long been a factor in shaping the Canadian urban environ-ment, even if they have only quite recently become forceful in reminding citizens of their patronage. Local governments similarly have helped broker public input, a local ameliorating effect, for the often extra-regional devel-opment capital that has set the pace for local landscape changes. Further, and perhaps ultimately, that localism is underpinned by notable physical and geographical contexts – mountain, sea, river, prairie, Shield (and climate) – that will always make Canadian urban places Canadian.

CHAPTER THREE

Measuring the Social Ecology of Cities

W.K.D. DAVIES AND R.A.
MURDIE

ORIGINS OF URBAN SOCIAL DESCRIPTION

In 1911 J.S. Woodsworth, Methodist minister, social reformer, and future CCF leader, published his seminal work, *My Neighbour*.[1] The book was an impassioned exposé of the social contrasts and the emerging squalor found in rapidly growing Canadian cities. Designed to alert influential members of society to the acute problems of homelessness, family breakdown, and social deprivation, it also contained a plan for reform. Heavily influenced by his religious convictions and social conscience, Woodsworth hammered home the theme that "the welfare of one is the concern of all," exhorting all urban citizens to improve their understanding of the social character, composition and life-style of people and areas around them.

Woodsworth's study stimulated popular interest in the social differentiation of Canadian cities. As such it was a Canadian echo – perhaps a muted and belated one in a world context – of rising concern about the urban consequences of Western industrialization. Such criticisms were not simply value judgments. From the mid-nineteenth century on, novelists sought to describe and interpret those conditions for an often sceptical and uninterested ruling elite. More explicit concern with illuminating and measuring the degree of social squalor motivated social commentators of the day, such as Mayhew (1862) in London, or the later Canadian examples, Ames (1897) in Montreal and Woodsworth (1911) in Winnipeg.[2] In time, the verbal descriptions of such commentators led to focused and systematic surveys of social conditions. This approach is best epitomized in the seventeen-volume work of Charles Booth (1902–3) who also provided descriptive commentaries on areas and trades and mapped – by individual street – the extent of poverty in London.[3] In addition he had previously proposed an index of social condition, based on six variables, which arranged each of the areas of London on a single scale of relative prosperity or status.[4] Booth designed a statistical instrument to measure what he called the new "terrae incognitae" of societies, for he believed that few urban residents understood the extent of poverty and the social variations in their home areas. Modern techniques applied to Canadian cities have similar objectives to Booth's index but include more input variables and more refined methods.

Contrasting with this gradual development of a quantitative, as opposed to qualitative and verbal description of areas, was the work of emerging social theorists such as Karl Marx.[5] He saw little point in wasting effort on empirical studies; the societal conditions causing these obvious problems needed to be changed, not studied. For Marx the salvation of society and cities could not be linked to the amelioration of conditions through gradual social reform, or through piety, as some penance toward a better life in another world; it could be achieved on earth only through the state acting as the agent for the people. Prescription through political action, not description, was the necessary approach; social reformers were dismissed as ineffectual dilettantes. Social reformers in Canada, such as Woodsworth, did follow the political route, but their Christian concern for the individual led them to eschew the dominant and exclusive role of the state as effective agent.

By the first decades of the twentieth century all these approaches were practised by different investigators, and such variety still exists in the contemporary literature. Unfortunately, most geographers ignored these origins until the 1960s, preferring to base their comparative studies of cities on a very different approach, the human or urban ecology school established at the University of Chicago in the 1920s.[6] Although the origin of the approach lies in a similar moral indignation about the social conditions of the new American cities, its principal advocates searched for explanatory principles borrowed from plant ecology to account for the great social contrasts in cities – but unlike the earlier theorists they did not emphasize the societal imperatives.

Perhaps the most famous example was E.W. Burgess's 1925 model of urban social areas, which identified a concentric sequence of social and land-use zones, arranged around the central business district.[7] By the late 1930s and early 1940s criticism of the utility of the concentric model had led other investigators to propose rather different spatial arrangements. Hoyt[8] argued that sectors were a more appropriate spatial generalization for housing and land-use changes, while Harris and Ullman[9] proposed a multiple-nuclei land-use pattern formed by the aggregation of uses around a series of separate, distinctive nodes. A rather different method was used by the sociologist Walter Firey.[10] In his study of Boston he developed a value-based approach which provided an important counterweight to decades of ecological interpretation emphasizing social or land-use distributions. In the latter context, J. Wreford Watson's study of Hamilton provided a pioneering Canadian example, a description of the social character of various areas using information from welfare agency case-loads to show the "remarkable concentration of unemployment, neglect, desertion and delinquency in the city's shatter zones"[11] – the area described by Burgess as the "zone in transition." Despite the usefulness of the many individual case studies of cities, rigorous measurements and tests of the utility of these alternative spatial patterns and their social relationships were rare. Mechanisms for change were assumed from the patterns, rather than being shown conclusively. The empirical emphasis also led to a marked lack of theoretical inquiry and a surprising lack of statistical sophistication, except in the study of ethnic variation, where segregation indices were used to measure the degree of separation of ethnic and occupational groups.

Reintegration came with the work of two sociologists, E. Shevky and W. Bell.[12] Instead of adopting the single-variable mapping of most ecologists, or combining variables into a single index of social conditions in the manner of Booth,[13] they argued for three basic sources of social variation in cities: social rank (economic status), urbanization (family status), and ethnicity. Each of these sources or dimensions of variation was derived from a broader theory of social change and measured by a series of census-based variables. Area scores on the first two dimensions were subdivided into four types that produced (when cross-classified) sixteen social area types. For the first time theory was linked to precise measurement in a social and spatial context.

MULTIVARIATE APPROACHES TO ECOLOGICAL DESCRIPTION

In retrospect, social-area analysis promised more than it delivered, as many critics[14] have shown. In particular, use of only three dimensions was found an inadequate description of the social mosaic of cities. Some problems were solved by the next phase of inquiry, factorial ecology,[15] which applied the multivariate procedures of factor analysis to a set of census variables for city sub-areas. This led to the use of bigger data sets, the derivation of dimensions from data rather than the imposition of them, testing of hypotheses as well as data exploration, and measurement and identification of both social dimensions and spatial patterns within one analysis.

Although many individual factorial studies of Canadian cities were undertaken in the late 1960s and the 1970s, many were repetitive, with few conceptual advances. None used comparable data sets. Nevertheless, the studies confirmed that the Shevky-Bell axes of differentiation represented initially useful, but not complete, summaries of the social variation of Canadian cities.[16] Two of them, Murdie's for Toronto, based on 1961 data, and Foggin and Polèse's for Montreal in 1971, are typical and used approximately the same number and type of variables. Both studies generated two economic-status dimensions. The first contrasted areas characterized by professional and managerial employees, high-income earners, and high levels of educational achievement with areas occupied primarily by manufacturing workers with much more modest incomes and educational attainment. This is the economic-status axis hypothesized by Shevky and Bell. In Montreal, this factor was further complicated by a strong relationship between economic status and ethnicity, with high-status areas occupied primarily by persons of British ancestry and low-status areas by those of French origin. The second economic-status dimension contrasted areas occupied by middle-class clerical workers and those containing low-income service employees, often living in housing that lacked basic amenities. Together, these two factors suggest that the social space of Toronto and Montreal can be differentiated by a hierarchy of four classes, a somewhat more complex picture than hypothesized by Shevky and Bell. Two family-status factors were also identified in each study; one contrasted areas occupied by family and non-family households, while the other distinguished areas on the basis of family type and residential turnover – younger families/high mobility and older families/relative stability. Two ethnic factors also emerged in both analyses, summarizing contrasts between the dominant ethnic groups, French in Montreal and British in Toronto, and two significant minority groups, British and Italian in Montreal and Italian and Jewish in Toronto. In Toronto, economic variables were also related to British/Italian contrasts, thereby mirroring British/French contrasts from the first economic-status factor in Montreal. Overall, the factor structures from these analyses were similar, suggesting that, aside from obvious differences in ethnic composition, there is strong structural similarity between Canada's two largest metropolitan centres.

Through time, factorial studies produced additional important findings for the study of urban social differentiation. First, the presence of additional axes was reported in many studies, illustrating that social-area analysis had only partially measured the social complexity of urban areas.[17] Second, they also resolved one of the dilemmas facing urban ecologists. Which spatial pattern was more representative in modern cities: the concentric zones of Burgess, the sectors of Hoyt, or the multiple nuclei of Harris and Ullman? Murdie's factorial study of Toronto demonstrated that all three spatial patterns could be identified – they were complementary, not competitive descriptions. He showed that each was linked to a particular dimension of variation in 1961: family status with concentric zones, economic status with sectors, and various ethnic clusters with segregated nuclei.[18] Third, the approach has been used to measure the variation in social areas in all cities in a country, rather than simply providing studies of individual or at most a few cities. The rest of this discussion illustrates the utility of the factorial approach in describing the social differentiation and selected spatial patterns of all Canadian metropolitan areas. Space constraints prevent discussion here of other major benefits, such as the integration of factor scores via cluster analysis to produce homogeneous social regions.[19]

As with all new methods, this multivariate approach to urban social differentiation has limits, many of them discussed elsewhere.[20] But two general issues do deserve some comment. First, the approach provides little explanation of why the social mosaic exists or changes. It makes no claims to do so – it is a method of pattern recognition and measurement, for identifying both social dimensions and their spatial patterns. There is nothing, however, to prevent its being used to contribute to or test theoretical explanations, leading to what Davies[21] anticipated to be a multivariate-structural phase of investigation. Second, the approach does not deal with the social character of areas as part of a lived-in experience, which demands very different philosophies and methods. Development of such concepts as territorial attachment, defensible space, and various forms of urban perceptions extends the concepts of sentiment and symbolism proposed by Firey,[22] and these are explicitly addressed in subsequent chapters (5, 7, 10, and 11).

AXES AND PATTERNS OF DIFFERENTIATION IN CANADIAN METROPOLITAN AREAS

Recent literature has shown new trends in Canadian cities, such as gentrification, increased long-term poverty, the emergence of empty-nesters, new concentrations of recently arrived ethnic groups and older people, greater numbers of childless families, and the presence of larger, more socially complex and multi-centred cities. Factorial studies that have adopted a wide variable set have also revealed these new axes of differentiation. It has been suggested[23] that these axes may be considered products of an emerging post-industrial society – the homogeneity of the traditional Shevky-Bell–type axes has been split by the growing social complexity of cities, especially with increasing differentiation of family-related groups. Moreover, these additional axes often exhibit particular spatial patterns – at least in metropolitan areas below a population of one million that are still dominated by one major downtown – and each pattern can be linked to a variety of spatial explanations of intra-urban differentiation. Most larger metropolitan areas have incorporated once separate satellite cities, of different origins and social character, producing multi-nucleated social mosaics not easily explained by a single pattern.

In an attempt to summarize the contemporary differentiation of social areas in Canadian cities, we carried out a new series of factor analyses of a 35 variable x 2981 area matrix of data. Its technical procedures are described elsewhere.[24] The variables were derived exclusively from the 1981 census for all census tracts (average population size 4,576)[25] in the twenty-four census metropolitan areas (CMAS). Incorporation of all the census tracts of the CMAS into a single data set permitted a joint analysis.[26] This ensured comparable measurement of the factors or social dimensions, which were derived from all the centres. The factor scores on every dimension for each census tract formed a series of common scales for census tracts in the CMAS on each of the factors. (Most previous factor results were city-specific because the dimensions and scores were individually scaled for each centre). The variables were representative of fourteen hypothesized sources of variation. Ideally it would also have been useful to have an indicator identifying public or social housing, but the census does not provide information on this source of differentiation. Only one ethnic axis was proposed in this study: since many ethnic types are concentrated in particular cities, only region- or city-specific axes would probably be identified. It seemed more logical to take for granted the need to provide supplementary evidence on the differentiation of ethnic groups, through ethnic segregation indices. Hence in this study one general ethnic axis was hypothesized, measuring the degree of deviation from the major ethnic type, those of British extraction – with 40.2 per cent of the population. The hypothesized axis was indexed by three variables: those of French extraction (26.7 per cent); immigrants – people who were born outside the country (16.1 per cent); and finally, all other groups whose ethnic origin was neither French nor British (33.1 per cent).

A nine-axis solution using the principal axes method, followed by oblique rotation of the axes, and accounting for 85.8 per cent of the original variance, was selected for interpretation.[27] This means that the variation in the thirty-five variables was reduced to nine axes with the loss of only just over 14 per cent of total variability. The loadings for each social dimension or factor axis, which identify how much of the variability of each variable is associated with the dimension, are shown in Table 3.1, together with a short title chosen to reflect the general character of the axis. Constraints of space mean that only some of the factor results can be presented here to illustrate the patterns.

As hypothesized, two dimensions of social rank emerged from the analysis. Table 3.1 shows that the first of these is economic status, with high loadings on the occupation, education, and income variables, providing the familiar contrast between areas occupied by high income, managerial workers and lower-income, blue-collar employees. Use of a joint analysis means that the relative importance of each census tract in all CMAS on each separate source of variation is provided by the factor scores. Table 3.2 shows that on average 11.2 per cent (5.1 per cent) of the CMA tracts had positive, or high-status scores above 1.0 (above 2.0). The average score would be 0.0 for all

Table 3.1
Factor titles and loadings for metropolitan-area analysis, 1981

Economic status		Impoverishment	
University educated	92	Female-parent families	99
Males in managerial occupations	86	Low-income families	72
High income families (> $40,000 p.a.)	78	Dwellings rented	52
Low education (< grade 9)	−60	Local movers (within census area)	45
Males in industry, construction	−93	Apartment dwellers	42
		Male unemployment rate	38
Family and age			
FAMILIES		*Early and late family*	
Persons per family ratio	75	Young children (under 6 yrs)	86
Children (0–14 yrs)	62	Adults (25–34 yrs)	79
Adults (25–34 yrs)	32*	Movers (in last 5 yrs)	53
Middle-aged (55–64 yrs)	−67	New housing (under 10 yrs old)	36*
Old age (> 65 yrs)	−82	Childless families	34*
Completed families (children not home)	−93	High income	−30*
		Singles (adults never married)	−32*
NON-FAMILY		Middle-aged (55–64 yrs)	−42*
Non-family/family ratio	87	Young adult (18–24 yrs)	−81
Singles (adults never married)	76	Early middle age (45–54 yrs)	−90
Divorced/married adults ratio	68		
Childless families	53	*Young adult*	
Male unemployment rate	36*	Adults (18–24 yrs)	−76
Apartment dwellers	30*	Females in labour force	−59
Children (0–14 yrs)	−42*	Childless families	−38*
		Persons per family ratio	−30*
Housing			
Poor housing (repairs needed)	76	*Migrant*	
Median length of occupancy	68	Distant migrants (out of province)	−73
New housing (under 10 yrs old)	−59	Local mover (within census area)	44
		French ethnic origin	37*
Ethnicity		Apartment dwellers	32*
Immigrants (born outside Canada)	95		
Non-British and -French ethnic origin	95		
Females in labour force	40*		
Male unemployment rate	−35*		
French ethnic origin	−73		

Note: All indicators are percentage values unless noted as a ratio, rate, or median. All loadings have had their decimal points removed: +0.99 is shown as 99. Variables with second- and third-ranking loadings on an axis are differentiated by asterisks.
Source: Davies and Murdie (1991).

the tracts together. The political and economic capitals of the country, Ottawa and Toronto, respectively, have proportionally more than double the tracts that could be expected in this high-status category, with most of the other larger centres, such as Montreal, and the provincial capitals also having more high-status tracts than average. The exceptions to these broad generalizations are Calgary and London, the first a centre of major oil-related

Table 3.2
High-status scores on the economic-status axis: CMAs, 1981

Percentage of tracts with factor scores		CMAs*	Highest-scoring areas for individual CMAs[†]
> 1.0	> 2.0		
35.8	10.8	Ottawa	Rockcliffe Park (3.84), Nepean/Graham Park (3.18), Kanata (2.36), Alta Vista (2.26), Glebe (2.22)
27.8	10.4	Toronto	Bayview/Post Road (4.03), Upper Forest Hill (3.93), Harbour Square (3.46), Lawrence Park (3.32), Rosedale (3.28)
26.0	8.6	Calgary	Mt Royal (3.02), Eagle Ridge (2.64)
22.8	2.8	Regina	Hillsdale (2.09)
19.6	3.2	Halifax	North West Arm (3.01)
19.3	4.4	London	Stoneybrook (2.23), Sherwood Forest (2.14)
16.8	5.3	Quebec City	Sillery (2.66)
16.3	4.2	Vancouver	West Vancouver/Sherman (2.77), University Hill (2.66), Shaughnessy (2.65)
14.6	4.4	Saskatoon	Wildwood (1.40), Nutana (1.32)
13.8	5.5	Montreal	Westmount (3.96), Westend Laval (3.36), Outremont (3.23), Mount Royal (3.22), Hampstead (3.01)
12.2	3.0	St John's	
11.3	0.0	Victoria	Oak Bay (1.78), McNeil Bay (1.68)
11.2	5.1	All CMAs	
10.4	2.2	Winnipeg	Silverheights (2.89), Riverheights (2.31)
7.7	1.4	Hamilton	Burlington/Lakeshore (2.17), Burlington/Harbour (1.47), Westdale (1.36)
4.9	0.0	Kitchener	Beechwood (1.78)
2.7	0.0	St Catharines	

* No other CMAs have a census tract with a score above 1.0.

† Census tracts are not co-incident with the communities named in each city; the name of the community nearest to the tract is given in this table. Where communities encompass more than one tract, the value of the highest-scoring tract is given.

growth in the 1970s, the latter a traditionally prosperous regional service centre in southwestern Ontario. Seven centres, mainly the smaller industrial or port cities (Chicoutimi, Trois-Rivières, Oshawa, Windsor, Sudbury, Thunder Bay, and Saint John) do not have comparable high-status tracts with scores over 1.0. High-status areas can be found in these centres but are not sufficiently prosperous or spatially extensive to be isolated as high-status areas when we use all Canadian census tracts. The table also shows the highest-scoring areas for individual CMAs, isolating the familiar high-status communities of Canadian cities. These areas may be intuitively known to those familiar with each metropolitan centre, but the technique shows how they can be measured on a single comparable scale.

Maps showing the distribution of factor scores for this axis for all twenty-four CMAs would be needed to provide a comprehensive summary of variations in economic status. Constraints of space make this impossible. Instead, some examples of individual cities are used to summarize the key features within all Canadians cities – an approach also adopted for all the other axes. Figures 3.1A and 3.1B show the basic variations in economic status for Quebec City and Vancouver.[28] In Quebec City there is clear contrast between high- and low-status areas: high-status areas characterized by managerial occupations and high-income earners hug the heights of the St Lawrence shoreline west of the historical Citadel and Plains of Abraham; the low-status areas that show strong low-income, blue-collar characteristics are in the lower town, behind the old harbour, and lie north and east of the approximate line of the CPR rail line to the west, along which newer industrial areas have developed. Despite some exceptions, the division of the Quebec CMA into two broad zones is clearly visible. Such contrasts can also be seen in Vancouver. Most of the North Shore is of high status, as are the west-side areas of the city of Vancouver, west of Cambie Street. To the east, as far as the Burnaby boundary, lies an area of mainly lower status, which continues in the southeast as far as New Westminister. The lowest status scores are found in this long-established industrial area and east of the downtown in the Hastings-Terminal area.

Given Vancouver's much larger size and its absorption of a series of satellite centres, it is perhaps not surprising that it has a more complex pattern of economic-status variation than Quebec City. All Canadian metropolitan centres display similar contrasts in economic status, and the difference is usually linked to the ability of high-status areas to establish and consolidate themselves on view lots or in environmentally attractive areas, leaving lower-status areas to develop and be perpetuated around the industrial areas. This zonal contrast demonstrates the continued utility of Hoyt's[29] suggestion that broad sectors characterize social differentiation in cities – at least for this particular axis and source of differentiation. Yet the location of the high-status areas is not necessarily determined by environmental features. Their hill or waterfront locations were capitalized upon by developers.[30] Although such sectors may be perpetuated, subsequent urban development often truncates their outward growth, preserving them as enclaves.

These and other examples suggest that Canadian cities now have four dominant types of high-status regions: inner-city areas which have maintained their traditional status; new suburbs, often associated with some exclusive recreational complex; newly gentrified neighbourhoods; and the redeveloped central-city areas. The latter are usually typified by condominium developments. Yet the similar status of these areas cannot disguise quite

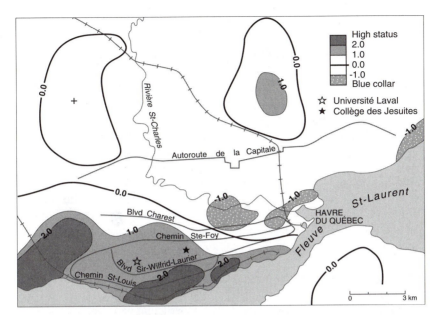

Figure 3.1A
Economic status (factor scores), Quebec City, 1981

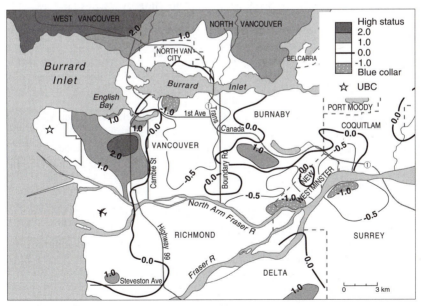

Figure 3.1B
Economic status (factor scores), Vancouver, 1981

different social worlds in terms of life-style and interaction patterns[31] (see chapter 11).

These broad contrasts in Canadian cities are not the only source of economic variation. Table 3.1 shows a second social rank factor, best described as impoverishment. It is different, however, from the low-status, blue-collar employment end of the major economic-status continuum. It reflects the intersection of low income, high local mobility, and substandard housing with high proportions of rental and apartment units that has been identified in many factorial ecologies in Britain and Canada, as well as in the United States, where it has often been associated with race.[32] Yet a high association with the growing numbers of single-parent female-headed families testifies to the greater importance of gender-related social differences.[33] For Canada as a whole in 1981, 9.3 per cent of families were headed by single female parents, many of whom now constitute the new and expanding pool of the impoverished. There is clearly a need to identify this growing sector of disadvantaged members of society left behind by the increasing affluence of the majority of the population – in addition to the homeless people rarely identified in census surveys.

This population is highly concentrated in particular parts of the city. Most of these tracts are in inner cities, particularly near downtown, in the older working-class districts that began as low-income areas and have suffered since from an absence of renewal and insufficient social welfare. Since the 1960s, federal-provincial schemes for housing improvement may have mitigated most of the worst examples of such stress in Canada – unlike the wholesale abandonment of many such areas in the United States – but such improvements and public housing have often further concentrated the lowest-income groups.

There are also significant between-city variations in the number of tracts with high values on this axis. Table 3.3 shows that for Canada as a whole, 14.3 per cent of the almost three thousand census tracts in this analysis have positive scores over 1.0 (4.5 per cent over 2.0). These are highly spatially concentrated. The five CMAs in Quebec and New Brunswick have a one-fifth (20 per cent) or more of their census tracts with scores over 1.0 on this scale of impoverishment: Saint John (37 per cent), Quebec (29 per cent), Trois-Rivières (28 per cent), Montreal (27 per cent), and Chicoutimi (24 per cent). This compares with only 9.7 per cent of the tracts having scores over 1.0 in Toronto, 6.6 per cent in Vancouver, and 2.6 per cent in Calgary. The percentage of tracts with people of low income and single-parent families reflects the relative post-war economic prosperity of these centres, but the high association with the Quebec locales and Saint John also seems to point to the greater incidence of single female parents, a key variable on this axis. In Saint John and Quebec, 15 per cent and 13 per cent of families respectively were headed by single parents in 1981.

Table 3.3
CMAs with the largest number of extreme scores
on the impoverishment axis, 1981

Percentage of tracts with factor scores		CMAs*
> 1.0	> 2.0	
37.1	25.7	Saint John
29.2	8.8	Quebec
28.0	12.0	Trois-Rivières
27.1	8.9	Montreal
23.5	0.0	Chicoutimi
17.1	8.2	Winnipeg
17.1	3.6	Windsor
17.1	0.0	Regina
14.3	4.5	All CMAs
9.7	3.1	Toronto
6.6	0.0	Vancouver
2.6	0.9	Calgary

* Toronto, Vancouver, and Calgary are shown for comparison.

Figures 3.2A and 3.2B show for Quebec and Winnipeg that the distribution of areas with these characteristics follows the familiar east-end-of-the city pattern, which has persisted for most of this century. Examples can be seen in part of the lower town called "Le croissant de pauvreté" in Quebec[34] and in the area north and east of North Main Street in Winnipeg – the zone of distress identified by Woodsworth in 1911. Such traditional areas of impoverishment and entrapment have experienced relatively little residential migration to other parts of the city. Elsewhere, large public renewal projects of the 1960s and early 1970s, such as Regent Park and Lawrence Heights in Toronto or Jeanne Mance in Montreal, also show high scores on this axis. In contrast, inner-city areas redeveloped to encourage greater social mix have less extreme scores. Toronto's St Lawrence neighbourhood, for example, is 57th on the rank-order list, and the tract incorporating the False Creek development in Vancouver is 211th. Indeed redevelopment of this latter area with a mix of tenure types and incomes gives the area relatively high economic status (Figure 3.1B).

Previous factorial ecologies and many studies of social change in inner cities have suggested increasing differentiation of contemporary urban society by family or family-related variables. The results of this study illustrate that the hypothesized family-status dimensions were summarized by four separate factors, measuring a variety of different features. This complexity,

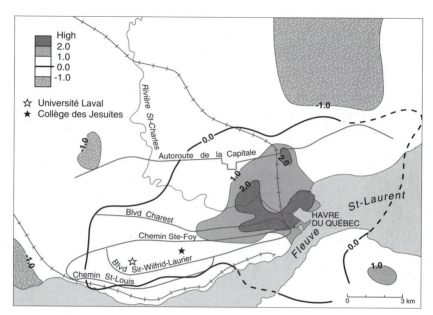

Figure 3.2A
Impoverishment (factor scores), Quebec City, 1981

derived from all twenty-four CMAs and not a single city, confirms that a single axis identifying family variations can no longer define the social variability of Canadian metropolitan centres. The traditional family-status axis was identified in this study, but its links with age differentiation make it more appropriate to refer to it as a family-and-age axis. This dimension separates the positive high loadings of variables linked to families in the child-rearing stage of the life cycle from the high negative loadings of indicators measuring the older-age and completed-family character of so-called empty nesters. In addition to family and age, the presence of a separate axis called early and late family is consistent with findings from similar studies of individual Canadian cities.[35] The latter dimension has high positive loadings for those indicators picking out young family heads and a large proportion of pre-school children from those variables indexing older parents and children in their late teens and early twenties which have negative loadings. The third family-status factor, young adult, also identified in studies of British cities, identifies the segregation of this group within Canadian metropolitan areas as a whole. This seems to be a growing phenomenon, given the spatial concentration of high-density apartment accommodation and social facilities for these individuals, but also post-secondary educational opportunities and office jobs in parts of the inner city. Finally, a non-family axis was characterized by high loadings for the variables measuring high proportions of single adults, childless families, and the ratio of divorced to married adults.

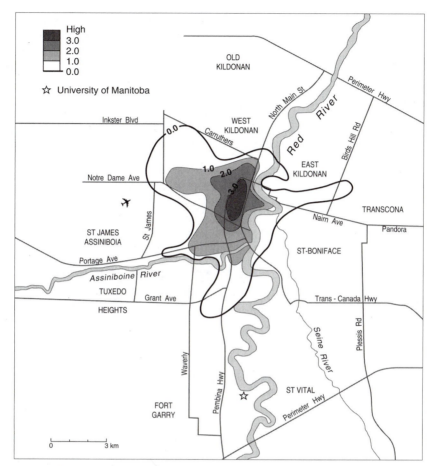

Figure 3.2B
Impoverishment (factor scores), Winnipeg, 1981

When these family axes of differentiation are mapped for individual cities rather divergent patterns emerge. Unlike with the economic-status axis, however, it is harder to obtain consistent spatial patterns across sets of cities. The same peculiarities caused by the accidents of local topography and in-dustrial zones that interrupt the continuity of economic variations are inten-sified by variations between areas caused by differential growth rates, the ageing of family groups in these communities, and the degree of family re-newal. Despite these often city-specific variations, many cities show broadly similar spatial patterns. This time, rather than choosing different city exam-ples, we illustrate them by the distribution of scores for all the family-related dimensions for Calgary (Figure 3.3). This metropolitan area was chosen be-cause it has grown primarily by concentric accretion on its edges, not by the

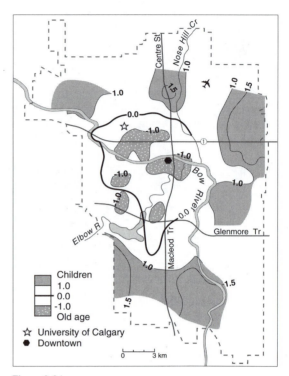

Figure 3.3A
Family and age (factor scores), Calgary, 1981

peripheral incorporation of large, older communities. Also its newness and explosive growth in the 1960s and 1970s mean that its neighbourhoods are more likely to highlight some of the emerging trends in the social mosaic of modern Canadian society.

The distribution of the axes measuring family and age scores displays the familiar broadly concentric pattern found in previous factorial studies of family status (Figure 3.3A). The negative scores, indexing areas of the old and middle-aged and completed families, are found in the inner city – approximately the area built up by 1960. The highest positive scores (over 1.0) show the 1970s growth areas dominated by their young families and children. Although the highest old-age characteristics are found west and southwest of the central business district, there is a discontinuous zone of scores over − 1.0 – indexing completed families – in the inner city, up to 4 km from the city centre. The presence of such high positive scores of young families around Mount Royal College and Sarcee Military Barracks in the southwest picks out the more youthful army and college population.

The distribution of scores on the early-and-late-family dimension shows similar concentricity, but without a simple city-suburb transition (Figure

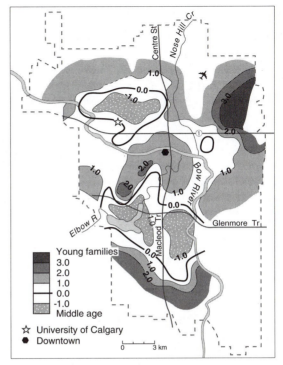

Figure 3.3B
Early and late family (factor scores), Calgary, 1981

3.3B). The general pattern is best summarized as a cross-sectional profile in the form of an open letter W – with local deviations. In the suburbs built to the northwest, northeast, and south in the 1970s, high scores index the youngest children and the 25–34-year-old age groups. Inner-city scores over 1.0 show that the area still has some younger age groups because of the larger number of cheaper apartment blocks and the gentrification movement in the 1970s and 1980s.[36] The highest negative values, picking out the communities with mature families, form a broken belt – split by major traffic arteries – between 4 and 7 km from the city centre in the northwest, southwest, and south-southeast. The ageing of those 1960s young-single-family communities has left these inner suburbs with a mature population in which many children have left or are leaving home: the empty nester syndrome.

The spatial distribution of the young-adult factor, associated with 18–24-year-olds and high levels of female participation in the labour force, shows that the separate presence of this axis is caused mainly by the increasing desire of young adults – who are able to afford houses on their own – to locate close to areas of work in the central city and to educational institutions or hospitals (Figure 3.3C). The highest scores are found south of Calgary's

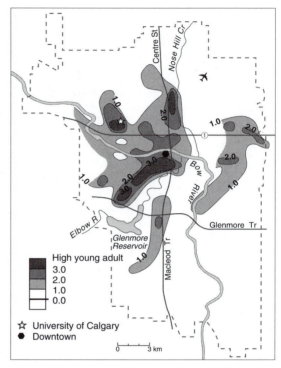

Figure 3.3C
Young adult (factor scores), Calgary, 1981

downtown, with outliers in the northwest, near the Southern Alberta Institute of Technology and the University of Calgary, as well as in those parts of the suburbs built in the 1960s close to the major industrial and commercial areas.

There are wide variations among Canadian metropolitan areas in the incidence of these axes.[37] One example is the extent to which young adults have established their own homes in particular areas in the city. For instance, there are twenty-six tracts in Calgary with high scores over 2.0 for the young-adult axis, whereas in Quebec City there is only one – in the Ste-Foy area west of the old city, between the Université Laval and the Collège des Jésuites. This contrast illustrates the variable presence of a separate young-adult axis in Canadian metropolitan areas; the CMAs of Alberta, with their heavy in-migration in the 1970s, show this pattern, unlike the cities of Quebec and the industrial centres of Ontario. Finally, Figure 3.3D shows the pattern of scores associated with the non-family axis. There are extreme values in the central city, extending through the apartment complexes to the southwest but attenuated to the south-southwest by the high-status area of Mount Royal and falling off rapidly outward from the central business district.

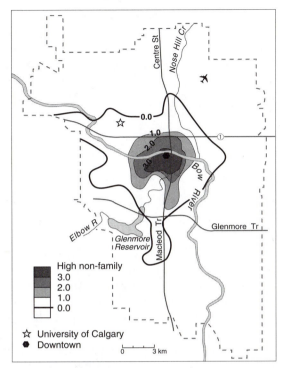

Figure 3.3D
Non-family (factor scores), Calgary, 1981

The remaining three axes shown in Table 3.1 are very different in type. A housing axis separates variables that measure tracts with a relatively large proportion of poorer housing in need of repair from those with predominantly new suburban housing. A migrant factor contrasts the indicators measuring a high proportion of out-of-province movers with those that show high mobility among people within the local municipality. Both show an inner-city–suburban contrast in the pattern of scores. These axes testify to the importance of both mobility and housing variations in the social differentiation of Canadian CMAs, although since the factor loadings are smaller and linked to only a few indicators, they are less important numerically than the other sources of differentiation discussed earlier.

The final axis of differentiation is ethnicity. It separates indicators of French ethnic origin from those that are of non-British and non-French origin, and from immigrants. A minor association with high female participation in the labour force is not unexpected, given the relatively high propensity among immigrant family members to enter the work-force.[38] Although extensive eastern European immigration at the turn of the century began reducing the basic French-British ethnic polarity in Canada, migration streams from southern Europe since 1945 have been more important in this respect, especially since the 1960s, while Asian and Caribbean immigrants have recently added to the cultural mélange of Canadian cities. Not all CMAs have shared equally in this emerging pluralism. Although the factor scores on this axis could be used to identify broad ethnic differences, greater precision in measuring the type of city-specific ethnicity and the degree of segregation is necessary.

ETHNIC DIFFERENTIATION IN CANADIAN METROPOLITAN AREAS

The index of dissimilarity has frequently been used to measure the segregation of specific ethnic groups and has been the subject of several recent Canadian studies.[39] As used in this study, the index measures the degree of spatial differentiation between a specific ethnic group and the rest of the population. The indices of dissimilarity for 1981, derived from several sources, are presented in Table 3.4 for all metropolitan areas and the larger ethnic-origin groups. These groups were selected to represent Native Canadians, the other "charter groups" (British and French), and the major streams of post-1945 immigrants.[40] In Table 3.4, both the metropolitan areas and ethnic groups have been rank-ordered by average segregation index (metropolitan areas ranked from high to low and ethnic groups from low to high).[41] A segregation index value of zero indicates that the group is distributed in the same way as the general population; a value of one implies complete segregation of the group in a single tract.

Metropolitan averages indicate considerable differences between centres. The two largest Quebec cities (Montreal and Quebec City) and the two other cities with a large French-minority population (Ottawa and Sudbury) have the highest average levels of ethnic differentiation, followed closely by industrial cities in Ontario, such as Hamilton, St Catharines, and Windsor, as well as Toronto and Winnipeg. These empirical findings relate to two features: first, the historical social and spatial divide between British and French in cities where either group represents a substantial minority; and second, the primary focus of post-war international migrant streams on the larger centres and the smaller industrial centres of southern Ontario. In contrast, the lowest average segregation levels are found in ethnically homogeneous centers such as St John's, Chicoutimi, and Trois-Rivières, as well as the four CMAs in Alberta and Saskatchewan. The latter, particularly Calgary and Edmonton, grew very rapidly during the 1970s, and in such a new and highly mobile environment recent migrants from different ethnic groups may not have had time to form separate community structures.[42] Moreover, many – particularly in Alberta – also had qualifications that gave them rel-

Table 3.4

Indices of dissimilarity for Canadian CMAs and selected ethnic groups, 1981

Census metro area (CMA)	British	German	French	Ukrainian	Polish	Italian	Native Canada	Black	Chinese	Indo-Pakistani	Greek	Jewish	Portuguese	CMA mean index of dissimilarity*
Montreal	0.458	0.407	0.478	0.479	0.442	0.565	0.454	0.463	0.605	0.614	0.656	0.831	0.600	0.573
Quebec	0.211	0.393	0.232	—†	—	0.389	0.610	n.d.	0.727	—	—	—	0.761	0.529
Ottawa	0.372	0.275	0.573	0.332	0.340	0.481	0.409	0.443	0.445	0.499	0.533	0.504	0.586	0.482
Sudbury	0.157	0.160	0.326	0.255	0.294	0.444	0.432	0.605	0.571	0.614	0.601	0.775	0.706	0.457
Hamilton	0.151	0.171	0.184	0.224	0.278	0.373	0.516	0.496	0.508	0.493	0.457	0.679	0.640	0.443
St Catharines	0.142	0.259	0.308	0.167	0.245	0.331	0.491	0.544	0.556	0.570	0.487	0.636	0.778	0.442
Winnipeg	0.208	0.200	0.390	0.279	0.279	0.338	0.492	n.d.	0.452	0.516	0.536	0.725	0.682	0.427
Halifax	0.075	0.117	0.117	0.422	0.368	0.401	0.468	n.d.	0.557	0.514	0.585	0.580	0.693	0.417
Toronto	0.263	0.194	0.198	0.344	0.388	0.505	0.447	0.385	0.452	0.409	0.461	0.741	0.632	0.413
Windsor	0.103	0.151	0.208	0.195	0.261	0.293	0.514	0.451	0.550	0.534	0.435	0.610	0.667	0.400
London	0.084	0.112	0.147	0.253	0.273	0.308	0.453	0.457	0.426	0.495	0.375	0.597	0.539	0.386
Kitchener	0.155	0.237	0.136	0.254	0.211	0.267	0.426	0.366	0.518	0.397	0.426	0.439	0.545	0.386
Victoria	0.113	0.119	0.188	0.127	0.250	0.285	0.398	n.d.	0.451	0.499	0.562	0.446	0.452	0.374
Edmonton	0.123	0.149	0.154	0.202	0.198	0.410	0.372	n.d.	0.362	0.481	0.512	0.655	0.633	0.365
Saint John	0.093	0.203	0.160	—	—	0.478	—	n.d.	0.625	—	0.652	—	—	0.362
Vancouver	0.174	0.159	0.211	0.158	0.209	0.448	0.395	0.325	0.509	0.389	0.482	0.559	0.589	0.362
Oshawa	0.084	0.118	0.159	0.264	0.274	0.238	0.307	0.421	0.424	0.434	0.489	0.496	0.517	0.354
Regina	0.117	0.128	0.109	0.116	0.143	0.311	0.376	n.d.	0.374	0.440	0.671	0.591	—	0.336
St John's	0.180	0.301	0.177	—	—	—	—	n.d.	0.610	0.464	—	—	—	0.346
Thunder Bay	0.118	0.115	0.113	0.203	0.132	0.205	0.374	0.551	0.550	0.542	0.544	0.726	0.549	0.346
Calgary	0.083	0.100	0.153	0.112	0.168	0.329	0.364	n.d.	0.300	0.418	0.537	0.489	0.556	0.322
Saskatoon	0.070	0.088	0.103	0.138	0.141	0.288	0.334	n.d.	0.238	0.520	0.606	0.510	0.809	0.321
Chicoutimi	0.160	0.451	0.140	—	—	—	0.368	n.d.	—	—	—	—	—	0.280
Trois-Rivières	0.195	—	0.173	—	—	—	—	n.d.	—	—	—	—	—	0.184
Mean index of dissimilarity	0.162	0.200	0.214	0.238	0.258	0.366	0.429	0.472	0.491	0.492	0.530	0.610	0.628	

Sources: All indices except for the black group are from Bourne et al. (1986: 58–9). Indices for blacks in Ontario cities are from Bourne, Baker, and Kalbach (1985: 32) and for Montreal and Vancouver from Balakrishnan and Kralt (1987: 151). The index for Ottawa is for the Ontario part of the Ottawa-Hull CMA. Bourne, Baker, and Kalbach (1985) and Balakrishnan and Kralt (1987) do not define black ethnicity identically.

* Based on twenty-four ethnic groups, excluding blacks. For details see Bourne et al. (1986: 58–9). Only thirteen of these groups are shown here.

† Index not calculated because group represents less than 0.05 per cent of CMA's population.

n.d. = no data available.

atively high-paying jobs and access to a home in the suburbs rather than the historical immigrant entry route via the lower-status inner city.

Previous studies have hypothesized positive relationships between ethnic segregation and the size and ethnic diversity of a city.[43] The argument concerning size is that ethnic groups in larger cities are likely to have sufficient members and therefore more opportunities to develop specialized institutions and neighbourhoods, while in cities that are more ethnically diverse residents are likely to be more aware of ethnic differences and have a higher level of identification with their own group. Calculations revealed that there is a modest rank-order correlation of +0.56 between population size and average segregation index, but no substantial relationship between segregation and ethnic diversity (a value of +0.02) or the proportion of the population non-British and non-French (−0.07).[44]

As expected, segregation by individual ethnic groups is generally related to both period of immigration and social distance from the "charter groups." Table 3.4 also reveals that the thirteen ethnic groups reported here vary in their degree of segregation not only absolutely in the mean values for the CMAs but also for each city. Those of British origin are, not surprising, the most scattered, with the lowest index of dissimilarity. Yet in Montreal and Ottawa-Hull the presence of distinct francophone and anglophone areas means that segregation indices are high. Of the non-British and non-French groups, western and eastern Europeans are the least segregated. These groups have been in Canada for a relatively long time and exhibit low levels of social distance from the "charter groups." In contrast, southern Europeans are more highly segregated than western and eastern Europeans. The Portuguese, in particular, who arrived relatively recently in Canada, have retained their linguistic identity and in many cities have established a high level of institutional completeness. Visible minority groups such as blacks, Chinese, Indo-Pakistanis, and Native Canadians exhibit lower dissimilarity indices than Greeks, Jews, or Portuguese. This is surprising given the relatively high levels of social distance between visible minorities and the "charter groups." Perhaps many of the visible minority groups are, in fact, more heterogeneous than southern Europeans, and therefore Indo-Pakistanis from different religious traditions, or Caribbeans from different island backgrounds, may settle in different parts of the city. Though census sources do not identify ethnic groups in sufficient detail to test this assertion, it is well known that in Montreal French-speaking blacks from Haiti have settled in the French-dominated east end of the city, and blacks from the British West Indies in the city's English-speaking west end.[45] Finally, the Jewish group has the second highest average dissimilarity index and a high level of segregation for all metropolitan areas. Much of the segregation today, however, is voluntary, related primarily to a conscious attempt by conservative groups to be within walking distance of a synagogue, although in some cities many

Jews are in high-income categories, and so their residential locations are linked to high-status areas.

SUMMARY AND CONCLUSION

Serious interest in identifying the social character of different areas in Canadian cities can be traced back to the turn of the century and has gone through a series of conceptual and methodological refinements. In recent years the availability of census tract data[46] and of more sophisticated statistical measures has meant that the focus on poverty that motivated many of the earliest workers has been augmented by a wider scholarly interest in the description and analysis of all types of sub-areas. The knowledge so gained may be directed toward prescriptive or planning ends, since factorial ecologies provide background information for the evaluation of social policy. Comparative measurements of poverty or ethnicity can act as sampling frames for more detailed quantitative or qualitative studies.

The empirical study reported here has been a cross-sectional analysis at one point in time, and it obviously cannot be used directly to trace temporal changes. Nevertheless, the results of this multivariate analysis of the social composition of Canadian metropolitan areas confirm their increasing complexity. The three-axis model proposed by early workers is simply not comprehensive enough to display the social variations in these areas. Ideally, this analysis needs to be complemented by parallel studies at previous census dates using identical data sets to trace and measure detailed patterns of change. Nobody has yet attempted this task for all Canadian CMAs, although glimpses into some major changes can be derived from longitudinal studies of change done for Montreal (1971–81), Toronto (1951–81), Hamilton (1961–81), and Winnipeg (1971–81).[47] Since the data sets are all different, ranging from twelve variables in the Hamilton study to fifty-nine (sixty in 1981) for Montreal, it is impossible to compare the results rigorously. Nevertheless, some general trends can be derived.

First, the major axes of differentiation that structure urban social space and the corresponding levels of explanation – in term of variance accounted for – have been relatively stable over short periods. Yet Murdie's study of Toronto showed a decline in the level of explanation from 73.5 per cent in 1951 to 65.6 per cent in 1981.[48] This suggests that over a longer period Toronto's structure has become more complex: the variables used now have more localized patterns that cannot be captured as general sources of differentiation.

Second, despite the continued importance of general axes, variables often change their associations with particular factors. For example, in Toronto in 1951 female labour-force participation emerged as the most significant variable on the family-status factor, but by 1981 it was much higher and more uniform through all social groups, and consequently this variable was not associated with any of the general factors. In Montreal somewhat similar findings were reported,[49] but over a shorter period. Also the single-parent variable is no longer associated exclusively with older age groups, a situation linked to marriage breakdowns and Canada's more liberal divorce laws since 1968.

Third, in contrast to the general stability of the axes of social differentiation, the spatial patterning of Canadian metropolitan areas has changed considerably. Many inner-city areas, for example, have shown higher economic status under the impact of revitalization, perhaps most apparent in Vancouver, Ottawa, and Toronto.[50] Taylor and Murdie have also shown how in the inner areas of Hamilton and Toronto, respectively, family status increased, thereby breaching the simple zonal pattern.[51] Both areas experienced recent European immigration, primarily Portuguese. Yet an increase in long-term poverty has also perpetuated impoverishment, particularly among the elderly and single-parent families. Areas of deprivation that were of concern to the turn-of-the-century reformers, such as Woodsworth and Ames, and later geographers such as Watson, are still largely problem areas, although the "gentrification movement" has revitalized some of these communities. Similarly, most of the traditional areas of high status have remained intact since the Second World War.

Fourth, ethnicity has displayed more complex variations because of differences in immigration levels and composition and the social mobility of individual groups. It is difficult to generalize for ethnic groups as a whole. Canada is now a more pluralistic country, with multiculturalism as official government policy. It is not surprising, then, to find considerable stability in average levels of ethnic segregation between 1961 and 1981. Studies have shown that ethnic segregation does not totally disappear with increased social mobility. Indirect evidence also suggests that although ethnic segregation is generally lower in areas of higher economic status, there is still considerable between-group segregation in more affluent parts of cities.[52] Thus, as most researchers have argued, differences in economic status between groups is not the only reason for ethnic segregation in Canadian cities. Instead, factors such as period of immigration, size of group, social distance, and language retention play a much more important role. Development of new concentrations of ethnic groups has taken place alongside the outward movement of post-war European immigrants, particularly Italians, to new housing in the outer suburban communities of the bigger cities.

Despite the large number of immigrants in Canadian society, most cities outside the three biggest centres do not have the extreme levels of segregation by ethnic and racial character found in American or European cities. There are certainly areas of concentration, some of which are growing rapidly. But if immigrants continue their process of assimilation into the host society as a whole, segregation levels may decline. Again it is difficult to produce a conclusive statement. It will depend on future levels of immigration, particularly from underdeveloped countries; the economic prosperity of the country; and continued commitment to multicultural equality of opportunity and expression. The experience of other Western countries shows that high levels of immigration of visible minorities with limited education, combined with restricted job opportunities, may create an increasingly large and distinctive underclass in society. This will be added to groups of long-term impoverished in our society – Native Canadians and the Canadian-born underclass. Such ethnic groups – like the growing numbers impoverished and of single-parent families – are very likely to be segregated and discriminated against, and this may mark a return to the conditions that Woodsworth found so intolerable in Canada at the turn of the century. Whether or not this scenario occurs, the increasing social complexity of Canadian cities merits serious attention if we are to understand what Booth referred to in 1893 as the "terrae incognitae" of our urban societies – the differential character of our urban social areas.

CHAPTER FOUR

Demography, Living Arrangement, and Residential Geography

J.R. MIRON

The past four decades have seen remarkable shifts in the demography of Canada's population, in living arrangement, and in the daily pattern of household life. Demographic change has contributed to alterations in Canada's residential geography – for example, metropolitan growth, the emergence of the automobile-oriented commuter suburb, the high-rise residential construction boom of the late 1960s and early 1970s, and the development of senior citizens' and other purpose-built housing. At the same time, new trends in residential geography were also shaped by changes in the economy and by social policies that included income stabilization, fiscal restraint, "targeting," and "deinstitutionalization."[1]

This chapter reviews concepts and evidence central to understanding these trends. First, it presents the principal data available on changes in demography, living arrangement, household life, and residential geography. Second, a micro-economic framework is used wherein living arrangement, household life, and housing consumption are seen to be outcomes of demand, supply, and policy constraints. Finally, the chapter interprets the effects of changes in demography, consumer income, and social policy on household formation and residential geography.

CHANGES IN DEMOGRAPHIC COMPOSITION

Let us consider important post–Second World War changes in the demography of Canada's population in four areas: fertility, marriage and divorce, migration, and longevity.[2] Let us summarize the changes briefly as a basis for speculation on the effects of shifting demography on the social geography of Canadian cities.

Fertility

One major and well-known demographic event of the past half-century was the post-war baby boom and subsequent baby bust. Prior to 1939, fertility had slumped.[3] Fertility rates gained momentum during the war, peaked around 1959,[4] fell sharply during the 1960s and early 1970s, and thereafter continued to drift down slowly.[5] The baby boom created a bulge in Canada's age pyramid evident in Table 4.1 as a peak near 1961 in the number of children aged under 6, near 1971 among 6–14-year-olds, and near 1986 in the 25–34 group.

Marriage and Divorce

Corresponding to the baby boom and bust were a marriage boom and bust that resulted in a sharp drop, then moderate increase, in typical age at first marriage. During the marriage boom, first-marriage rates (conventional marriages only) increased for men under about 30 years of age and for women under about 25.[6] The median age at first marriage for women dropped from about 23 years for cohorts born between the 1880s and 1920s to 21 years for the cohort born in the 1940s;[7] among females born in the 1950s, the marriage bust is evidenced in a higher median age (21.5 years). Using a marital life-table approach, it has been estimated that, between 1971 and 1981, the propensity to ever marry among women declined from 0.92 to 0.88, and average time lived as never-married increased from 25.0 to 29.8 years.[8]

To what extent was the decline in first marriage offset by an increase in cohabitation, or common-law marriage? The 1981 census was the first in which opposite-sex common-law couples were separately identified.[9] Counts from the 1981 and 1986 censuses do indicate that common-law marriage became more popular (especially among persons under 35).[10] Nonetheless, the total percentage of young adult men and women enumerated as married (both conventional and common law) fell between 1981 and 1986. Cohabitation had only partly offset the decline in conventional marriage rates.

Also important during the marriage bust was the rise in divorce rates and the decline of remarriage. Using a marital life table based on marriage, divorce, and survivorship rates for 1980–82, it has been estimated the 29 per cent of marriages end in divorce, up from 19 per cent a decade earlier; that the average duration of marriage declined by about four years over the same period; and that 69 per cent of divorced women remarry, down from 79 per cent a decade earlier.[11]

Table 4.1
Population and families by living arrangement, Canada, 1941–86

	1941	1951	1961	1971	1981	1986
	(thousands of persons)					
Total population (usual residents)*	11,490	13,984	18,238	21,568	24,203	25,207
In private dwellings						
Family members						
Living with spouse[†]	4,432	5,923	7,600	9,184	11,222	11,763
Lone parent	309	326	347	479	714	854
Child[‡]						
Under 6	1,245	1,470	2,662	2,198	2,076	2,109
6–14	1,909	2,254	3,447	4,090	3,251	3,141
15–24	1,539	1,820	1,669	2,567	2,926	2,770
Over 24 years old	451	423	371	334	414	559
Non-family individuals[§]	1,237	1,384	1,659	2,323	3,195	3,578
In collective dwellings	368	384	484	393	406	434
	(thousands of families)					
All families[#]	2,525	3,287	4,147	5,076	6,325	6,735
Primary families["]	2,333	2,967	3,912	4,915	6,133	6,534
Living alone	–	–	3,263	4,286	5,556	5,939
Others present	–	–	649	629	577	596
Secondary families**	192	321	235	161	192	201

Sources: Calculated from published reports of the Census of Canada, various years.

*   Columns may not total because of rounding. – Indicates data not available.

†   Since 1981, common-law couples have been enumerated as marrieds. In earlier censuses, where such couples chose not to list themselves as married they were counted as either non-family individuals (if no children present) or lone-parent families (if children present).

‡   1941 estimates were prorated from published data for the under–7 and 7–14 age groups; 1951 data interpolated from "Total under 15."

§   Includes individuals whose family status could not be ascertained.

1981 and 1986 census counts exclude families in collective dwellings. Families in collective dwellings are included in earlier counts.

"   Under the 1941 census definition of a household, there could be two or more households per dwelling. Compared with the subsequently used definition which assigns only one household per dwelling, the number of primary families was overstated, and the number of secondary families understated, in 1941.

** Since the 1981 census, primary and secondary statuses have depended on whether the "household maintainer" (i.e. the person chiefly responsible for financial maintenance of the dwelling) was resident. Where a family lived alone but was financially supported from outside, there was no primary family. Prior to 1981, a family living alone was always enumerated as a primary unit. Thus censuses since 1981 overcount secondary families relative to earlier censuses.

Table 4.2
Immigrant population by period of immigration, Canada, 1986

	Total immigrants (persons)	Period of immigration				
		Before 1945 (%)	1945–66 (%)	1967–77 (%)	1978–82 (%)	1983–86 (%)
All Canada	3,908,145	10	40	31	12	6
Three-CMA total	2,084,435	6	38	35	13	7
Montreal	459,495	6	41	31	14	8
Toronto	1,233,090	5	38	37	13	7
Vancouver	391,850	10	34	36	14	7
Other Canada	1,823,710	15	42	26	11	5

Sources: Calculated from *1986 Census of Canada*, Cat. 93–156, pp. 10.7, 10.17, and 10.8, and Cat. 93–155, pp. 1.6–1.9.
Note: Consolidated CMA data employed. Subject to round-off error, percentages sum to 100 for each row.

Longevity

Longevity increased steadily among Canadians over the past half-century because of a drop in infant mortality and improved chances of survival at older ages. The probability of surviving to one's first birthday increased for males from 0.913 around 1931 to 0.989 around 1981; for females, from 0.931 to 0.991. Remaining life expectancy at one's first birthday increased for males from 64.7 to 71.1 years over the same period; for females, from 65.7 to 78.5 years.[12] Against a backdrop of increasing nuptiality in older age groups, the persistent and growing gap between males and females meant that a rising proportion of women experienced widowhood and that they could typically expect to spend more of their life in that state.

Migration and Immigration

The history of immigration into Canada has been one of peaks and troughs. From the low levels of wartime, the rate of gross immigration peaked in 1948 (at 10 immigrants per 1,000 population), then again in 1951 (at 14), before reaching a post-war high of 17 in 1957. After a slump during the recession of the late 1950s, gross immigration rates varied but generally fell from about 10 in the mid-1960s to about 4 by the 1980s.[13]

The cumulative effect of this immigration has been substantial. In the 1986 census, almost one person in six in Canada reported having immigrated from abroad. Further, much of this immigration, especially after about 1960, was into Canada's three principal census metropolitan areas (CMAs) (see Table 4.2). Further, the ethnic mix of Canada gradually shifted.

Table 4.3
Percentage distribution of population by mobility status, Canada, 1961–86, showing breakdown by three CMAs and rest of Canada, 1986

	All Canada					1986	
	1961 (%)	1971 (%)	1976 (%)	1981 (%)	1986 (%)	Three CMAs (%)	Other Canada (%)
Population aged 5 or older	100	100	100	100	100	100	100
Non-movers (same dwelling)	55	53	52	52	56	53	58
Movers (changed dwelling)	45	47	48	48	44	47	42
Non-migrants (same CSD)	25	24	24	25	24	26	23
Migrants (different CSD)	20	24	25	23	19	21	19
Same province	14	14	16	15	13	15	17
Different province	3	4	4	5	4	3	4
Abroad	3	4	3	2	2	4	1

Sources: Calculated from *1986 Census of Canada*, Cat. 93–108 (Table 1) and Cat. 94–128.

Notes: Mover is a person who, at census date, reported living in a different dwelling five years previous. Migrant is a mover who reported that dwelling five years previous was located in another census subdivision (CSD). A CSD is an incorporated city, village, township, other municipality, or Indian reserve.

From largely British, French, and aboriginal origins (84 per cent of the 1921 population), Canada became increasingly diverse; by 1986, these three groups accounted for only about 71 per cent of the total population.[14]

Mobility has also helped shape the spatial distribution of population. Various quinquennial censuses since 1961 have found that just under half of Canadians report living in a different dwelling from five years previous (see Table 4.3). Also, mobility rates were higher in the three principal CMAs than elsewhere.

The effects of the baby boom, immigration, and improved longevity on Canada's total population were unmistakable. From 1946 to 1986, Canada's population more than doubled. However, this growth was not spread evenly; principally because of immigration and migration, the populations of the three largest CMAs increased faster than the rest of the country.

CHANGES IN LIVING ARRANGEMENT

The marriage and baby booms increased the incidence of people living in census families (a husband-wife couple with any never-married children or a lone parent with one or more never-married children who share a dwelling; anyone else is a non-family person).[15] As seen in Table 4.1, the number of

Table 4.4
Census family children as percentage of all singles (never-marrieds) and as percentage of
total population by age group, Canada, 1951–86

	1951 (%)	1961 (%)	1971 (%)	1981 (%)	1986 (%)
15–19-year-old children as					
percentage of singles aged 15–19	80	85	86	90	91
percentage of total population aged 15–19	76	81	83	86	89
20–24-year-old children as					
percentage of singles aged 20–24	83	78	78	64	68
percentage of total population aged 20–24	51	43	43	40	47

Sources: Calculated from published reports of the Census of Canada, various years.
Notes: For 1961 and 1971, numbers of children interpolated from counts of persons aged 15–18 and 19–24. For 1951, numbers interpolated from counts of persons aged 14–17 and 18–24.

non-family individuals increased less rapidly between 1941 and 1961 than did the number of persons living with spouses or as census children. However, with the onset of the baby and marriage busts, non-family individuals increased as a proportion of the total population, even though, during the 1970s and 1980s, the baby-boom generation swelled the age groups most prone to family formation.

Another change in living arrangement has been an increased modality in age at home-leaving among young adults (Table 4.4). With more young adults completing high school, the percentage of young singles remaining in the family home (and hence remaining census family children) through about age 18 or 19 increased steadily. However, until around 1981, young singles more and more tended to move out of the family home on completion of high school or soon thereafter. In 1951, 83 per cent of 20–24-years-old singles were children in census families; the figure dropped to 64 per cent by 1981. What happened to this group between 1981 and 1986 was something new. Suddenly, the percentage enumerated in a parental home began to increase. Young single adults again became more likely to delay leaving home.

A third major change in living arrangement over the past half-century was the increasing tendency of families and non-family individuals to maintain a dwelling. To understand the concept of "maintaining a dwelling" and its changing interpretation over time, we must first define the concepts of household head, person 1, and household maintainer. In Canadian censuses up to 1971, head of household was defined to be (1) a person living alone, (2) the parent in a lone-parent family living by itself, (3) the husband in a husband-wife family living alone, or (4) any person in a group of people sharing a dwelling as partners. Subsequent censuses used a similar definition

Table 4.5
Percentage distribution of women by household status and (for private households) relationship to person 1 by age group, Canada, 1971 through 1986

Percentage of age group

	15–24				25–34				35–44				45–54				55–64				65–74				75 or older			
	71	*76*	*81*	*86*	*71*	*76*	*81*	*86*	*71*	*76*	*81*	*86*	*71*	*76*	*81*	*86*	*71*	*76*	*81*	*86*	*71*	*76*	*81*	*86*	*71*	*76*	*81*	*86*
Person 1 or spouse	30	33	33	29	87	89	90	87	92	94	95	94	91	93	94	94	86	90	92	90	78	83	86	87	58	61	64	59
Daughter	57	57	57	62	6	6	6	8	3	3	2	3	2	2	2	1	1	1	1	1	0	0	0	0	0	0	0	0
Mother (in-law)	0	0	0	0	0	0	0	0	0	0	0	0	1	1	1	2	3	3	2	4	9	6	5	6	17	13	11	19
Sister (in-law)	2	2	2	2	1	1	1	1	1	1	1	1	1	1	1	1	3	2	1	1	4	3	2	2	3	3	2	2
Daughter-in-law	1	1	0	0	1	0	0	0	0	0	0	0	0	0	0	0	0	0	0	0	0	0	0	0	0	0	0	0
Granddaughter	1	0	1	1	0	0	0	0	0	0	0	0	0	0	0	0	0	0	0	1	0	0	0	0	0	0	0	0
Other relative	1	1	1	1	0	0	0	0	0	0	0	0	0	0	0	0	0	0	0	0	1	1	0	0	1	1	1	1
Non-relative	6	5	4	5	4	3	2	3	2	2	1	1	2	1	1	1	3	2	1	1	3	2	1	1	3	2	1	1
Other arrangement	3	2	2	1	2	2	1	1	2	2	1	1	2	2	1	1	3	3	1	1	6	5	5	3	17	19	20	18

Sources: Derived from the 1971, 1976, 1981, and 1986, Census public use samples (individual files). Calculations by the author.

Notes: "Other arrangement" includes people in collectives, temporary or foreign residents, or people in households abroad. Columns sum to 100 per cent, subject to round-off. In 1971 and 1976, relationship shown is to household head rather than to person 1. Treatment of common-law marriages has changed: see notes, Table 4.1.

except that either spouse could be household head in case 3, and the name was changed to "person 1" in 1981. In 1981, the concept of household maintainer was introduced – defined as the person (not necessarily resident in the household), or a person, responsible for paying rent, mortgage, taxes, electricity, and so on, for that dwelling. The person who is household head or maintainer, and the family, if any, that includes the household head, or the non-family individual that is head, are said to be a primary individual or family: i.e. to maintain a dwelling.[16]

Table 4.5 details changes in the living arrangements of Canadian women from 1971 to 1986. From 1971 to 1976, females in all age groups shown became more likely to be person 1 or spouse of person 1 – i.e. to be a primary non-family individual or part of a primary census family. They became less likely to be mother, mother-in-law, sister, sister-in-law, or other relative of person 1 and to be unrelated to person 1. In these respects, they showed less tendency to be secondary non-family individuals or members of secondary families. Except for women aged 75 or older, they also became more likely to live in a private household.

However, the trend toward being person 1 or spouse was arrested or reversed after 1976, most notably at the ends of the age spectrum. Among women aged 15–24, a larger proportion remained in a parental home in the early 1980s; among women 75 or older, the big increase was in living with a son or daughter's family.

There is additional evidence of the growing propensity of Canadians to maintain a dwelling. Table 4.6 presents data on families maintaining a dwelling or living alone, on individuals who are non-family, on non-family people living alone, and on households with one or more lodgers (related or unrelated, paying or free). These data indicate a substantial increase between 1941 and 1971 in propensity to maintain a dwelling. However, this trend was arrested between 1971 and 1981; between 1981 and 1986, it was even reversed in some cases.

The key change in living arrangement from 1951 to 1971 was that more Canadian families and non-family individuals tended to live alone; a trend evidenced in Table 4.6. Explanations might thus focus on why families and individuals sought to exclude others from their dwellings. Admittedly, a rising propensity to maintain a dwelling, when universally applied, implies that more families or individuals live alone. However, since the propensity to live alone did not increase substantially among primary families and non-family individuals, it is helpful to think instead in terms of why families and individuals increasingly chose to maintain a dwelling of their own.[17]

This trend was also arrested between 1981 and 1986. For some reason, individuals and families did not become more likely to live alone during this period. What caused this change?

Table 4.6
Indicators of undoubling, Canada, 1941–86

	1941 (%)	1951 (%)	1961 (%)	1971 (%)	1981 (%)	1986 (%)
Families						
Percentage of families maintaining dwelling*	92	90	94	97	97	96
Percentage of families living alone	–	–	79	84	88	88
Individuals						
Percentage of individuals who are non-family[†]	14	13	12	13	15	16
Percentage of non-family persons who live alone	–	14	20	30	47	47
Private households						
Percentage with lodgers[‡]	–	–	19	16	12	12

Sources: Calculated from published reports of the Census of Canada, various years.

Note: – indicates data are unavailable.

* See the last two notes to Table 4.1. In 1981 and 1986, the numerator excludes a small number of families maintaining a dwelling that consists of two or more families.

† Includes persons living in collective dwellings, regardless of family status.

‡ Includes any household not consisting of a person living alone or family living alone.

CHANGES IN THE DAILY PATTERN OF HOUSEHOLD LIFE

How did the daily pattern of household life (i.e. activities in or around the home) change over the past half-century?[18] Were there changes in the kinds of activities undertaken at home? Did the nature of these activities change? Did the time typically spent at each activity change? Although we lack good empirical data, several studies have identified qualitative changes in household life using popular literature such as advertisements, advice books, magazine articles, and novels.[19]

There is powerful, even if indirect or partial, evidence of change in the daily pattern of household life over the past half-century. One piece of evidence has been the increasing participation of wives in the paid work-force. Reviews of empirical studies of household time budgets suggest that, among husband-wife families, total time spent on housework declined with wives' increased participation in the labour force.[20] Similar evidence is found in the rising incidence of the one-person household. Persons in the paid work-force and living alone have less flexibility in offsetting the demands of work and household life; hence, they typically cannot spend as much time at home as can others. People living alone are also less frequently able to remain at home when they need medical care.

Other evidence of change is found in the rise in expenditure outside the home on substitutes for household activities. The amount that consumers spent on restaurant meals, for example, rose over the past half-century,[21] as did spending on accommodation other than principal residence (e.g. second homes, cottages, student and staff residences, hotels, and time-sharing units). With an increasingly metropolitan population and attendant suburbanization, the time (and money) spent commuting may also have increased, further reducing the time that individuals spent at home.

CHANGES IN THE RESIDENTIAL GEOGRAPHY OF CANADA

How has the residential geography of Canada changed in recent decades? This is a broad question. Let us here focus on just three changes that affected the quality of life: expansion of the housing stock, the rise of metropolitan Canada, and change in the social, ethnic, and income mix of households within neighbourhoods.

Expansion of the Housing Stock

An important aspect of the changing residential geography of Canada over the past half-century has been the overall growth in the amount, and improved quality, of housing. At one level, we can think simply of the net total new stock added. Between 1941 and 1986, the stock of private, occupied dwellings rose from 2.6 to 9.0 million units. That this total increase is net of any losses (e.g. demolitions, conversions, and abandonments) understates the gross total of units added.

Further, the quality of the typical new addition to the stock (e.g. floor area, lot size, quality of construction, and quality and amount of built-in appliances, fixtures, and fittings) also increased. Table 4.7 shows the new construction component of business residential gross fixed capital formation (BRGFCF) for selected years from 1951 to 1989.[22] Also shown is the implicit price deflator for BRGFCF (1981 = 100) from which can be estimated real new construction in 1981 dollars.[23] By dividing the latter by the number of dwelling completions, an average quality estimate can be obtained (bottom row of Table 4.7) of the typical investment in a new dwelling.[24]

Average dwelling quality increased from 1951 to 1956, and again from 1971 to 1986. However, the average quality of a new dwelling fell during the 1970s – surprising given the growth and robustness of the Canadian economy during that boom period. However, consider the effect of the changing mix of dwelling completions. As late as 1961, there were about three completions of single, detached dwellings for every apartment completed. With the high-rise apartment boom of the 1960s, there were more completions of apartments than of single, detached dwellings by the end of the decade. Apartments, typically smaller in floor area and hence less expen-

Table 4.7

Annual investment in housing, Canada, selected years, 1951–89

	1951	1956	1961	1966	1971	1976	1981	1986	1989
Housing investment ($billion)	1.1	2.2	2.2	3.2	5.6	14.1	20.6	30.8	48.0
New construction ($billion)	0.7	1.6	1.4	2.1	4.1	9.5	11.?	15.4	24.4
Implicit price deflator (1981 = 100)	23	25	26	31	39	70	1C	122	155
Real new construction (1981 $billion)	3.1	6.6	5.6	6.9	10.3	13.5	11.2	12.6	15.7
Total new completions (000s)	81	136	116	162	201	236	175	185	217
Average quality (1981 $000/unit)	38.1	49.0	48.3	42.4	51.2	57.0	63.6	67.9	72.4

Definitions and sources: Housing investment is business residential gross fixed capital formation in current dollars; it excludes government investment in housing and includes new construction, alterations and repairs, and transfer costs. Data taken from *National Income and Expenditure Accounts: Annual Estimates, 1926–1986*, Cat. 13–531; and 1979–1980, Cat. 13–201, Statistics Canada, 1988; and 1992.

New construction is the component of business residential gross fixed capital formation that includes investment in new buildings as well as conversions of existing residential structures.

Implicit price deflator for business residential gross fixed capital formation. Data taken from Statistics Canada, 1988 and 1992 (as above).

Real new construction is the new construction component deflated by the implicit price deflator for business residential gross fixed capital formation.

Total new completions is an annual total as estimated from *Canadian Housing Statistics*, CMHC, various years.

Average quality is the ratio of real new construction to total new completions.

sive per dwelling to construct, pulled down the average investment per completion. The boom in apartment construction ended suddenly in the early 1970s, succeeded by a shorter and less intense boom in row housing and the re-emergence of completions of single detached dwellings. Average investment per completion then increased.

Some additions to the stock also came about through conversions – as in the transformation of an existing building that contains one dwelling into two or more dwellings, usually by creating a separate upstairs or basement apartment.[25] Offsetting these additions were deconversions such as those that sometimes occurred during "whitepainting" when a multi-family house was converted back to a single-family dwelling.

As well as new additions to the stock, the quality of existing stock changed. Census data suggest a considerable upgrading in the existing stock in terms of heating, plumbing, electrical systems, and installed appliances.[26] The alterations-and-improvements component rose from one-quarter of BRGFCF in 1951 to one-third in 1986.[27]

Significant changes in tenure also took place in this period. Overall, the percentage of Canadian homeowning households declined during the 1960s and recovered only partially after that. However, overall percentages may be misleading because metropolitan areas, which historically tended to have lower rates of homeownership, grew more rapidly than other parts of Canada, and the kinds of households – for example, those that do not contain a husband-wife family and/or have low incomes – less likely to be homeowners proliferated. In other words, it is perhaps surprising that the incidence of homeownership did not fall faster than it did.

In addition, tenure options increased: new forms of owning and renting emerged (particularly condominium ownership), and property rights associated with existing forms were altered. Other modes of tenure also became more prevalent, including equity and non-equity co-operatives and mixed owned-rented schemes (most widely used for owned mobile homes on rented sites or lots). With increased regulation of land, building, and development, the property rights of land-owners changed. In many provinces, security-of-tenure legislation was also introduced to protect certain rights for tenants.[28]

Metropolitan Growth

Over the past fifty years, the population grew more rapidly in metropolitan areas than elsewhere in Canada.[29] By 1986, the number of CMAs in Canada had increased to twenty-five. This includes settlements as small as Saint John, with just 121,000 people. Still, there were three CMAs with over one million people in 1986. In 1941, only 22 per cent of Canadians lived in the three largest CMAs, compared with 30 per cent in 1986.[30]

Unfortunately, we have few precise data about how the residential environments of Canadians were altered over the past half-century as a result of metropolitan growth. In terms of the well-being of residents, important changes were rising standards in the provision of public services such as sewage, water, roads, sidewalks, and street lighting; increased homogeneity in residential experience; and, at the same time, a greater diversification of neighbourhoods (e.g. ethnic areas; social, income, and demographic groups; and adult, senior, and other purpose-built housing).

In what respects did the residential environments of many Canadians become more homogeneous? Although good data are lacking, I suspect that Canadians increasingly found themselves living in either a high- or a low-density cluster of similar residential buildings. Three post-war trends probably caused this situation: the emergence of a large-scale land-development industry able to assemble, design, construct, and sell "planned neighbourhoods"; greater control by public planners over suburban development;[31] and heightened demand by consumers for such development. Many consumers found their housing geographically separated from other land uses, especially industrial uses. Also, local and regional shopping facilities became commonplace, standardized, and similar across the nation. Finally, access to regional and national newspapers, television networks, and national chain stores became widespread.

Metropolitan concentration often also implies social diversity. By virtue of their size, big cities can have the "critical mass" for many kinds of activities, especially when there is an accompanying social differentiation of neighbourhoods. Such differentiation and the activities that require critical mass are interrelated. Many consumers want access to these activities, which in turn need sufficient consumers nearby to provide the critical mass. The rapid growth of the three principal CMAs in Canada has thus enabled a kind and scale of neighbourhood differentiation that might not otherwise have occurred. (see chapter 9).

Social, Ethnic, and Income Mix

The past half-century has witnessed changes in the social, ethnic, and income mix of particular neighbourhoods. Examples abound of neighbourhoods that were eliminated by bulldozers, that have been allowed to deteriorate slowly over time, or have been improved, regenerated, or "whitepainted." Social scientists have come to appreciate the variety of processes and factors that change or preserve neighbourhoods. Unfortunately, the concept of neighbourhood is also elusive. There is no simple definition of neighbourhood for which historical data have been systematically and comprehensively assembled that would tell us overall how particular kinds of neighbourhoods have changed over time in Canada.[32]

What has been studied at length, as chapter 3 outlined, are changes in the characteristics of census tracts. In the last few decades, it has become possible to make comparisons across censuses. A review of studies of ethnic segregation reports mixed findings; in some cities and for some ethnic groups, segregation appears to have increased since 1961, while in others it has apparently declined.[33] Nonetheless, whatever the change over the past few decades, neighbourhood differentiation by ethnic, social, or income groups is an unmistakable fact of the contemporary residential geography of urban Canada.

Some of this differentiation was made possible by the spinning off of non-family individuals and families from shared accommodation. As time passed, for example, widowed grandmothers became more likely to live alone in a separate dwelling and less likely to live in a child's family home. Some could then live in a housing project or neighbourhood where they would be clustered with others like them. Similar arguments can be applied in the case of many young singles, low-income couples, and lone-parent families.

A THEORETICAL PERSPECTIVE

Microeconomic demand theory can help to explain shifts in demography, living arrangement, household life, and residential geography, as well as their interrelationships. How, with whom, and where people live can, for many of us, be thought of as a choice among alternatives. In a demand-theoretic perspective, choice is a function of the income of the consumer; the prices of alternatives, substitutes, and complements; and the preferences of the consumer.

A microeconomic perspective is useful because it provides testable hypotheses that bear up to empirical scrutiny and that integrate explanations of seemingly disparate phenomena. It has two principal limitations, however. First, it is a clumsy tool for considering effects brought about by changes in supply (e.g. with respect to technology or factor prices) or by public policy. Second, and not unrelated, some consumers have little or no choice; for them, with whom, how, and where they live are shaped principally by that which is available. However, it is beyond the scope of this chapter to do more than mention briefly the effects of supply changes, housing availability, and public policy (see chapters 9 and 15). A demand perspective helps provide insights about changes in living arrangement, household life, and residence that have been identified so far.

It is helpful to begin by imagining three separate, but related, markets. The first is the market for residential structures. On the demand side are would-be landlords and homeowners; on the supply side are builders and existing landlords and homeowners. The second is the market for housing services: landlords combine structures with heat, water, grounds maintenance, and other services which are then rented to a tenant; homeowners in contrast both supply and demand housing services. The demand for residential structures is also linked significantly to the supply of housing services. The third market is the market for household production: each household both supplies and demands its own household production.

Microeconomics of Household Production

Beginning with the work by Becker and others in the 1970s, a useful body of economic theory has been developed to explain living arrangement and household production.[34] Central to this theory is the notion of a household as a source of both demand and production of certain household services. This theory is derived from two antecedents: the idea that consumption requires time as well as goods, and the idea that the benefit of consumption arises from attributes of the consumer good, rather than simply the quantity of the good itself.

In the economic theory of the household, members contribute time and goods to domestic production which directly provides utility, and that either cannot be purchased in the market or is a substitute for market-supplied goods.[35] There are a variety of domestic production activities: from reading a newspaper, preparing a meal, and gardening, to the provision of affection and childbearing. Some domestic production activities have close substitutes in the marketplace (e.g. restaurants and professional gardening), while others do not. The time that members of a household spend in activities that might be described as household life depends on the preferences of the household (determined culturally and by household composition), on the opportunity cost of time, on the prices and availability of substitutes and complements, and on the household's income.

Viewed from this perspective, living together offers advantages and disadvantages.[36] Economies of scale arise in shared use of facilities: for example, all members of the household derive the same benefit from having a front door lock, regardless of whether the dwelling contains one person or more. However, as the number of people sharing a dwelling increases, so does competition and congestion in the use of common facilities such as bathrooms and stoves.

At least in theory, living together permits a better division of household labour in the split between outside (paid) work and domestic (unpaid) work. Someone who lives alone has to undertake all the required domestic and paid work. When two adults share a dwelling, they can have different combinations of domestic and paid work that better use their comparative abilities.[37] This can be especially helpful if one of them has, for whatever reason, limited access to paid work. Household members, in sharing domestic work, may, at least in principle, be able to specialize in household duties which they enjoy or at which they are best.

Living together can permit substitution of market-supplied goods in favour of household production (e.g. home meals over restaurant meals, at-home recreational pursuits over outside paid entertainment, at-home care during illness versus hospital/clinic care). Living together may also enable complementarities: i.e. to encourage residents to consume several commodities domestically, where they are by their nature consumed jointly and where at least one is best supplied domestically − for example, the larger domicile that becomes affordable/necessary for a bigger household makes it possible to entertain others at home and to have a garage in which to store things or to make repairs or a basement for hobbies.

There can also be substantial changes in the rate of domestic production over one's lifetime. For example, individuals may wish to engage in more domestic production during childbearing and childrearing, in part because of new complementarities that arise in the consumption of domestic production or substitutions made possible by an accompanying change in labour-force status. Domestic production also changes partly because shifts in the composition of the household alter preferences and hence demand patterns. Standard microeconomic treatments assume that a consumer (an individual or a household) have an unchanging set of preferences over time. However, even if individuals' preferences remain the same, those of the household (however these may be collectively defined) can be expected to shift with the arrival of new household members or the death or departure of members.

Microeconomics of tenure

Consumers expect their real income to vary over a lifetime. Typically, their incomes rise through middle age, then decline in old age. Such changes in income are not necessarily tied to offsetting shifts in household composition and spending needs. As a result, consumers find it advantageous at some points to save and at other points to deplete their savings.

In this context, microeconomic models have been developed to explain tenure choice.[38] In practice, they model simply the choice between owning and renting a home; ignoring the distinctions between condominium and fee-simple ownership and between co-operatives (equity or non-equity) and other types of tenure. They ignore as well the effects on tenure choice of changes in the property rights of both renters and owners in recent decades. Most such models treat the household as an entity whose income may change over a fixed lifetime but that otherwise remains the same. They also usually assume that the household must save for a down payment to make homeownership "affordable,"[39] that the household will remain a renter until then, and that the household will switch to homeownership only where it is financially advantageous. Such models typically solve for the age of household head at the time of the switch to homeownership and the size of down payment. Among the explanatory variables are expected future incomes, tax rates, mortgage rates, opportunity costs of capital, prices of owned homes and expected capital gains, and depreciation rates.

These models, however, do not typically take into account that the composition of the household – and hence its preferences with respect to housing – change over time.[40] Consider as one indicator of household composition the effect of an increase in census family size on tenure choice. Larger families might strongly prefer the security of tenure that comes with homeownership.[41] But large families, especially those of modest income, also feel greater pressure on their budget in terms of other essentials (e.g. food and clothing) and thus may find it more difficult to accumulate the downpayment required to make homeownerhip feasible.

Microeconomics of Location and Neighbourhood

Starting with the seminal works of Alonso and Muth in the 1960s, much attention has been paid to the development of microeconomic models of residential location.[42] These models focus on the trade-offs that consumers make in choosing between a central site (with good access to worksites but typically high rent) and a suburban site (perhaps less accessible but with lower rent). In their simplest form, these models assume that the consumer chooses a location (hence a commuting cost), an amount of land (Alonso model) or housing services[43] (Muth model) to be rented, and a quantity of other goods to be consumed that maximizes utility subject to the constraints of a dollar budget, and sometimes a total time budget. They also assume a large group of identical consumers.[44] However, heterogeneity of consumers can be built into the model by having more than one class. In monocentric city versions of these models,[45] classes of households less sensitive to the marginal cost of commuting locate toward the urban periphery.

One interesting speculation has to do with the effect of rising real incomes on the density of land use. In monocentric models where there are only monetary costs of trip making, a rise in average real income leads to a lower-density, more dispersed city wherein average commuting distance is greater than before. However, the scene is muddied when we introduce a time budget in addition to a money budget. As the incomes of consumers increase, so too does the opportunity cost of their time (whether in terms of paid work or of domestic production forgone); some consumers may choose to undertake less commuting (and therefore live at a higher density) as their incomes increase.

EXPLAINING THE CHANGES

What caused the changes in living arrangement, daily pattern of household life, and residential geography observed at the outset of this chapter? To what extent, in particular, were these shifts brought about by a new demography, rising affluence, changing technology, or other factors? To what extent were these changes themselves interrelated? These are substantial questions, for which there is room here only to sketch the basic components of one kind of explanation. Let us now briefly outline aspects of these arguments, divided into demand (demographic change, change in real income), supply (changes in availability, amenity, technology, and affordability), and public policy initiatives.

Demographic Change

Greater longevity, the baby boom, and extensive immigration caused the number of families and individuals in Canada to more than double in the past fifty years, thus propelling the demand for housing and the spread of post-war suburbs. Against this backdrop of overall increase, however, were significant changes in aggregate number and propensity. Four demographic shifts substantially affected living arrangement and the residential geography of Canadian cities: the marriage boom and bust, the persistent and growing differential in longevity between males and females, the baby boom and baby bust, and the changing pace of immigration.

The marriage bust of the 1970s and 1980s as noted above contributed to the substantial growth of non-family households. Whether through never marrying, or through increased rates of separation or divorce, more adults were tending to live alone or without family, a trend offset only in part by the rise of cohabitation. The numerical importance of the non-family individual was most marked in Canada's three principal CMAs. In 1986, 17 per cent of all people resident in the three principal CMAs were non-family individuals, compared with 13 per cent in the rest of Canada.[46]

How did the marriage bust manifest itself "on the ground"? While there was some new construction geared to the needs of a "singles" or "adult only" population, much of the housing built in the 1970s and 1980s continued to be aimed at the family household. New households formed out of the marriage bust – singles and lone parents – coped typically through adaptation of the existing stock. The high-rise stock, whether rental or condominium, built up during the great apartment boom was attractive for many singles, sometimes because of a central location, always because of the convenience of tended facilities. That rents increased less than inflation added to the attractiveness of such housing and facilitated this living arrangement.

During the past half-century, women increasingly outlived men. Both sexes were living longer than before. However, the trend to lengthier widowhood was only partly offset by the fact that, since males were also living longer, wives could expect to have their husbands around for more years. The number of elderly widows increased rapidly, and this sex differential in survival rates helped fuel the demand for purpose-built senior citizens' housing. Such residences, mainly small, inexpensive dwellings with grounds-keeping and maintenance services, reduced the cost and physical effort of household maintenance for many of the elderly. Much of this type of housing was constructed by government agencies and non-profit enterprise; these organizations quickly found that economies of scale in production and operation led to larger and larger projects. Although a few projects mixed different kinds of households (i.e. elderly and non-elderly), many isolated the elderly physically from the rest of the community, and this segregation itself has recently become a public concern.

In the late 1940s and the 1950s, the baby boom also enlarged completed families. As it aged, the baby-boom generation swelled the ranks of young adults, beginning in the late 1960s, with subsequent effects on aggregate household formation. Changes in the nature and timing of life cycles over the past half-century affected the desirability and extent of domestic production. The advent of the baby bust brought rising childlessness. The number and duration of domestic production activities associated with, or complementary to, childbearing and childrearing presumably declined as a consequence.

Accompanying the baby bust was a compression of the age interval among women normally thought to constitute the childbearing years. The post-war baby boom was manifested partly as a decline in the typical age of women at first birth, a trend reversed in the baby bust. At the same time, there was a steady and marked decline in fertility among women over forty years of age during the past few decades. Thus the past fifty years saw a shift from more births dispersed over a woman's childbearing years to fewer births concentrated narrowly within the childbearing years. If one held other things constant, one might expect a family to adjust its housing consumption over its lifetime in response to changes in family size and composition. A drop in completed family size should have reduced the housing consumption of a typical family, and the compression of childbearing should have shrunk the period during which the family's space needs were at their peak.

Immigration has also affected living arrangement and residential geography. Immigrants tended to be poorer, on arrival, than their Canadian-born peers in terms of both wealth and income. In addition, some immigrants were ineligible for certain social assistance programs – for example, not all elderly immigrants were eligible for Old Age Security (OAS) and/or Canada/Quebec Pension Plan (CPP, QPP).[47] Living with relatives or friends was one way of coping with inadequate income; it could also be a hedge against economic uncertainties and provide ease of access to job opportunities and other employment information, as well as social contacts and support.

Changes in Real Income

Overall, Canadians have experienced considerable prosperity since 1945. In 1946, per capita personal disposable income was $728. By 1989, it had increased to $16,664. Even after accounting for inflation, real per capita personal disposable income rose by 180 per cent.[48] Between 1946 and about 1976, incomes grew faster than did Statistics Canada's overall consumer price index (CPI) generally, or its shelter-cost component specifically. In this sense, housing became more affordable. Consumers were able on average to purchase bigger and/or better housing. And until the mid-1970s, more of them were able to afford separate housing than had been possible before. The experience from the mid-1970s through much of the 1980s was reversed, however, when individual and family incomes, especially of male family heads, typically failed to keep up with inflation generally, and shelter prices in particular (see Figure 4.1).

The growth of real incomes differs depending on the base used – income per household or per capita. Because the number of people living in a typical household fell, income per household rose more slowly than income per capita. The increase in per capita income and the decline in household size are linked: increased prosperity enabled a spinning off of non-family individuals and families from shared accommodation, which in turn reduced household income. Rising prosperity may well have increased the demand for smaller dwellings, the kind that would be occupied by the newly formed small households.

Metropolitan growth also affected incomes. Canada's three main cities grew during this period partly because of the scale, agglomeration, and urbanization economies they made possible; in addition, the new jobs and the high incomes that they offered helped attract migrants. At the same time, metropolitan residents typically spend more time and money commuting and on shopping and recreation trips. In other words, as gross incomes increased, so did pressures on their money budgets and constraints on leisure time.

Offsetting the effect of smaller household sizes, and also contributing to a rise in household incomes, was the entry of wives into the paid work-

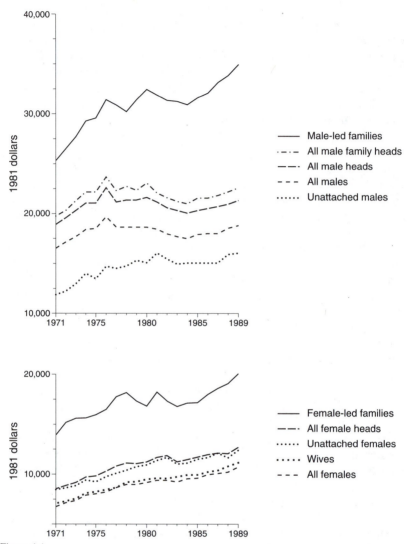

Figure 4.1
Real incomes of families by sex of head and of individuals by family status, 1981 dollars, Canada, 1971–89. "All female heads" includes three categories that follow and others (e.g. daughters).

force. Couples with two income earners made up just 20 per cent of Canadian husband-wife couples in 1961; by 1986, the figure had risen to 54 per cent. This rise was a consequence in part of the growth of the service sector, which included many jobs that had traditionally been female-dominated, and of changing attitudes that saw an increasing number of women in what had been male-dominated occupations.

There was less time spent at home because more women were working, in paid jobs. This had several implications for household life: the extent of lodging decided, and households producted fewer services for themselves (e.g. meals/food prepared at home) and bought more services (e.g. house-cleaning, nanny, concierge). For women who had not previously entered the paid workforce, the new income and independence may also have reduced the need for marriage.

Childbearing also brings with it demands for childcare services. To the extent that this reshapes the allocation of labour within the household, it also creates the possibility of – or (with the resultant drop in income) necessitates – increased domestic production. Also, childbearing decisions may well be linked to decisions abour formal education in particular and to human capital formation (which also includes on-the-job experience and skill acquisition) in general.

Boarding and lodging had traditionally been services offered in the home by wives that contributed to household income. Increased work outside the home made it increasingly difficult, or undesirable, for families to take in boarders or lodgers. Viewed another way, work outside the home was a sub-stitute for commercial work inside the home. Also contributing to the rise of household incomes since about 1965 has been teenagers' increased labour-force participation, much of it in the form of part-time work by students. Prior to about 1965, work-force participation among teenagers had been slowly declining. The growth of the service sector, the rising costs of post-secondary education, and changing attitudes toward part-time employment helped make part-time paid work more common among young adults.

Together, the increased work-force participation of wives and students substantially changed the daily pattern of household life. Households' in-comes grew, but members had less discretionary time in which to engage in domestic production and spent more on commercial products such as pre-pared foods.

How have rising real incomes affected the choice of location within cities? Have cities become more spatially dispersed as would be predicted by the simple microeconomic model outlined above? There certainly was evidence during the 1950s and 1960s of lower densities in new suburban de-velopments. However, with the booms in housing prices that have swept the markets for owner-occupied stock in various locales since the early 1970s, increased regulation of urban sprawl, and changes in household composi-tion, and hence preferences, the role played in subsequent changes by rising incomes alone is unclear. In particular, the higher number of working wives and the increasing number of people living alone probably reduced the time available for home-oriented maintenance and hobbies, and declining fertility reduced the need for children's yard space; all favour the higher densities in new residential development that began to emerge in the 1970s and 1980s.[49]

Rising real incomes have also influenced tenure choice. There is evidence that, during the 1970s, the likelihood of ever owning a home increased with the growth in real incomes.[50] In addition, households tended to become homeowners at ever-younger ages. This trend appears to have slowed during the 1980s, however, as incomes failed to keep pace with inflation generally.

There is also indirect evidence of a growing volume of income/wealth transfers among households, particularly between generations.[51] Part of the wealth accumulated by parents was passed on to their adult children to assist in making rent payments (e.g. for student housing) or down payments on a home purchase. Also time and effort were passed between generations, as in childcare services provided by grandparents. In other words, part of the growing affluence of some households was passed on to other households by cash and in-kind transfers, enabling the formation and survival of households whose incomes would have been too low to maintain a separate dwelling by the standards of time past.

Changes in Availability, Amenity, Technology, and Affordability

Until about the mid-1970s, the cost of a standard unit of housing services[52] increased less quickly than inflation generally, and much slower than per capita income. Whatever the cause for this, rational households might be expected to consume more housing, because of both the income effect and the substitution effect. At the same time, less-expensive housing would induce more consumers to form separate households.[53]

It is surprising that, with metropolitan growth, housing costs did not rise faster. One might have expected that, as more Canadians moved into the three principal CMAs, they faced the prospect of paying higher costs for land, and hence housing. While housing costs were somewhat higher in these large centres than in nearby communities, the increase in shelter costs attributable to metropolitan growth was, at least for renters, modest. Table 4.8 presents estimates of the monthly shelter cost (rent plus utilities) for a rental unit of standard quality in each of fourteen urban areas across Canada between 1953 and 1988.[54] Also shown are the population-weighted average shelter costs for samples of small and large CMAs as well as for the three principal CMAs. Average shelter cost in the case of the three principal CMAs was held down by sluggish increases for Montreal. Nonetheless, shelter costs were higher overall in the three largest CMAs, although smaller, more remote communities such as St John's also had high shelter costs. Thus, for some migrants at least, housing costs did not increase (or at least do so substantially) when they moved to the larger cities.

Owner-occupied housing may also have become more affordable to the extent that liquidity in the housing market grew. Housing has traditionally

Table 4.8
Average monthly payment (rent plus utilities) for standard-quality dwelling in fourteen Canadian urban areas, 1953–88

Year	Metropolitan				Large cities							Smaller cities					
	Mtl ($)	Tor ($)	Van ($)	Avg ($)	Cal ($)	Edm ($)	Hal ($)	Ott ($)	Que ($)	Win ($)	Avg ($)	Reg ($)	St Jn ($)	Ssk ($)	St Jn's ($)	Thn By ($)	Avg ($)
1953	94	124	125	111	106	109	115	104		96	104	104	91	99	132		107
1958	102	142	140	124	112	116	129	123		107	116	115	108	109	139		118
1963	107	143	141	127	113	117	136	129		112	120	120	114	113	143		123
1968	117	160	156	141	125	129	144	138		120	130	130	122	123	154		133
1973	130	185	179	161	145	149	170	170		139	154	144	141	136	180		151
1978	194	282	285	247	244	245	242	255	198	225	235	237	207	210	281	233	235
1983	300	409	422	369	386	392	357	377	321	339	364	353	301	327	428	335	352
1988	379	534	460	463	389	407	418	471	391	406	416	424	357	386	494	388	413

Source: Estimated by author; see text for details.

Notes: Column-label abbreviations: Montreal (Mtl), Toronto (Tor), Vancouver (Van), Calgary (Cal), Edmonton (Edm), Halifax (Hal), Ottawa (Ott), Quebec City (Que), Winnipeg (Win), Regina (Reg), Saint John (St Jn), Saskatoon (Ssk), St John's (St Jn's), and Thunder Bay (Thn By).

been illiquid in that it is tied to land – it requires foundations, water and sewer connections, and other physical and social infrastructure. It is also difficult or expensive to convert a residential structure to another use, to move it elsewhere, or to tear it down at one location and build anew or reconstruct elsewhere. Hence, investment is subject to the vagaries of the factors that determine the demand for housing at that specific location. In larger cities and towns, there has also been a relatively well-developed market for housing throughout the past fifty years. In rural areas, and smaller and remote communities, the market was not as well developed. Improved liquidity came about partly because of the geographic shift in population into larger cities and towns and because liquidity increased in the housing markets of smaller cities and towns, although geographic differentials still persist. Greater liquidity makes households more willing to risk homeownership and may have increased housing demand overall.

Over this period, housing and neighbourhoods have been increasingly subject to planning regulation. As a result of higher standards for public facility and infrastructure provision, housing and neighbourhoods became healthier, safer, less prone to fire or flooding, and more efficently serviced. There was also growing uniformity in the nature and extent of regulation and in the standards employed. The proliferation of regional governments, especially within metropolitan areas, also enabled specialization in the production of public services (e.g. education, health, police). Much of this homogenization was a consequence of the post-war proliferation of suburbs. When the infusion of new residents began in earnest after 1945, many cities experienced growing pains, including suburban spawl, traffic congestion, costly lot servicing, water and sewer trunk capacity limits, and storm flooding. Municipalities and provinces responded with new programs, policies, and regulations designed to ensure that new residential development was permitted only when certain minimum design, construction, and servicing standards had been met.

Finally, the past half-century has witnessed major changes in the technology of household production. New materials and products have been developed – for example, automatic clothes washers and dryers, dishwashers, frost-free refrigerators, self-cleaning ovens, microwave ovens, vacuum cleaners, wax-free floors, durable broadloom, wrinkle-resistant clothing, and prepared foods – that, at least in principle, enable household members to spend less time in domestic production. New services aid or substitute for household production – for example, condominiums, home service clubs, childcare centres, home nursing programs, and "meals on wheels." In a similar vein, group housing and other special-needs housing combine private accommodation with certain enabling services. This changed technology made it feasible for individuals to maintain a living arrangement that would otherwise simply not have been possible.

The Effects of Public Policy

Over the past several decades, increased expenditures by Canadian governments on a variety of social programs had implications for choices of living arrangement and rates of new household formation.

Among social programs, changes were made to the principal income maintenance programs: old age security (OAS), guaranteed income supplement (GIS), Canada/Quebec Pension Plan (CPP/QPP), family allowances, and unemployment insurance (UI). In addition, each province operates a health insurance plan, jointly funded with the federal government, and a variety of social assistance schemes, also jointly funded, under the Canada Assistance Plan. These programs have all been modified in the last few decades. Changes include extensions of UI coverage, indexing of pensions against inflation, introduction of spouse's allowances, elimination of family allowances, and introduction of a refundable child-tax credit. These changes boosted and helped stabilize the incomes of some of Canada's poorest households (at the expense of wealthier households). As a result, some individuals and families can afford living arrangements and forms of housing that would earlier have been out of reach.

New initiatives were introduced throughout the 1960s and 1970s, followed by a scaling back of some programs in the 1980s. Some program benefits that became fully indexed against inflation in the 1970s were partially de-indexed in the 1980s. Eligibility requirements were tightened. Some benefits to wealthier households became subject to partial or total "clawback," and others were replaced by tax credits. A number of programs were changed in the name of targeting – i.e. to direct benefits at those most in need. In other cases, the need for fiscal restraint simply made it impossible to continue prior levels of benefits. It is too early to say how this scaling back will affect future living arrangements, daily patterns of household life, and residential geography.

Some of these same initiatives may well have reduced the demand for owner-occupied housing. Traditionally, owner-occupied housing served as a "nest egg" against both unemployment and poverty in old age. UI and welfare assistance now provide an alternative form of protection against unemployment; while OAS, GIS, and the CPP/QPP offer some income security for the elderly.

Varied policy initiatives in the housing area have also affected living arrangements and housing consumption. Five sets of tools have been used by governments in Canada: direct subsidies, tax expenditures, loans and loan guarantees, regulation, and crown corporations. *Direct subsidies* include operating grants, interest differential subsidies, and capital gifts to low-income, non-profit, limited-dividend, co-operative, and special-needs housing; examples are the Assisted Home Ownership Plan (1973–78) and the Residential Rehabilitation Assistance Plan (1973–   ). *Tax expenditures* include the absence of taxation on the imputed rents or capital gains of homeowners, property tax credits, and the recent policy to permit RRSP funds to be used for home purchase, as well as depreciation and other expense allowances on rental property; one example is the Multiple Unit Rental Building (or MURB) Program, 1974–79 and 1980–82. *Loans* include direct lending by CMHC and NHA-insured mortgages for new construction, existing housing, and repairs. *Regulation* includes all national and provincial building codes, home-warranty programs, landlord-tenant and condominium legislation, rent regulation, and the extensive network of controls exercised over land subdivision, zoning, and development. The role of *crown corporations* includes the dissemination of research and information on housing-market activity and new technologies by Statistics Canada, CMHC, and the National Research Council, as well as initiatives in market development, such as mortgage-backed securities (1986–   ).

Each of these activities may have directly or indirectly affected the price of shelter and hence the living arrangement choices of individuals and families. In some cases, as in the subsidy programs, these policies reduced the price of shelter, making housing more affordable and possibly increasing the income available to consumers to spend on other goods. In other cases, such as the increased regulation of new construction, the result may have been a better stock of housing, but also one that is more expensive than would otherwise have been the case.

CONCLUSION

Unquestionably, changes in demography (i.e. in fertility, nuptiality, longevity, and migration) have helped shape living arrangement, the daily pattern of household life, the quantity and forms of housing consumed, and the residential geography of Canadian cities. This chapter summarizes evidence of change in each of these areas. It also shows how a microeconomic perspective can help explain living arrangement, housing consumption, residential location, and tenure. Such a perspective helps us to understand how demographic change operates, in conjunction with shifts in consumer income, the prices of housing alternatives, and public policy, to determine residential geography.

CHAPTER FIVE

Urban Social Behaviour in Time and Space

D.G. JANELLE

Time is an increasingly central factor in the emerging geography of Canadian cities. This chapter describes recent changes in the temporal order of social and economic relationships, explores their implications for the behaviour of urban residents, and considers their importance to the spatial structure of urban life and city form. A general conceptual discussion is followed by a survey of time use and space-time behaviour in several cities and by a more detailed profile on the time-geography of Halifax.

THE TEMPORAL ORDERING OF URBAN LIFE

In the social geography of cities, minutes count: the need to mesh individual life with the time demands of social and economic obligations helps establish the temporal ordering of life in Canadian cities.[1] Spaces within cities, whether for individual establishments, such as households or shops, or districts, such as industrial or shopping zones, are scheduled for specific events and patterns of activity, for opening and closing, for entering and exiting. So too, linkages within and among these spaces, represented by flows of people, goods, money, and information, are controlled by the clock and by transport limits to the speed of movement.[2]

Individual parts of cities often assume their own tempo of events, with rhythms that give them distinctive character according to the time of day.[3] Compare the solitude and emptiness of Bay Street in Toronto at 0300 hours with the air of expectation and the bustle of people at 1500 hours. Changes in the demographic composition of such spaces throughout the day create a (re)cycling of behavioural settings, each associated with a cognition and symbology of appropriate behaviour.[4]

The temporal character of individual urban spaces is bound up in a web of timed linkages with other parts of a city, with events elsewhere in Canada and the world, and with the needs of individuals. Thus, for example, food distributors begin rounds to the shopping districts long before customers arrive, stock brokers in St John's adjust their business hours to synchronize with events on Wall Street and Bay Street, and all activities ought to respect the individual's biological needs for food and rest. Even essential airline links with other parts of the world are governed by the need to suspend airport traffic during normal sleeping hours.

The temporal order of social and economic life has always reflected needs to co-ordinate different levels of activity and different scales of societal organization. However, the most basic units of social structure (the individual and the household) must co-ordinate at the broadest levels. Not only must they respond to the temporal structure of their economic lives, but they must adapt their programs to the social needs of family, friends, and neighbours, to institutional requirements such as those relating to schools, and to the temporal prescriptions of behaviour that accompany cultural and religious observance. Behaviour at the individual and household levels best reflects the impress of temporal controls on the social geography of cities. Individual decisions on career, places of residence and work, accommodation type and neighbourhood, marital status, and mode of transport all impinge on one's time demands.

Other chapters in this book identify significant recent changes in the social character of Canadian households and its response to rapid changes in employment conditions.[5] The wage economy is based, in part, on the monetary value attached to a unit of work (usually the hour). In this context, the household seeks satisfactory levels of comfort and enhancement of its purchasing power. However, since more than money is at stake in the allocation of time to employment, society is continually re-evaluating its legislative regulations and norms over the timing of work and other activities. Examples include concerns about acceptable working hours (flexitime, staggered work hours, shift work), working days (compressed work weeks), and minimum wages; acceptable times for shops to remain open (twenty-four–stores, Sunday shopping); ages for children to enter formal education and for adults to begin and end their work careers; and the timing of statutory holidays and the scheduling of the school year.

As throughout the past century, these matters are related to labour-management relationships, changing interactions between religious and other cultural values, and a range of other dimensions. However, the intensity of these concerns in the period since 1950 is unparalleled – the increasing complexity of meshing individual needs with household and societal demands has encouraged changes in the ways in which children are raised, food is prepared and served, and work and play are defined and carried out.

This environment of change has been accompanied by an array of technologies to squeeze the most from each minute, whether measured in profit or in pleasure. Microwave ovens, telephone answering and fax machines, electronic mail, cellular phones, vcrs, digital watches with alarm and time-lapse capabilities, and a profusion of time management courses are all seen as enhancing mastery over the timing of events in one's life. Even recent land use practices reinforce this intensification in the exploitation of time – quick-stop shopping now prevails for such basic services as laundry, eating, groceries, banking, video tapes, and donation boxes for recycling used items. Things are increasingly available not only when we want them, but where we want them.[6]

In spite of increased mobility and access to a range of time-saving innovations, Canadians report that their lives are increasingly "rushed."[7] The sense of being rushed is shared across various forms of household organization – but particularly among members of dual-career households with children and for single parents. And within any household, women report more pressure on their time than men. While the general problems associated with these groups are discussed in other chapters, changing temporal patterns in social and economic activity help explain the nature of the problems and identify solutions. The flexibility provided by such innovations as adjustable work hours and banking machines may answer some needs, but entirely new social services are also being invoked – especially, extended public and private provision of child daycare services for working parents. These and other concerns over the timing of life in Canadian cities are seen through a brief review of recent reports in the press.

CONSCIOUSNESS OF TIME-BASED SOCIAL AND ECONOMIC ISSUES

Headlines from selected newspapers in 1986–88 reveal lively debates over time-related concerns for structuring social relations and patterns.[8] Issues are organized around three dominant themes: the structure of work, work time and family life, and changing temporal patterns in shopping. They reveal problems that mark the current political agenda of federal, provincial, and municipal governments in Canada and that pervade the everyday consciousness of residents.

The temporal structure of work is marked by awareness of a growing dichotomy between full-time and part-time jobs, the lack of adequate social benefits for part-timers, and the inability of many to move into the full-time work-force. Some see compressed work weeks as a basis for sharing scarce jobs, while many full-time workers express interest in trading work time for more free time, even at lower incomes. But shift work, in particular, is seen as interfering with family activities. Conditions of employment impinge most directly on the social life of Canadians, through their effects on families.

Topping the press coverage of all time-related issues in the 1986–88 survey was the provision of daycare service for children of working parents, prompted by the rapid movement of women into the labour force, because of economic necessity or a desire for a career outside the home.[9] Many dual-income and most single-parent households see the availability of reasonably priced daycare as essential to their economic livelihood. In most cities a growing consensus has formed on the need for this service, and business participates increasingly in providing child care, either at or near work sites.

Meanwhile, recent polls show that even when women are co-equal household breadwinners, they retain an unequal burden of household responsibility, including weekly shopping.[10] In this context, too, shopping malls have become focal points of urban life, and heated debate has emerged in some provinces over the appropriateness of Sunday shopping. On the surface, the latter debate is a clash of religious, union, and family-life values with economic interests. However, extension of the shopping week is seen by some as a safety valve to relieve the pressure of cramming all shopping events into already crowded evenings and days off. In another sense, this colonization of time[11] by commercial activity may be seen as intensified exploitation of the time dimension to facilitate a more rapid turnover of capital.[12]

TIME USE IN CANADIAN CITIES

Any focus on time as a factor in urban social geography must consider the use of time – how people package their days with discrete events of specific duration and sequence. Canadian researchers have been particularly active in time-budget research since 1970. Interest in monitoring the quality of life and the need for indicators of social trends inspired research undertaken in Halifax, Vancouver, and Toronto.[13] The 1971–72 survey in Halifax, to be described in more detail below, was the most extensive study of its kind in Canada and set a standard for comparison with the Multinational Time Use Study.[14]

The first survey of comparative time use for different locations across Canada was conducted in 1981. The Canadian Time Use Pilot Study included eleven cities and three rural areas with sample sizes of 105 to 197 persons for each, except for Halifax with 495.[15] The larger sample for Halifax was to permit evaluation of a separate panel survey of some of the original participants in the earlier 1971 study. Only a selected portion of information for the ninety-nine activity types used in the study (mean number of minutes per activity per day for respondents who are at least fifteen years of age) is shown for the fourteen places in Table 5.1.

Table 5.1 suggests that commuting time increases with city size, that general conversation ranks unusually high in Quebec centres, and that leisurely home meals and crafts are more popular in rural than in urban places. Dif-

Table 5.1
Time use across Canada

Places	Average minutes per day on selected activities in cities and rural areas						
	Normal work	Trav. to/ from work	Meals home	Meals rest.	Sports & exer.	Home craft	Talking/ conver.
St John's	180	17	62	10	10	14	28
Charlottetown	157	11	72	17	2	28	24
King's Cty, PEI	190	16	67	13	2	25	29
Halifax	173	18	53	15	7	17	25
Saint John	161	17	57	9	3	17	29
Sherbrooke	144	13	58	21	11	13	75
Brome Cty, Quebec	185	12	70	13	5	17	32
Montreal	188	26	61	14	11	7	52
Toronto	208	32	56	16	7	10	21
Sudbury	154	16	62	14	13	10	24
Census div. 18, Manitoba	143	13	76	19	7	20	19
Regina	205	20	55	16	6	13	24
Calgary	180	24	47	18	11	6	25
Vancouver	154	17	50	20	4	11	30

Source: Kinsley and O'Donnell (1983: 141–97).

ferences in demographic composition, economic function, and cultural background may account for some of the variability among places, but geographical variation in Canadians' use of time is poorly understood.

In their Halifax panel survey of the same 453 respondents in 1971 and 1981, Harvey and Elliott report an average reallocation of 110 minutes over the ten years.[16] Significant declines occurred in sleep, home care, and recreation, and notable gains in passive leisure activities such as watching television and reading. And, at the expense of time for home and family care and for free time, there were sizeable increases in paid work for women who entered the work-force after 1971. These findings complement Michelson's studies on shifting patterns of time use by women in Toronto.[17] His interpretations illustrate the importance of separating out major structural changes in the economy and in household organization.

For most people, the allocation of daily time is focused on the workplace, home, and the transportation link between them. Therefore changes in commitments of time to households, employment, and travel will impinge on the range of activities in which people may engage and influence the types of services and infrastructure needed to accommodate them. To understand

these relationships, we should look at alterations in conditions of employment and changing patterns of household formation.

CHANGING CONDITIONS OF EMPLOYMENT

Economic statistics document the growth of part-time at the expense of full-
time jobs, work-sharing schemes, longer periods of unemployment, trends
to earlier retirement, and difficulties for some age cohorts in finding work.
Reasons cited for these changes related to substitution of capital for labour,
the internationalization of economic activity, and rapid growth of a highly
competitive service sector.[18] Through their influence on individuals' time
commitments and incomes, these employment trends directly affect the allocation of time and activity patterns of urban populations.

Although part-time work may be seen as a trade-off of income for time,
and although several recent public opinion polls suggest that many workers
would prefer more free time,[19] there is evidence that for many people part-
time work is involuntary. For example, 14 per cent of men and 13 per cent
of women part-time workers in British Columbia preferred full-time work in
1975; but, by 1985, those preferring full-time jobs increased to 40 per cent
for women and 45 per cent for men.[20] Accordingly, more than half a million
Canadian part-timers (about one-third of the total) were underemployed in
1985.[21] Growth in involuntary non-work time also occurs because of longer
average periods of unemployment – from 15 weeks at the peak of an economic recession in 1981 to 20 weeks in a period of prosperity in 1987.[22]
And early retirements, whether forced or voluntary, have reduced the proportion of men over 65 years of age in the work-force, from more than 56
per cent in 1931 to less than 12 per cent in 1987.[23]

While these trends in the structure of employment are likely to affect significantly the activities of urban populations, workers also have considerable
control over their time at work. For example, a Quebec opinion poll in 1984
showed that two-thirds of workers followed a rigid time table and that more
than one-third worked evenings, nights, or shifts.[24] This situation affects
people's engagement in other important activities during the day or week
and their discharge of household responsibilities. A study in Toronto, for instance, showed that one-fifth of full-time and one-fourth of part-time employed mothers were working outside normal hours (0800 to 1700). And,
among dual-career households with children, 42 per cent had partners working different shifts.[25]

Since the early 1970s, more of the labour force has benefited from staggered and more flexible work hours.[26] Staggered hours stage the check-in
and check-out times of employees to spread the pattern of traffic over a
longer span of time; the intention is often to relieve congestion on streets and

transit systems. Because of the ease of predicting traffic flow based on known numbers of employees, the staggered scheme has been a preferred low-cost strategy for traffic management by transport planners in major cities. In contrast, flexible work hours (flexitime) give workers freedom to decide when to be at work during designated flexible periods, usually two-to-three-hour blocks at the start and end of the work day and lunch periods. Employees can more easily plan daily schedules to accommodate both personal and household needs and job obligations, and the scheme is particularly suited to routine office work.

D.C. Bowman attempted to assess the behavioural effects of flexible working hours on the employees of fourteen Ontario firms in London, Markham, Toronto, and Woodstock. She hypothesized that new flexitime workers would try to create larger blocks of time for personal use, engage in a greater variety of activities, and travel greater distances to take advantage of alternative recreation, shopping, and other opportunities.[27] Ninety per cent of respondents reported taking advantage of the new freedom. Many were saving time by travelling to work in periods of less congestion, spending more time with the family, and adapting work schedules to their personal biological clocks. Significantly more time was devoted to household maintenance activities, shopping, household and non-household recreation, and socializing; only 7 per cent took on a second paid job.[28]

Part-time jobs, staggered work hours, and flexitime are only three examples of altered schedules. Others include compressed weeks (for example three days of twelve hours and four days off), eight-day weeks (four on and four off), and a variety of formats for shift work. Although there is little research to document the behavioural responses to different schemes, the range of potential opportunities would probably vary considerably. Computerized simulation models support this contention. Based on a mathematical conceptualization of Hägerstrand's time-geography model for an idealized urban environment, Burns demonstrated how the introduction of flexible work hours expands the range of potential activities – more so than costly transportation projects to speed up traffic flow.[29] An empirically based simulation of activity patterns in the greater Toronto region yielded a similar conclusion – lessening constraints on timing of work liberated people's movements and activities more than did transport-system changes.[30]

TIME CONSTRAINTS AND CHANGING HOUSEHOLD NORMS

Recent alterations in conditions of employment have been responses, in part, to the changing character and needs of the Canadian work-force, the most dramatic being the rapid growth of female participation. An increasing diversity of female roles is based on choices of whether or not to participate in the wage economy (either full-time or part-time) and on variations in marital and parental circumstances, economic well-being, and stage in the life cycle.

Following Hägerstrand's time-geography perspective, Palm and Pred illustrated the extent to which North American cities pose problems for women in carrying out typical daily routines.[31] These included difficulties in co-ordinating work with childcare obligations and other household functions, the segregation of work from residential environments, the isolation of many suburbs from basic services, the timing of opening hours for facilities that preclude their use by many, and the overall design of cities which forces dependence on automobiles. Although they do not adopt an explicit time-geography perspective, Wekerle and Rutherford document the concerns identified by Palm and Pred.[32] In a study of 8,222 employed individuals in Scarborough, Ontario, they show how the distance separation of home and work and the spatial segregation of suburban land uses interrelate with changes in households and employment to pose time penalties and spatial constraints on the activities of respondents, particularly women, with the most severe penalties on single-parent working females.

In comparison with married mothers, lone-parent women in Scarborough have higher job participation rates, greater orientation to full-time jobs, lower household incomes, and much lower rates of automobile ownership – only 46 per cent of those with children under the age of twelve own cars, compared with 96 per cent for married women. Not surprising, they are disproportionately represented among captive riders of the transit system. Coupled with the time inflexibility associated with full-time work and the inability to share household and childcare responsibilities with a spouse, single parents also spend significantly more time commuting to work. Transit users spent 55 minutes getting to work compared with 23 minutes for those using automobiles. Though female single parents live on average closer to jobs than married men, their average one-way journey to work was 7 minutes longer (39 v. 32 minutes). Wekerle and Rutherford suggest that the trade-off for women in working closer to their homes is lower access to better-paying jobs, a view reinforced by Villeneuve and Rose's consideration of male and female employment patterns in Montreal.[33]

The insensitivity of both community land use and work environments to the effective use of time, and to providing access to basic services and other opportunities, is central to much of the feminist critique in urban geography. Women, more than men, have been disadvantaged by suburbanization, job decentralization, land use segregation, and automobile-oriented transportation planning.[34] However, Isabel Dyck has illustrated a response by women in Vancouver's suburbs to assert control over neighbourhoods. Dyck conducted an intensive participant-observation study of activities for twenty-five mothers with at least one child under the age of ten in two-parent households.[35] The women established co-operative arrangements among neighbours to share information, to pool resources, and to provide time-saving services for one another. In childcare, strategies were used to appropriate control over neighbourhood environments, to provide security and opportunities for children, and to permit some mothers to meet employment obligations and career ambitions without jeopardizing families. The incorporation of community support networks to co-ordinate complex patterns of activity reinforces the space-time context of household behaviour.

Research on changing spatial behaviour in Canadian cities points to an incidental rather than focused concern among scholars for the space-time dimensions of activity patterns in cities.[36] Nonetheless, issues of social and economic restructuring, which underlie much of social geographical, feminist, and time-budget inquiries, provide a context for developing an explicit time-geography of urban social patterns. The rudiments of such an approach are outlined next, in a review of research in Halifax, Nova Scotia.

TIME GEOGRAPHY IN A CANADIAN CITY

Knowledge of the time behaviour and time geography of Halifax-Dartmouth residents ranks among the highest for all cities in the world. An unusually extensive one-day space-time diary survey for more than 2,100 randomly selected residents of the region took place in 1971–72, as well as a follow-up time-use survey in 1981.[37] Coding of activities for both surveys followed the typology used in the International Time Budget Survey,[38] but the 1971–72 study was supplemented with a geocoding of respondents' locations to the nearest one-tenth of a kilometre throughout the diary day. This scheme offers unusual flexibility for aggregating information at any spatial or temporal scale, for observing activity patterns of individuals or groups, and for creating a census for any time of day.

Since much of our understanding of urban social geography has been based on census data, it reflects a bias toward night-time patterns of residence. In contrast, a time-geography perspective, in conjunction with the Halifax surveys, provides insight into the dynamic daily cycle of urban social patterns. Table 5.2 describes one attribute of space-time paths for Halifax respondents – their average speeds of travel for all trips on diary day, regardless of mode.[39] To provide social context, respondents are grouped into single-variable sub-populations and multi-variable role groups. Role groups are defined by combinations of personal characteristics that are indicative of the types of constraints that people face in daily life relative to job, childcare, and spousal responsibilities and to mobility.

The rank-ordering of sub-populations by average speeds highlights the significance of gender, employment, marital status, and car ownership on

Table 5.2
Average trip speeds for Halifax sub-populations
and role groups

Sub-population/ role group	Symbol for category	Speed km/hr	Sample size
All respondents		8.4	1,091
Male	M	9.6*	532
Homeowners		9.5*	616
Employed	E	9.4*	791
Autos in household	A	9.3*	917
Married	W	9.2*	886
Children at home	C	8.9**	730
No children at home	N	8.0**	438
Renters		7.6*	569
Female	F	7.5*	675
Single	S	6.7*	312
Unemployed	U	6.4*	416
No autos	P	5.8*	227
MWCE		11.0*	22
FWCE		9.0	104
FWNE		10.6*	71
MWNE		10.5*	97
MWCU		7.8	22
FWCU		6.6*	236
FSCE		9.8	33
MSCE		7.9	26
MSCU		6.2**	23
MSNE		6.8**	57
FSNE		6.3*	105
FSNEA		6.5*	31
FSNEP		5.3*	48

* Significant at 0.05 level
** Significant at 0.01 level

A two-sample *t*-test was used to test the null hypothesis that a sub-population or role group and the remainder of the sample population were drawn randomly and independently from the same normally distributed population. For further details, see Janelle, Goodchild, and Klinkenberg (1988: 895 and Tables 2 and 5–7).

this simple indicator of movement behaviour. However, as shown by role-group averages, different combinations of attributes significantly affect the results – compare averages for men and women with similar marital, child-care, and employment circumstances. A gender bias in the life-cycle patterns of employed males and females is suggested in Figure 5.1. This

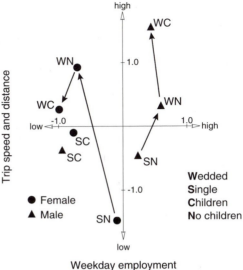

Figure 5.1
Life cycle changes in mobility and daily time fragmentation
for employed males and females in Halifax. Modified and
reproduced from Janelle and Goodchild (1983: 334) with
permission of V.H. Winston and Son.

diagram identifies the positions of respondent groupings by standardized
factor scores on two constructs of temporal structure in daily life – block
time and mobility. High block time is associated with uninterrupted attention to individual activities and is highly related to the extent of weekday
employment; low block time implies fragmentation of the day into a large
number of short episodes. High mobility relates to high average speeds for
trips and high average distances for individual travels; low mobility implies
short, slow trips. Marriage and childcare obligations have no apparent influence on levels of block time and employment time for males but significantly increase the distances and speeds of travel. In contrast, whereas
marriage results in large increases in average travel speeds and trip lengths
for women, both marital and childcare responsibilities drastically fragment
their day. Most disadvantaged of all are single parents (SC), male or female,
who have below-average mobility and highly fragmented time.

In response to functional specializations and role needs, city residents
move about the local environment. This movement affects the concentration
of particular demographic and social groups in different areas at various
times. Such changes are shown in Figure 5.2 for the distribution of unmarried respondents over an average weekday. Concentrations are highest on
the central peninsula, near universities (Dalhousie and Saint Mary's), but

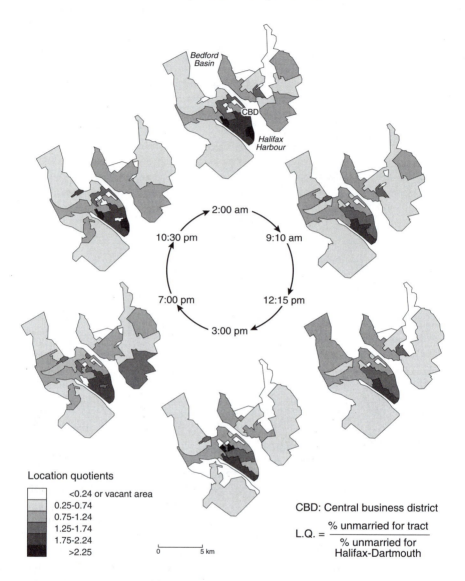

Location quotients

	<0.24 or vacant area
	0.25-0.74
	0.75-1.24
	1.25-1.74
	1.75-2.24
	>2.25

CBD: Central business district

$$L.Q. = \frac{\% \text{ unmarried for tract}}{\% \text{ unmarried for Halifax-Dartmouth}}$$

Figure 5.2
Diurnal variations in spatial concentration of unmarried respondents in Halifax-Dartmouth.
Reproduced with permission from Janelle and Goodchild (1983: 412).

the pattern varies through the day. In this example, the location quotient is used to measure for each of six times the degree to which the proportion of single people in a tract either exceeds or falls below the average for the Halifax-Dartmouth region as a whole.

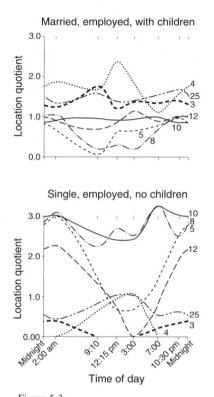

Description (based on 1971-72 evidence)

3   Affluent and middle-income suburban residential areas; yacht club

4   Middle-income suburban residential; undeveloped land; Mount Saint Vincent University

5   Mostly residential units; young population; lower-middle income

8   Affluent and stable residential area, mostly single family housing; Saint Mary's and Dalhousie universities; Point Pleasant Park

10 Parts of Halifax central business area; public institutions; commercial docks; middle-to-high-density residential areas; low to middle income; student accommodations

12 Mixture of income groups, subdivided houses, apartments, and restored homes; port facilities, container terminal, Canadian National Railway marshalling yards

25 Dartmouth central business area; shipyards, ferry terminal; rental flats and single-family homes; lower-middle income

Figure 5.3
Diurnal variations in spatial concentrations of female role groups in Halifax-Darmouth. Modified with permission from Janelle and Goodchild (1983: 421).

The location quotient is also used to compare changing daily concentrations of single employed females with employed married mothers. Pronounced daily shifts in location for the single group, in comparison with generally less varied levels of tract concentration for married employed women, is shown in Figure 5.3. Employed single women show exceptional concentration throughout the day in the central peninsula near the central business district (CBD) and universities (tracts 10 and 8 respectively). In contrast, tracts 5 and 12 show prominent cycles, as women leave these areas during the workday and return in the evening. The high concentration of single women helps account for their particularly low average scores on trip distances and speed of travel (Figure 5.1). Apparently, marriage results in considerable diffusion of women (and men) throughout the city and greater reliance on cars.

The previous examples confirm the dynamic quality of urban social spaces. An extension of this idea uses the composite profile of activity patterns by all diary respondents to interpret the city's changing social ecology. Cyclic (re) mixing of people and activities over space during the day gives way to greater homogeneity in social spaces and activities by late evening. By monitoring the movements of approximately 1,500 randomly selected people in and out of thirty-two subregions, we created census-like data for six different times of day. We then treated each subregion as six separate cases and collapsed variables for the resulting 192 space-time units (6 times multiplied by 32 spaces) into their underlying dimensions (factors) by a principal-axis factor analysis.[40] The forty variables, all derived from the diaries and from an accompanying survey, included fourteen socio-demographic traits (e.g. age and income), six general activity categories (e.g. home-centred discretionary), three travel-purpose indicators (e.g. to or from non-employment obligatory activities), nine site characteristics (e.g. in eating and drinking locales), and eight social-contact descriptions (e.g. with child of household or alone).

Figure 5.4 plots two of the eight factors derived from the analysis as bar graphs of the standardized scores, by time of day for each tract. The factors are defined by correlations (factor loadings) with the original variables. The presence of a factor in a tract is indicated by the length of bars above or below the baseline, which represents the average for all 192 space-time units.

Employment-related variables define the most significant factor, which accounts for 29 per cent of the explained variance. Thus work provides the dominant rhythm to the city's weekday routine. Areas with high daytime scores include the CBDs of Halifax (tracts 10 and 18) and Dartmouth (25), shopping centres (13, 17, and 27), industrial and military establishments (9, 19, and 28), and hospitals (15). Movement in and out of these work zones is conspicuous by early morning and evening fluctuations in the scores, but the hospital area and the Stillwater Air Base (in 28) are busy work environments at all times of day. Employment is also associated with regions that boast evening-oriented events, such as the Exhibition Grounds and the Halifax Forum (16 and 1, respectively) and night courses at Mount Saint Vincent University (tract 6).

Contrary to the employment factor, the non-home-social-and-leisure factor explains (at 7 per cent) little of the variation in the city's space-time ecology. Nonetheless, the pattern is distinctive. Morning and midday visits to neighbours and nearby recreation sites account for above-average scores (tracts 6, 13, 20, and 26), and specific recreation facilities yield some of the

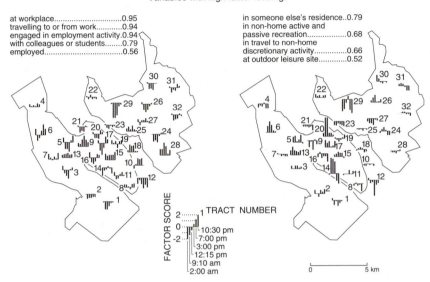

Figure 5.4
Ecological structure in the time geography of Halifax. Reproduced from Goodchild and Janelle (1984: 814–15) by permission of Pion Ltd.

more prominent peaks (golf clubs in 13 and 26; an arena in 20; and boat clubs in 14, 22, and 26).

This exploratory and partial review of weekday patterns in the ecological structure of Halifax reveals temporal cycles shaping the city's social geography.[41] In general, social obligations (particularly at the household level), biological needs for rest and sustenance, and the employment function set the time commitments and movement behaviours for individuals. But, because of links established by specialized roles and because of distances that must be overcome for essential exchanges, the variable behaviour of individuals contributes to a social geography with distinct daily rhythms and temporal cycles.

CONCLUSIONS

This chapter has attempted to give a space-time view of cities and of behaviour in cities by pulling together separate threads of research concerning the restructuring of economic and social life, time-budget surveys, and time geography. While the time dimension has been secondary to research on changes in conditions of employment and norms of household structure, these changes are associated integrally with social and economic issues over the timing of activities and services and over the autonomy that people have in their lives. All of these are major public issues in Canadian cities.

Time is not just a container of activities; it functions "as a social resource (commodity) to be administered and exchanged, and as an organizer and regulator of different areas of activity."[42] Geographically, the opening and closing hours of activity sites and the time required for travel affect both the spatial range of mobility and the activity choices of people. However, two aspects of daily allocation of time stand out from the individual's perspective – work time and childcare – imposing a rigidity on the daily programs of many.

Political struggles to introduce flexibility in work timing and to expand childcare services reflect broad social awareness concerning the resource value of time. Outcomes of these efforts impinge directly on individuals' livelihoods, their responsibilities as parents, and their ability to meet social and work obligations and use opportunities provided in local environments. Given the centrality of these issues to gender equity and labour-management relations, it is surprising that geographers and other social scientists in Canada have been slow to address the structural underpinnings of the social geography of time use – the ability of business and government to control when activities and places are scheduled. Power relationships that surround the timing of places and events, and the subsequent space-time patterns of people's behaviour, need investigation.

A strict physical interpretation of time as sequential chronology cannot explain complexities of urban social behaviour and temporal changes in the character of Canadian urban spaces. Describing behaviour via time-budget data or space-time paths is insufficient for understanding behaviour. As Rose observes,[43] there are non-linear variations in the perception of time. Its resource value may vary among social classes. But, even for the individual, its value will change according to the needs associated with different time horizons (the daily, weekly, yearly, and life-time levels of time consciousness). While these complexities may preclude development of a general dynamic model of urban social geography, they should not discourage attempts to understand more fully the time-geography of our cities.

PART TWO

Contexts: Social Structure and Urban Space

CHAPTER SIX

Migration, Mobility, and Population Redistribution

E.G. MOORE AND
M.W. ROSENBERG

Changing residence is a pervasive form of behaviour in North American society. The average Canadian can expect to move about a dozen times during a lifetime. Although most moves cover relatively short distances within the same local jurisdiction, they serve an important role both for individuals and for the larger society. For individuals, they facilitate an ongoing adjustment to shifting needs and desires in relation to housing, employment, and access to amenities and to social and economic environments. For the larger society, they both reflect collective values of freedom of choice in a world bound by many constraints and, as an aggregate outcome of flows and counter-flows at all scales, redistribute the population and its characteristics. Concerns about individual welfare and aggregate redistribution are central to discussions of the current state and future prospects for Canadian cities, and it is therefore appropriate to consider the role played by mobility.

In this chapter, we shall consider four main topics: the evolution of conceptualizations of the mobility process in the social sciences, the main dimensions of mobility and their magnitudes, links between mobility and the changing social geography of Canadian cities, and possible directions for future research.

THE NATURE OF MOBILITY

Mobility in general is defined as any change in the usual place of residence,[1] whether a move from one apartment to another in the same building or a relocation from Halifax to Vancouver. It is also common to distinguish between local moves, or residential mobility, and long-distance moves, or migration. Residential mobility usually involves housing adjustments within local areas with little effect on the spatial organization of other activities; migration, in contrast, implies shifts between spatially distinct labour markets and thus major disruptions to established patterns of daily activities and social networks.[2]

Views about the nature of mobility have changed considerably over the years. For early students, mobility came to symbolize the "restlessness of American life" and the "source of freedom and opportunity";[3] it reflected the belief that upward social and economic mobility was inextricably linked with willingness to move physically from one place to another. Links were drawn between mobility and frontier growth, but, as the expansionary period drew to a close at the turn of the century in the United States, other concerns arose. Mobility did not decline and, particularly in the city, came to be seen as reflecting an inability to develop ties to community. In the years between the two world wars, the sociological perceptions of rootlessness and anomie among recent arrivals in cities led to an identification of mobility as a pathological condition which required public intervention to promote stability, particularly at the scale of the neighbourhood.[4] These two perspectives of progress and pathology represented popular polar positions with no great empirical support: many moves take place without clear social advance, while much evidence exists to counter the negative image of mobility by showing that the majority of movers are more content after they have moved.[5]

More sophisticated conceptions of mobility have recognized the essential link between changing residence and adjustments in housing and labour markets. In his classic work on residential mobility, *Why Families Move*, Rossi argues for the primacy of the adjustment process in housing consumption to changes in housing need as households grow, shrink, age, or accommodate altered social and economic circumstances.[6] At much the same time, migration was conceptualized in terms of employment-related adjustments involving moves between spatially separate labour markets. The emphasis was on setting the perceived long-run benefits of new opportunities against the short-run costs of moving (termed investment in human capital).[7] Such a perspective implied that older individuals should be much less likely to migrate: their costs of moving would be higher, reflecting both longer-term investments in the current location and lower benefits, as the expected time over which benefits would accrue would necessarily be shorter.

Different disciplines added their own flavour to these basic ideas. Economists saw mobility as a basic mechanism matching housing or labour supply and demand in a competitive market. Initially it was seen to produce a system in equilibrium in which no one could become better off without making someone else worse off (the condition of Pareto optimality).[8] More recent conceptualizations recognized that real markets do not attain states of equilibrium; mobility reflects adjustments to markets in disequilibrium, and each move tends to take the system toward an equilibrium that is constantly shifting, as external economic, social, and political conditions change.[9]

Perhaps the most important contribution from sociologists was that of Michelson, based on a detailed study of environmental choice and residential satisfaction in Toronto.[10] He argued that the traditional views were essentially static, presenting adjustments as ad hoc at any given time. In reality, for many households, successive residential choices represented steady progress toward specific housing goals; thus a move to a larger apartment may be made as a rational but temporary adjustment on the way to a more permanent housing solution in the future. Michelson's view refocused the basic question from "why do people move?" to "why do people make the residential choices that they do?" His work was highly suggestive of why people chose particular residential environments with very specific social and locational characteristics at various times in the evolution of the household. Perhaps more than any other work, his has brought together the sociological pespectives on individual and household behaviour with spatial views of the nature of residential environments and the organization of urban residential patterns.

A critical element in Michelson's view was the status of homeownership. It represented a stable state appropriate to the rearing of children, most often in a suburban location with access to private space. The perspective contained a substantial empirical truth for the bulk of the study population, which was unabashedly middle class, and thereby both circumscribed the generality of the findings and perhaps undeservedly diminished some of the real value of his view of a dynamic process. It does raise, however, questions about the role of access to homeownership in current mobility patterns, to which we return below.

Geographers have made considerable contributions within the frameworks established by both Rossi and Sjaastad. Place utility – the perceived costs and benefits associated with a given alternative location relative to the current location – has been an important component of models of individual movement decisions.[11] Many questions arise about measuring place utility, particularly in regard to social and cultural as opposed to economic components, but it forces us to recognize the richness and complexity of movement processes in a spatial context.

The conventional view from all disciplinary perspectives has stressed a voluntaristic, demand-oriented perspective on mobility. Households became dissatisfied with current conditions; they sought out alternatives in a systematic fashion and selected that one which best suited their choice criteria, defined in terms of short-run or long-run goals. This view has received considerable criticism in recent years. First, fundamental changes in the nature of the Canadian family[12] and in the role of women have undermined the evolutionary model of stages in the life cycle in which a stable family with children in an owner-occupied dwelling is seen as the logical culmination of a sequence of demographic events.[13] Second, many empirical studies have demonstrated that a significant proportion of moves occur under circumstances that can hardly be decribed as voluntary, because of eviction or disaster or forced job relocation, or as a consequence of marriage, separation, divorce, drastic income losses, or rent increases.[14] As a corollary, economic conditions and poor information or overt discrimination severely constrain many choices.

Much of the criticism of conventional approaches has been played out in the context of the tension between choice and constraint. Some have argued that the constraints arising from both public action – regulation of the housing stock and management of public housing – and private-sector decisions – access to finance, lease arrangements, and the influences of real estate agents within a more general "manipulated city" environment[15] – best explain mobility behaviour.[16] Yet few constraints are truly coercive, and thus understanding the spatial and social order that arises out of individual decisions requires notions of preference and choice for all but the poorest and most disadvantaged segments of the community.

The relation between choice and constraint does lead to more specific concerns with the effects of significant changes in the Canadian housing market over the last fifteen years. Depending on how prices and income are measured, prices of owner-occupied houses on average have increased more rapidly than incomes, at least in the larger cities, making the transfer to homeownership more difficult for most and impossible for some, although the effects are primarily on the socio-demographic composition of owner-occupation rather than its level.[17] At the same time, a decline in rental construction and the imposition of rent controls in several provinces have tightened the rental market in virtually every major city. Vacancy rates in the largest metropolitan areas fell to less than 1 per cent in the mid-1980s and were regarded as too low to ensure proper flexibility in the market. More important from a social perspective, certain groups become marginalized in the search for adequate and affordable housing. As the rental market tightens, landlords exert greater control over who gains tenancy in their properties. Groups that are perceived to be "poor tenants," particularly single parents and those on welfare, find their choices restricted.[18] The mobility experience of these groups becomes significantly different from that of the majority, predominantly middle-class Canadian. Although we do not have comparable data in Canada, research in the United States has revealed a growing differentiation between renters and owners in terms of the proportions that reflect traditional "housing choice" reasons for moving. A recent study of mobility in four US cities showed that whereas about 60 per cent of owners moved voluntarily to improve their housing conditions, the comparable figure for renters was less than 30 per cent.[19] For renters, a much higher proportion consisted of non-voluntary relocations associated with evictions, household dissolution, and rent increases too onerous for the current occupants.

Recognition of the role of specific market relationships in structuring mobility raises a number of problems which are fundamentally geographical in nature. Whereas the socio-demographic changes and shifts in the housing needs of households that underlie desires to move are quite consistent from one locale to another, the factors that control the supply of housing tend to be much more place-specific (see chapter 9). They reflect the historical character of the stock – for example, the significantly higher rental component of the stock in Montreal compared with other cities – as well as specific mixes of local regulation and development controls which interact with more general economic and programmatic trends. The student is faced with what Entrikin so aptly describes as the "between-ness of place."[20] The account of mobility must reflect the specific context in which the behaviour occurs, while we struggle to identify more general relations between the attributes of context and their effects on behaviour. However, to develop the latter, we need to be able to consider the relation between behaviour and context in several different places within the same analysis, which imposes great strains on our research resources unless suitable secondary source data are available. In the United States, the American Housing Survey is beginning to generate such research,[21] but there is no comparable source in Canada. We must assume that federal or provincial housing programs directed at such activities as the promotion or regulation of access to homeownership or to affordable housing must generate differential effects in different cities, depending on local conditions. Unfortunately, to demonstrate this and to provide appropriate input to program design requires analyses that current data do not permit.

MOBILITY AND REDISTRIBUTION

Along with demographers and regional scientists, geographers have also been concerned with the larger-scale consequences of mobility. In aggregate, flows of movers from one place to another are redistributive. The specific patterns of arrivals and departures generate changes in the social, economic, and even political character of areas from the scale of the neighbourhood to entire provinces and regions. In this context, who does not move is as important as who does. The departure of the young leaves behind the old; that of the more affluent, the less affluent; that of those with high skills, the low-skilled; and that of the Canadian born, those born elsewhere. Such exchanges are not without social and economic costs. In a society that values freedom of movement, we must be aware that such freedom also imposes differential costs and burdens on communities. These costs and burdens need to be recognized in the development of public policies and programs directed toward both equity and equality.[22]

Although the logical relation between redistribution and mobility clearly involves the net effects of inflows and outflows with different magnitudes and characteristics, the theoretical and empirical links are less well defined. We have a number of conceptual models of redistribution, including those developed with respect to suburbanization of middle-class families, gentrification of inner-city neighbourhoods, and ethnic neighbourhood succession and white flight. But the role of mobility is often implicit rather than explicit. In almost every case, including quite rapid racial change, the gross flows are many times larger than the resulting net effects (the difference between gross in-flows and gross out-flows), implying much more complex processes than might be expected from the simple representation of the net effects alone.

Unfortunately, it is difficult to tackle the redistribution problem from the theory of gross population flows. While many models are well developed, often representing flows in terms of the relative attractiveness of origin and destination and mediated by the distance between them,[23] they cannot provide reliable estimates of net change. The error in the estimate of net change is defined by the sum of the errors in the constituent in-flows and out-flows, and so the errors in estimating net change from gross flows are often greater than the estimate of change itself.

As a consequence, quantitative analysis of redistribution has been treated largely as an accounting problem.[24] This requires that we establish the complete set of flows into or out of an area, perhaps classified by age, sex, or other socio-demographic attributes (the accounts). Such flows are then modelled in terms of the rates at which particular flows are generated from a given origin group and area to a corresponding destination. The linking of accounts with more general redistributive concepts is a slow process, as it is difficult to develop suitable data. However, it is a critical issue if we are to avoid over-reacting to the emotive power of some of the earlier generalizations. For example, significant debates arose in the 1970s, especially in the United States, over the potential of neighbourhood rehabilitation programs to displace low-income residents, yet when detailed data are available, the magnitude of the problem appears small and highly localized.[25] In contrast, recent detailed analyses of demolition, conversion, and renovation have pointed to substantial loss of rental units in inner areas of Canadian CMAs, with consequences for the profiles of both the housing stock and its occupant households.[26]

The relation between redistribution and mobility is a function not just of those who move but also of those who stay. Certain types of neighbourhood require relatively high levels of mobility in order to maintain their character, particularly those that cater to apartment living among young singles and couples. Lowered mobility may lead not only to rapid ageing of a population but also to notable shifts in patterns of demand for health and social services.

If moves to family-oriented housing are postponed because of increasing costs of access, presence of more children imposes new pressures on daycare, nursery, and primary school facilities. In a similar vein, the low mobility of older individuals, particularly in areas of high ownership, can produce rapid ageing in the population. In the major metropolitan areas in Canada, the fastest-growing elderly populations are in the early post-war suburban areas, as the first wave of residents are approaching retirement.[27]

Finally, there is a fundamental link among residential mobility within the city, external migration flows to and from the city, and overall rates of growth and change for the city as a whole. The simple concentric ring model of Burgess, so beloved of urban geography texts, contains a very strong dynamic, involving the in-migration of low-income individuals to the centre and resulting outward movement or replacement of existing households.[28] That simplistic view has largely been rejected, but internal and external relations remain important. The external relations for Canadian cities are crucial. Not only does the structure of inflows and outflows with respect to other metropolitan and non-metropolitan areas reflect the changing economic viability of cities and regions, but the strong metropolitan focus of immigrants is a major force in the evolving internal residential patterns of the city. The interplay between external and internal flows deserves much more attention that it has received to date.

MOBILITY IN CANADA

Although the concept of mobility as the change of "usual residence" from one location to another may seem quite amenable to measurement, a number of difficulties arise. Is the appropriate unit of observation the individual or the decision-making unit of the family or household?[29] However, since changes often occur in these aggregate units in association with the move itself, the structure of mobility is more appropriately tracked through observations on individuals, with reference to family or household structure defined as a characteristic of the individual.

Observations on mobility are themselves problematic. Some individuals move several times in a year, yet, without a mandatory registration system such as exists in Scandinavia, documenting such frequent relocations is very difficult. Even in surveys, highly mobile individuals are consistently underenumerated; in longitudinal panels in which the same individuals are observed repeatedly over time, movers are the most likely to be lost from the panel as it ages. Overall levels of mobility are then almost inevitably underestimated.

Given the complex temporal structure of individual movement behaviour, approximate measures are the norm. The Census of Canada asks "where did you live five years ago?" and records "transitions" between the beginning and end of the period rather than total moves in this time. Census measures do not consider multiple moves within the five-year period and, even more important, define movement only retrospectively. In other words, "risk factors" which define conditions prior to the move are not available – only the state of the individual at the end of the period.

The coding of moves is defined with respect to administrative boundaries, for moves are categorized in terms of whether they cross a census subdivision, a census division, or a provincial or a national boundary. The formal distinction between migration and local mobility is therefore arbitrary, with, for example, a move from Ottawa to Hull being both an interprovincial migration and a move within a local labour market.

Recently a wider range of possibilities has emerged for data analysis. In Canada, most notably, migration data based on administrative files, especially taxfiler records, have seen some limited application in the academic literature.[30] There is, however, nothing comparable to the richness of the American Housing Survey, which documents annual US housing and mobility experience both nationally and for individual metropolitan areas; the survey forms the focus of a growing body of both empirical and theoretical research.[31] The best alternatives in Canada, which contain perceptual and attitudinal variables as well as standard socio-demographic variables, are the Quality of Life Surveys from York University's Institute of Social Research (ISR) and the Current Population Profiles, which are constructed in conjunction with Statistics Canada's Labour Force Survey.

MOBILITY EXPERIENCE

The primary source for comparative statistics on mobility is the census. Comparison of the five-year mobility rate from 1961 to 1986 shows an overall decline in mobility that is partly a function of the ageing population. However, we also see a recent decline in age-standardized mobility (Table 6.1). This recent drop has been discussed by Rogerson in relation to Easterlin's theories of cohort size.[32] Easterlin suggests that large birth cohorts (particularly those associated with the post-war baby boom) place greater pressures on available opportunities in the workplace; fewer vacancies mean lower mobility in the job market, which may well translate into lower residential mobility. However, it is difficult to separate these effects from those associated with cyclic shifts in the economy in general as well as the tightening of housing markets. Even if the phenomenon does not have a clear explanation, the overall decline in mobility is in fact quite pervasive, applying across age groups and even across cities.

Mobility rates vary between cities, although not as dramatically as in the United States, where mobility in fast-growing cities of the southwest, such as Phoenix, is three times that of the old industrial cities of the northeast

Table 6.1
Changes in the five-year mobility rates in Canada, 1961–86

Census year	Total mobility rate*	Mobility rate (age adjusted)†
1961	44.2	45.5
1971	50.0	51.9
1976	50.4	51.2
1981	48.6	48.6
1986	44.5	44.7

Source: Calculated by the authors from censuses of Canada, 1961–86.

* Proportion of the population aged over five years in the census year who had moved in the five years prior to the census year.

† Age-adjusted values use the 1981 population of Canada as the reference population.

such as Detroit and Philadelphia.[33] In Ontario, for example, one-year mobility rates for renters in 1985 varied from 28 per cent in Hamilton to 34 per cent in Ottawa and 41 per cent in London.[34] The link between internal and external relations for cities is clear: the primary correlates of these mobility rates within cities are overall rates of employment and housing growth, as new opportunities have multiplier effects, producing additional local residential adjustments. Moore and Clark demonstrated these relations for selected US cities, while Bourne has noted the same effects for Canadian cities, with Calgary and Edmonton having particularly high rates during the "boom" years of the 1970s.[35]

The standard tables provided by Statistics Canada fail to deal with critical variables underlying mobility behaviour. Detailed characterizations of mobility thus require access to the Public Use Sample files, which give data for a 2-per-cent sample of individuals. Using the file for 1986, we are able to specify five-year rates for samples from major cities cross-classified by variables central to the movement experience: tenure, age, and household composition. In Table 6.2, we provide summary data for moves between 1981 and 1986 for Toronto and Montreal. The clear dependencies of mobility on age, tenure, and household status are shown; note, however, that status is reported after the move. Thus high mobility for owners aged 25–29 indicates that they probably moved from renting to owning in the prior five-year period. As might be expected, young singles are more likely to have moved, whether they are owners or renters, while lone parents are more likely to move locally as opposed to longer distances.

However, while the magnitude and structure of mobility provide a context for further discussion, these numbers give little idea of the meaning or im-

Table 6.2
Probabilities of having moved between 1981 and 1986 for Toronto and Montreal by tenure, age, and household type

	Owner-occupants				Renters			
	% movers		% migrants		% movers		% migrants	
	M	T	M	T	M	T	M	T
By age and tenure								
5–19	31.0	39.5	14.5	16.7	63.0	63.5	23.7	29.4
20–24	28.0	34.1	12.8	16.0	81.7	84.1	37.4	47.4
25–29	61.3	66.1	33.8	30.1	83.1	85.9	33.0	41.4
30–34	59.8	64.8	31.4	29.3	75.7	75.9	29.1	34.1
35–44	35.0	46.2	16.5	20.0	61.4	62.0	22.1	27.9
45–54	21.2	25.4	9.4	10.3	49.5	53.2	15.8	21.7
55–64	16.6	18.4	8.0	8.6	42.0	43.6	11.3	17.9
65–74	14.3	16.3	7.2	8.0	38.7	36.0	10.6	13.1
75+	13.2	15.8	6.7	7.3	29.8	30.2	7.7	11.7
By household type and tenure								
Hus./wife no children	34.6	36.0	18.4	16.8	64.0	69.0	25.5	32.8
Hus./wife children	31.0	37.5	14.9	16.5	57.2	63.9	21.9	31.4
Lone parent	32.3	35.9	13.4	13.2	66.2	61.9	21.3	22.7
Sing. per. < 40	71.1	72.6	28.9	32.8	84.1	79.5	35.3	37.4
Sing. per. > 40	22.6	17.2	8.8	5.8	43.3	38.0	10.6	11.1
Non-family	42.3	54.0	19.2	23.5	75.2	83.1	32.0	46.9

Source: Tabulation by authors from the 1986 Public Use Sample tape, Census of Canada.
Notes: M = Montreal; T = Toronto; migrants are those who lived in a different census division in 1981 than in 1986 or who lived overseas in 1981. Movers are all those who lived elsewhere in 1981.

pact of the phenomenon. In the following section we consider the effects of both migration and local mobility on the Canadian urban system.

MIGRATION IN THE URBAN SYSTEM

Migration within Canada reflects changing opportunities for employment and for retirement. People move in response both to changing opportunities in their current locations and to perceived opportunities elsewhere, although costs accrue to refusal of a move as well as its pursuit. The cyclic structure of regional economic development in Canada changes the patterns of internal migration, thus affecting rates of growth and internal mobility in individual cities.

The central role of Toronto, Montreal, and Vancouver in the migration behaviour of Canadians has long been recognized.[36] This was still true in the

Table 6.3
Net changes attributable to migration between CMAs, 1981–86, by birthplace

Birthplace	Montreal	Ottawa-Hull	Toronto	Winnipeg/ Thunder Bay	Edmonton/ Calgary	Vancouver/ Victoria
Canada	−8,760*	14,940	47,430	−3,885	−27,635	15,245
Elsewhere	−10,080	1,630	17,295	−2,775	−4,140	4,695
USA/UK/N. Europe	−5,310	1,150	5,740	−875	−2,800	3,295
S. Europe	−590	20	1,845	−40	−815	5
Asia	−1,975	15	6,545	−1,520	−100	880
Caribbean	−1,025	170	1,305	−130	5	15
All other	−1,205	320	1,880	−250	−470	545

Source: Moore, Ray, and Rosenberg (1989); data were based on a special tabulation obtained from Statistics Canada for the 1986 census. This tabulation combined designated CMAs in order to reduce the number of cells in the table to a manageable size.
* The net effect of immigration to Montreal from other CMAs and out-migration from Montreal to other CMAs among Canadian-born resulted in a loss of 8,760 Canadian-born for the Montreal CMA.

1980s, though with somewhat differing force for different sub-groups of the population. For 1981–86, the relative likelihood of moving between CMAs in the five-year period varies considerably with the relative size of places and the distances between them. Toronto and Vancouver are clearly involved with the largest flows, primarily as a function of their overall size. However, a recent study, which compared rates (rather than flows) by calculating the relative likelihood of selecting different destinations for moves from a given origin, produced a more interesting structure.[37] Not unexpectedly, younger Canadians are more likely to move to Toronto than the average, while older Canadians favour Vancouver. More dramatic differences emerge for groups defined in terms of birthplace and language. In this period, earlier immigrants, particularly those from the developing countries of Asia, Africa, and Central and South America were significantly more likely to move to Toronto than those born in Canada. Immigrants have less time to establish themselves in a particular place and therefore face lower social costs in moving; also, the metropolitan locus of many immigrant groups produces social networks that attract immigrant movers to the larger cities in general and to Toronto in particular.

This focus on Toronto is reflected in the net gains and losses attributable to migration during the years 1981–86. Table 6.3 summarizes the outcome of migration between CMAs for different population groups. For example, the net outcome of all flows of Canadian-born into and out of Montreal to other CMAs is a loss of 8,760, while for the Asian-born there is a loss of 1,975 to Montreal but a gain of 6,545 to Toronto. In fact, Toronto gains from every other CMA, while Vancouver gains from every other CMA but Toronto. These net gains and losses, however, must be put in context. For immigrant groups in most CMAs, internal migration between censuses was only a modest fraction of the total effect from direct immigration from overseas. In Toronto, for instance, the net gain was 17,300, while direct immigration generated an increase of approximately 99,000. In Montreal, the net loss of 10,000 is set against a gain of 40,500 from new arrivals from overseas. For Vancouver, internal gains were 4,700, while new immigrants contributed 34,200. Net migration effects for the Canadian-born population comprised gains of 47,400 for Toronto and 15,200 for Vancouver and a loss of 8,800 for Montreal. In other words, the net effects of internal migration of immigrants and direct immigration from overseas were substantially greater than the effects of net migration for the Canadian-born.

Migration flows (both inter-regional and international) are also very sensitive to comparative economic conditions. Human capital theory provides a rationale as to why potential migrants respond to differences in economic opportunity (especially in destination areas), and migration to Ontario was certainly sensitive to the rapidly improving conditions following the recession of 1981–83. These patterns contrast starkly with the migration experience of the previous five years, when the major attraction for migrants was the economic opportunities generated by the oil boom in Alberta.

The attraction of metropolitan areas for immigrants is somewhat in contrast to the shift in migration by the Canadian-born population over the last twenty years. A significant increase in migration has been noted from major urban centres to smaller centres and to rural non-farm areas, especially near large urban areas, reflecting differentials in housing costs and increasing preferences for less stress.[38] Relative rates of growth of urban and rural populations have altered, although these are mediated by economic conditions. In the Atlantic provinces and Quebec, which had the weakest economies, rural populations grew much faster than urban populations. In Ontario, with better overall economic conditions, rural growth rates were just a bit higher than urban ones, while the west had much greater urban growth.[39] There is evidence of another cycle, since 1984: a new spurt in the growth of southern Ontario cities and an increase in their attractiveness to migrants, coincident with the general upturn in the Ontario economy followed, in 1988, by a sharp decline in net migration to Ontario and a resurgence of migration-induced growth in the west, notably in British Columbia.

Ontario's experience emphasizes the limitations of the type of data available from the census. During the period 1981–86, the economy went through recession and recovery, associated closely with decreasing and then increasing mobility rates. These changes were paralleled by shifts and counter-shifts in the spatial patterns of migration, with moves during the low point of the recession being more localized and tending to avoid Toronto. The speed of response of migration to economic change, particularly among the most vulnerable groups – those in the early labour-force years – is very rapid and makes it difficult to generate forecasts of future distributions of these groups for any part of the country where migration is significant.

LOCAL RESIDENTIAL MOBILITY

Local mobility within urban areas has received much less attention in Canada than in many other countries, partly because of the paucity of suitable data for analysis. The main dimensions of the dependence of local mobility on age, tenure, and family status were illustrated in Table 6.2. Addressing the more specific issues of why individuals move and with what consequences requires access to data from special surveys.

In general, the dimensions of individual experience in Canada parallel those in the United States. The primary reason for moving is still the need to adjust housing experience to changing housing needs, whether the latter are instigated by shifts in the demographic structure of the household or changing economic circumstances. Research in the United States illustrates the growing role played by divorce in stimulating relocations, particularly from owning to renting, a transfer frequently associated with moves from family-oriented suburban environments to central cities.[40] The growth in Canadian divorce rates mirrors those in the United States, and one might expect similar mobility responses in this country, resulting in greater demand for smaller rental apartments and condominiums in more convenient, central locations. "Empty-nester" households, adjusting housing as their children become adults, increase this pressure.[41]

Do people exercise choice in improving their housing circumstances? To date, much of the evidence is indirect. For example, the incidence of unexpected moves is high. In a recent study using the ISR Quality of Life surveys for the period 1979–81, Crossman showed that, while almost 10 per cent of owners moved who had expected to stay, 50 per cent of owners who had expected to move did not do so.[42] The comparable figures for renters were 32 and 40 per cent. While "unexpected movers" include the notion of "displacement," "unexpected stayers" are more common and include what Michelson refers to as "blocked" moves.[43] The latter are particularly important for public policy: they suggest that many households at a variety of income levels become "trapped" in their current dwellings because of a lack of suitable, affordable alternatives elsewhere.

In tight housing markets with high prices and few vacancies, a situation typical of much of metropolitan Canada in recent years, those at the low end of the market become marginalized. Access to rental units is frequently decided by the landlord's perception of who constitutes "good" tenants (in the sense of being unlikely to generate problems), with serious consequences for specific sub-groups, such as lone parents.[44] Further, in such markets, flows of information are fragmented and the choices encountered in any search are highly constrained. This situation reduces the chances of finding accommodation that meets the desire of the searcher, thus increasing the likelihood of housing-related stress in the future.[45]

For more affluent groups, the transition from renting to owning is a major event in the life of the individual: it represents not only a drastic change in commitments and responsibilities but also a shift to a status with a much lower likelihood of moving. The contrast between the annual mobility of owners (5–8 per cent) and that of renters (20–30 per cent) is perhaps the most fundamental dichotomy in mobility behaviour. Canada is a nation of homeowners (67.1 per cent of people in households lived in owner-occupied units in 1986). Access to homeownership reflects not only economic conditions, defined by price, mortgage conditions, and alternative investment opportunities, but also the cultural values attached to homeownership. There are also significant differences in both homeownership and mobility for different ethnic groups, with southern Europeans having higher levels of ownership and lower mobility than other groups.

The ability to make the transition to ownership, however, is a function of individual economic circumstances and the entry price. In a recent survey by the Ontario Ministry of Housing, more than 30 per cent of all renters (and more than 35 per cent in Toronto and Ottawa) said that they did not plan to purchase a home because it was too expensive.[46] This same renter group, which contains high proportions of single persons and single parents on welfare, also has a high incidence of blocked moves. Thus, residential mobility may not be as effective in improving housing experiences as conventional consumer-choice theories of mobility would have us believe.

Mobility both maintains and alters residential patterns within the city – the growth of suburban or exurban populations or the shifting locus of ethnic neighbourhoods *must* involve moves by individual households. Although moves in any year are numerous, with as many as 10 per cent of the population in any given city moving locally, the intrinsic order in these moves and its sensitivity to economic, social, and political events are known only in the most general terms. Distance is still a major constraint, reflecting the nature of access to information and the desire to minimize disruption of estalished activities. The transfer from urban core to suburban periphery still dominates in shifts from renting to owning, although such outward movements are not associated with the socioeconomic and racial disparities that characterize so many US cities. In fact, Filion points to the increased demand for housing in core areas through the 1970s and 1980s as reflecting both the perceived quality of life in Canadian cities and the changing composition of households with decreasing emphasis on traditional family values.[47]

The contrast between US and Canadian cities in terms of their internal and external dynamics deserves continued research. It has been pointed out above that the general relation between overall growth and mobility holds for Canadian cities, though not as dramatically as south of the border. However, the strong spatial outcomes associated with racial segregation and central-city concentrations of low-income households find only limited parallels in Canada.[48] The differences stem in part from the ability of Canadian cities to retain high-quality and high-priced residential units in the central areas as the demand for such units remains high. Low-income housing, particularly public housing, is relatively dispersed, reflecting federal and provincial land purchases and development strategies.

Major economic and demographic changes over the last decade raise significant questions about the future effects of mobility on the residential structure of Canadian cities. The amendments to the Immigration Act implemented in 1978 substantially altered the composition of Canada's immigrants. Family-reunification and refugee categories grew to comprise 67.8 per cent of immigrants in 1986, which also meant a decline in educational attainment and average real incomes and more immigrants from "non-traditional" sources – in 1986, 69.2 per cent of immigrants came from Third World countries, with the majority being defined as "visible minorities."

The overwhelming focus of Third World immigrants within the three largest cities – Toronto, Montreal, and Vancouver – has been primarily on the central part of the urbanized area. In Toronto, this pattern of arrivals was associated with net departures of predominantly Canadian-born to surrounding regional municipalities of Durham, Halton, Peel, Waterloo, and York. Similar patterns of arrival and dispersal are evident in Montreal. At present, the residential patterns of low-income groups and visible minorities are not as concentrated as in many US cities. But with continued increases in immigration to Canada, which rose from 88,000 in 1984 to 213,000 in 1990, and with so many Third World immigrants choosing the three largest metropolises, we need to assess the effects first on residential patterns and then on education, social welfare, and other human services.

FUTURE MOBILITY RESEARCH IN CANADA

An assessment of the status of research on mobility, migration, and redistribution in Canadian cities reveals many more questions than answers. A study of mobility for its own sake provides few insights about the essential dynamics either in individual lives or in the communities in which it occurs. Thus studies of its frequency, the characteristics of movers, and the declining propensity to move with distance have high empirical content but unclear social meaning. Mobility has an important, two-fold function: it allows adjustments in individual lives which may improve individual welfare, and, in aggregate, it redistributes the population and its characteristics, thereby affecting planning and policy in both public and private sectors.

We still know little about whether mobility allows individuals to achieve appropriate adjustments in the housing market. Some US evidence has suggested a growing disparity in adjustment outcomes between owners and renters. Owners tend to behave in a more traditional, voluntaristic fashion; renters are more likely to be induced to move by adverse changes in housing or economic conditions and may find their opportunities for alternative housing circumscribed, particularly in the larger cities, where the costs of buying the first home form a major obstacle. Does inefficiency in the matching process merely slow individual improvement in the relation between housing needs and consumption, is the disruption of moving welfare-neutral and are there cumulative negative effects arising from forced or blocked moves?

The links between the demographic structure of cities and mobility raise interesting questions for future research. The strong decline in mobility with age is well established empirically and has solid theoretical foundations, both in human capital theory[49] and in the deterrent effect of the disruption of established social networks.[50] Canada is an ageing society with the largest "baby-boom" cohorts among all the industrialized countries. The current "baby-boomers" are between 25 and 45 and, with high female labour-force participation rates and dual incomes, are competing aggressively for access to homeownership. As their mobility declines with advancing age and within increasing homeownership, national mobility rates may decline still further. At the same time, the pressures on urban housing markets from the smaller cohorts behind may decline, particularly in urban areas that attract few immigrants. Such declines will change the real estate industry, but much remains to be done in developing the underlying demographic explanation which is even more powerful in Canada than in the United States.

A similar concern arises with the growing population over age 65, which has many implications for services and public expenditures.[51] Despite the popular perception of massive migration to amenity areas at retirement, the primary impetus in the geographically differentiated growth of the elderly is the current distribution of the population aged 45 to 65, which has both high ownership rates and low mobility. We need to give more attention to the implications of the concentration of this group in the older suburbs of Canada's major cities. Not only will it rapidly expand the proportion of elderly in these areas, but it may also form a demographic barrier to outward movement associated with pressures arising from increased immigration to central areas and to the outer suburbs.

Immigration itself raises a broad range of questions which affect the study of mobility. Clearly, if immigration continues its metropolitan focus of the last decade, pressures will build which must generate redistributive responses, either from existing residents or from immigrants. The problem is perhaps most pressing in Montreal, which accounts for well over 90 per cent of all immigrants to Quebec. While public officials would like to see a greater dispersal of immigrants from Montreal, it is not at all clear how this might be accomplished, given that new immigrants must be given the same rights of freedom of movement as Canadian citizens. Montreal is the hub of economic opportunity in Quebec as well as the overwhelming focus of its multicultural society. The task of creating comparable attractions in other communities outside Montreal in problematic.

From a geographical perspective, the relation between the general forces inducing mobility at the individual and societal levels and the character of individual places continues to pose fascinating questions for conceptualization and analysis. The evolution of the housing stock and its current management under public regulation and private initiative impose place-specific constraints on relocation decisions in any given city. Individual decision-making seeks to resolve both short- and long-term goals with regard to housing and employment, although such goals themselves are modified and social values change. A better understanding of how these forces play themselves out in Canadian cities may, among other benefits that it creates, improve the design of public intervention, which aims to facilitate and constrain action rather than coerce it.

CHAPTER SEVEN

The Emerging Ethnocultural Mosaic

S.H. OLSON AND
A.L. KOBAYASHI

Visitors are surprised by the way Canadian cities differ in their ethnocultural composition. A "composite" ethnicity contributes to their character, and the city of the senses varies from one city to the next and from one district to another. Marcuse once accused modern cities of having no erotic properties – no room, that is, for the senses, for emotion, sexuality, or love, for the exploration and exuberance of self. If these properties are present in the public sphere of Canadian cities, it is in large part through the vitality with which we retain and express our "old world" values and through the music we make together from our rooted identities. The ethnocultural dimension of city life is hard to define, volatile, and contentious, but it is always present. Its shifting nature contributes to a sense of dynamism as well as to perennial tensions. As generations succeed one another, as they plant and build and renovate, open shops and paint signs, they introduce new features into the urban landscape. As new streams of immigrants become part of the cultural mix, Canadian cities become more visibly diverse, noisier, more expressive, and more distinct, each one more "itself".

The present is charged with the past. Groups vary in their ways of expressing their identities, even in the conviction with which they assert ethnic traits. Individuals within a group vary in the importance they attach to their affiliation. Some deny it, some seek a low profile, while others celebrate it as a measure of pride, defiance, or sentiment. This variability makes generalizations about ethnic groups difficult, even dangerous. Three things can nonetheless be said. First, ethnicity is a shared social activity. Second, whether individual affiliation is strong or weak, we are all "ethnic". Third, Canada is characterized by cultural diversity and the absence of a single dominant cultural group. Since the arrival of Europeans in the sixteenth century, Canadian ethnic diversity has been continually regenerated through processes of immigration and city-building. We shall look first, therefore, at those twin processes, then turn to the ethnic communities that have

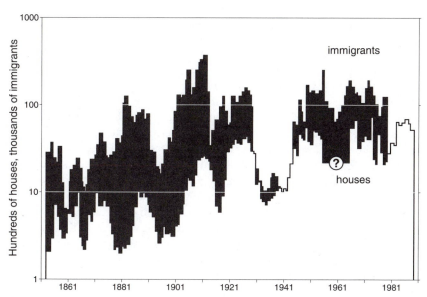

Figure 7.1
Rhythm of immigration and city-building, Montreal, 1851–1986. Note the strong
relationship between immigration waves into Canada (from annual reports of the
Department of Citizenship and Immigration since 1851) and building permits issued by
the city of Montreal.

emerged, to the expression of cultural diversity in the urban landscape, and,
finally, to the current process of communication between cultures.

IMMIGRATION AND CITY BUILDING

Immigration has occurred as waves breaking over the continent. As shown
in Figure 7.1, the swells of immigration synchronized with boom periods of
city-building.[1] The figure shows annual construction permits for Montreal,
but other Canadian cities shared the same peaks and troughs, sometimes
cresting even higher. Each construction boom has renewed the demand for
labour, and each wave of immigrant workers has added to the demand for
housing. Despite the attraction of farmland and official attempts to disperse
or direct them, immigrants have, in every wave, concentrated in metropolitan areas. it is therefore in the large cities that we see the major effects of
the rhythm of immigration.

Because each wave has comprised a distinctive mix of people, attracted to
successive regions of growth, shifts in patterns of immigration explain a
great deal of the variation in the ethnic composition of Canada's cities.
Some cities reflect the impact of only a few waves of immigration. Saint

John, for example, has not received a major influx of immigrants since its founding by Loyalists in 1783. Quebec City, despite massive arrival of Irish refugees from famine in the 1840s, is today predominantly French. At the other extreme, Montreal and Toronto reflect the influence of many successive waves. Against a backdrop of predominantly British origin in Toronto and French origin in Montreal is highlighted an array of communities with their own languages, religions, institutions, and cultural styles. In different proportions, they include well-established minorities of Italian, Jewish, and German origin, more recent communities of Greek and Portuguese origin, and newcomers from southeast Asia and the Caribbean. The landscapes of both cities display the effects of such shifting currents of immigration; witness the bulky loft buildings, which house the garment industry, along boulevard St-Laurent in Montreal or Spadina in Toronto (see chapter 10). Their labour force a century ago was heavily male and Jewish, from Russia, Poland, and Romania; by the 1950s it had become predominantly female, from Italy; today, women from Portugal, Asia, Latin America, and the Caribbean occupy these jobs, threatened by competition from new shirt factories in their home countries. Further west, the manufacturing cities of southwestern Ontario attracted specific European immigrant groups, Italian to Hamilton, German and Dutch to Kitchener. Populations of the prairie cities were established in the boom eras of the wheat economy, before the First World War with arrivals from Scandinavia, Poland, Germany, and Ukraine.[2] On the west coast, Vancouver has the highest level of ethnocultural diversity and the largest proportion of Asian Canadians, while Victoria is as "British" as Saint John.

Canada's immigrants arrived under a variety of pressures: a "push" provided by a pogrom in Russia, a junta in Greece, or repressive police in Haiti; the "pull" of jobs in construction or nursing in Canada; the drive of individuals and the enthusiasm of uncles and sisters already here. Under these pressures, members of particular communities tended to concentrate in certain economic niches, so that we observe some rather specialized cultural economies. "Chinese" restaurants, for example, and Italian-Canadian construction companies found in every metropolis reflect the intricacies of extended kinship networks and chain migration, as well as prejudices that have pushed these groups away from economic participation in other fields. We enjoy Greek pastry and Vietnamese springrolls from coast to coast, because these two peoples have exercised exceptional entrepreneurship and shared their cuisine beyond their own communities.

The distinctive skills and training of people of different backgrounds also allow creative teams to transcend their problems of communication. In one small sausage-making firm in Montreal, the foreman comes from Italy, the machinist from France, the helper from Russia; the owner, considerably older, is of Russian Jewish origin. The ingenuity of their machines for reconstructing sausage-casings from sheep gut is recognized in New Zealand and Australia; they blend and bundle spices for larger sausage and pickle packers in Pittsburgh, Cincinnati, and St Louis, to satisfy regional tastes derived from yet other streams of immigration.

Immigrants of each wave have been characterized by a particular balance of social classes and a certain "baggage" of political experiences and economic ambitions.[3] They have their own expectations of Canada and of themselves and their own perceptions of the values critical to the survival of their communities and the education of their children. These concepts change: in Montreal, for example, differences of religion were a strong barrier to marriage in the nineteenth century,[4] while today linguistic polarization seems stronger than the religious. If we penetrate any one community, we discover nuances, even antagonisms. Hungarian immigrants, for example, of Catholic, Jewish, and Protestant faiths, came in several waves. Those who arrived in the 1950s came from different regions of Hungary, with different skills, attitudes, and schooling from those who had come in the 1930s. Many Portuguese Canadians, concentrated around the parish church of Santa Cruz in Montreal, retain close ties with a home region of mainland Portugal, while others come from villages of the Azores, and a small group in a South Shore suburb comes from Angola, the remnant of Portuguese colonization in Africa. Immigrants of Chinese ethnicity re-create the divergent political opinions of the People's Republic of China, Taiwan and Hong Kong; while those of South Asian background are divided along ethnoreligious lines – Tamil, Sikh, Hindu, Muslim – as well as by place of origin.

The conditions under which migrants are received help define the sense of ethnic community that emerges, since each wave of arrivals is a new encounter, provokes a need for new forms of communication, and raises questions about existing privileges. In each generation Canadian society is rebuilt and the "place" of immigrants in Canadian urban society is redesignated. They become embroiled in the unresolved conflicts of earlier groups and subject to social definition within the dominant ideology of the time.

Until the Second World War, the dominant notion of "Canadian" was based on a stereotype of the "white Anglo-Saxon Protestant" (WASP). In some regions French Canadians were acknowledged as "belonging," but nearly all those of other backgrounds were viewed as "strangers." Even those advocating tolerance toward immigrants did so against a normative definition of "Canadian."[5] A normative or value-laden vision has affected policies concerning who will be welcomed to Canada and what sort of behaviour is acceptable. It influences the level of discrimination that has been directed at groups not seen as conforming to the Canadian norm.

Such notions have been supported by an assumption that "others" enter Canada at the bottom of the social and economic scale. In the large cities, many ethnic neighbourhoods were founded by new immigrants, initially men seeking accommodation close to their employment. In the late nineteenth century, Asian groups were used as strikebreakers in the primary industries of British Columbia, and labour disputes expressed divisions along both class and ethnic lines. As late as the 1920s, "navvies" and "coolies" were contracted like gravel and stone for canals, railways, and tunnels.

Norms of acceptability have had notable influence on immigration policies. In the 1890s one man, Clifford Sifton, was responsible for a radical shift in Canada's demographic make-up when he initiated a program to bring farmers from eastern Europe to settle the prairies.[6] These immigrants, referred to as "Sifton's Sheepskins," were shunted away from the eastern cities. Today their descendants are the established citizens of Winnipeg and Edmonton. In the next generation, Earl Grey, governor-general of Canada from 1904 to 1911, argued that Canada could attract gentlemen from Britain only by ensuring a sufficient number of Asian immigrants to act as their servants.[7] Others, viewing the "yellow peril" as competition for jobs, campaigned for the complete exclusion of Asians. The dominance of the latter view resulted in blatantly exclusionary laws which banned Chinese immigration almost completely from 1923 to 1947.

Manipulation of the labour force continues to take new forms. In the 1950s, domestic servants were personally recruited by Canadian tourists in the Caribbean. In 1987, an individual travel agent established a conduit of refugee claimants from rural Turkey to Montreal. In this kind of situation, many new arrivals start on the bottom step, and some individuals – like the Russian geologist who spent his first year in Canada sweeping floors – find themselves on a different step in the new world from where they stood in the old. Elaborate equations now attach arbitrary "points" to certain skills in economic demand, and special paths are provided for those with family sponsorship, church sponsorship, or the ability to bring in "suitcases of money" for a new enterprise.[8] Refugees from Vietnam, under the coercion of immigration points and church sponsorship, were initially assigned to rural areas, but a decade later they have recongregated in distinctive urban neighbourhoods.[9]

Immigration continues to have a significant effect on Canada's demographic structure. Total immigration in 1990 exceeded 200,000, slightly less than 1 per cent of the Canadian population,[10] but migrant streams were concentrated, as in the past, in a few cities. Immigrants experience pressures associated with concentration. Every day in any large Canadian city, the attentive observer can report incidents of racism – a landlord's abrupt response to a "foreign" accent, a taxi driver's unwillingness to stop for a dark-skinned couple, an employer's objection to turban or hairdo, or – with greatest risk – a tense encounter with police. The issues are often subtle. A visitor to Vancouver, for example, may hear immigrants from Hong Kong accused of defacing the residential landscape by building oversized houses, driving up real estate prices, or, by their children's diligence, raising the standard of education so that others cannot compete. These impressions usually have little factual basis, or have been distorted to suit certain views, but nonetheless create the pressures that arise as Canadians continue to negotiate place.

ETHNICITY AND COMMUNITY BUILDING

Because immigration is only a first stage in the process of ethnic group formation, we need to look at the process in the second and third generations. We usually find that particular groups occupy definite niches in the structure of status, class, and power in Canadian society, with some degree of conflict. Terms of negotiation are often drawn along ethnic lines. In that context, we need to consider the internal organization of the ethnic groups; to discover the ways in which a traditional heritage is maintained, transformed, denied, or used to jostle for position within the larger society; and to identify factors that influence the rate of retention of ethnocultural traits.

Little has been written by geographers and urban historians about the specific ways in which ethnocultural groups have redefined Canadian urban character. An exception is a collection of articles on ethnic enclaves in prewar Toronto, which treats the ethnic neighbourhood as "ambience" and seeks to "comprehend the group's sense of group" and to assess the concepts of neighbourhood and ethnic networks in terms of "the changing significance, for the immigrant and each succeeding generation, of various community institutions."[11] A neighbourhood is made up of corner stores, factories, houses, churches, schoolyards, and playgrounds: "Each was in a different spatial and psychic relationship to the 'little homeland,' the ethnic neighbourhood and the ethnic group. They were places where men and women had to negotiate their ethnicity, make those constant adjustments of style and thinking which were the milestones in the process of learning to live within the larger North American urban setting."[12] The neighbourhood became the focus of immigrant life and the centre of hope for future generations. It was also a refuge from which the rest of the city was sometimes felt to be foreign, even frightening: " Foreign immigrants and their children rarely developed a balanced map of the whole city. An understanding of its other people, of the use of its other spaces, or its history developed slowly. The first *Bulgaro-English Dictionary* included phrases on how to take the streetcar to King Street East "where the Macedonians live," side by side with how to find the Wabash ticket agent for trains to other Macedonian settlements near Chicago or St. Louis. It had no phrases about the Ward or Rosedale."[13]

Zucchi describes the beginnings of Toronto's "Little Italy" in boarding houses established along Elizabeth Street in the 1870s to provide winter housing for navvies returning from summer work sites on the railroads. Grocery stores and boarding-houses were nodes around which a large community developed an impressive institutional network.[14] Out of its cohesion has come genuine economic and political empowerment.[15] In contrast, Anderson, from an outside perspective, interprets the formation of Vancouver's "Chinatown" as an expression of racial discrimination. She argues: "Chinatown" was a social construct that belonged to Vancouver's 'white' European society, who, like their contemporaries throughout North America, perceived the district of Chinese settlement according to an influential culture of race. From the vantage point of the European, Chinatown signified all those features that seemed to set the Chinese irrevocably apart."[16] The contrast between these two accounts of "ethnic" neighbourhoods reflects differences in the ways in which groups view their own and others' neighbourhoods and hints at the tension that occurs on the streets as the spatial meaning of the city is negotiated. To its residents, a neighbourhood may convey the meaning of home, security, and opportunity, while outsiders may designate it a "slum" or a landscape of heathen degeneracy, in order to affirm their own sense of superiority and to reinforce an ideology of difference.

The example of Vancouver's "Little Tokyo" suggests that both interpretations express a dimension of the total experience and that our interpretations of the meaning of place need to be further adapted as circumstances change. The neighbourhood of Powell Street developed as a collection of boarding-houses for sojourners working in the Hastings Sawmill in the 1890s.[17] Owners of the boarding-houses were simultaneously labour contractors or bosses, translators, immigration counsellors, outfitters, and suppliers. Over four decades the area became a self-contained neighbourhood of Japanese immigrants and their children, with a full range of commercial and institutional services. That community too became a focal point, a place set aside. The white population perceived "Little Tokyo" as a ghetto, but in the minds of Japanese Canadians, scattered in other Vancouver neighbourhoods and in agricultural, logging, or fishing communities, Powell Street was a symbol of shared community, a source of traditional services and commodities, and a setting in which their common culture could be maintained. Economically, it was the command centre that controlled most of the contract employment throughout British Columbia. The fate of this community represents a tragic example of deliberate manipulation of the landscape in order to remove a designated group. The dispossession and incarceration of Japanese Canadians during the 1940s attest to the strength of hostility that can arise against a group whose presence is viewed as a "threat."[18] Powell Street became an empty shell of a place, where derelict buildings and wisps of newspaper behind cracked and dusty windows were poignant reminders of a community once vibrant and full of purpose.

How well do residential patterns in cities reflect those social dimensions and social distances? Traditional studies in urban geography have tended to take for granted a process whereby immigrants enter a country as poor and ill-educated, seeking security, kinship, and mutual services among their fellow émigrés in the low-cost housing of the inner city, close to industries employing "cheap labour." One model depicts the city as a series of concentric zones where affluence, assimilation, and acceptability increase with distance from the centre. Through a process of invasion and succession, it is claimed, migrants move into the centre and then outward, "escaping" from one zone to the next, thereby improving their lot in life.[19] The evidence does not always support such a theory, and actual patterns of distribution of ethnic communities defy simple explanation. The Jewish community of Montreal, for example, has achieved high levels of income and university education but remains strongly segregated within established suburbs. In Toronto, several populations of Caribbean ancestry (among whom there are high proportions of immigrants) are not as strongly concentrated as some of the third- and fourth-generation ethnic groups. A partial explanation for this pattern may lie in Toronto's high land costs.[20]

These brief sketches indicate something of the complexity and dynamism of the process by which distinctive ethnocultural communities are formed. A question that continues to puzzle urban geographers is whether strong patterns of concentration result from the segregation of a group from the mainstream, from positive choices to enjoy and maintain tradition and solidarity, or from a combination of both. We do know that such patterns exist today in Canadian cities,[21] but why do some groups concentrate in specific areas of the city while others are widely dispersed?

The forces of discrimination have been much in evidence. Prior to the Second World War, Jewish and Asian Canadians, among others, found refuge in segregated communities in the face of discrimination that reached violent proportions.[22] In areas such as West Vancouver's "British Properties," restrictive by-laws excluded certain groups from residence and sharpened the lines of segregation. At a more subtle level, segregation occurs because lack of economic opportunity restricts the purchasing power of certain groups, and therefore their residential choices. Studies have shown that the tendency for ethnic segregation goes beyond simple economic distinctions and beyond the power of class to define residential patterns.[23] Just as difficult to specify are the expressions of empathy, or of antipathy, by which groups sort out their spatial relations, or the intricate values by which they designate some neighbourhoods as accessible, some groups as welcome, others not.[24]

Other theories, focusing on the internal structure of the ethnocultural community, claim that "institutional completeness"[25] and ethnic segregation occur together. A study of six groups in Winnipeg showed that some groups (Jewish, Ukrainian, Polish, and French) have maintained community as well as institutional completeness over several generations. In the French-Canadian case, as described by Bernard Thraves (Figure 7.2), cultural sur-

Figure 7.2
Ethnicity expressed in the landscape, St-Boniface. French Canadians in Winnipeg sustain
their cultural identity through institutions such as the Cathedral of St-Boniface, built
inside the shell of its earlier burnt-out building. Close by are a college, museum, cultural
centre, library, hospital, schools, and convents, many with the same stone construction. The
neighbourhood has well-defined boundaries – the Red River, rail yards, and an industrial
zone – and a shared history is also expressed in street names. *Photo*: courtesy Bernard
Thraves.

vival has been assured by a critical mass of population and institutions of
group identity and mutual assistance. In the Jewish case, residents carried
their institutions with them as they migrated from the central city to the suburbs, while other communities, including German and Scandinavian, have
largely lost their residential and institutional distinctiveness.[26]

We do not know why this variation occurs. Is the retention of cultural
traits a statement of the strength of original traditions and values or a legacy
of past discrimination? Is the strength of institutions positively motivated, or
is it a defence against exclusion? Two principles seem to help answer these
questions. First, we need to look not only at the strength of institutions
within ethnocultural communities but also at the extent to which external notions and stereotypes of those groups (prejudices) have themselves become
institutionalized.[27] Second, all our observations tell us that, despite the role
of external conditions, and the interrelated factors of class formation and
prejudice, groups are still different from one another, and they create within
the city communities that are in many respects unique. This means that we

need to understand ethnocultural groups within the contexts that they themselves create.

The challenge is to make sense of this diversity and complexity, to comprehend the strength and durability of tradition even in times of rapid social change, to recognize the ways in which people invoke their ethnocultural heritage in building cities. As buildings are put up, altered, and torn down, a landscape is reworked as evidence of fundamental social change. At some stage in our investigations, therefore, we usually turn to the visible landscape, for it is there that daily lives are lived out and social meanings are expressed and interpreted.

CULTURAL LANDSCAPES IN THE CITY

How have people of different cultures modified Canadian urban landscapes to express their identifies? What spaces are recognized as having ethnic qualities? To address these questions, we turn to the streets of Montreal.

It has been argued that architecture itself is "ethnic." Of "French Canadian" and "Italian Canadian" architectural features identifiable in Montreal, not all would be recognized by a European visitor as "French" or "Italian." The modern cement church of Santa Cruz recalls Portuguese tradition by its exterior simplicity, its flower boxes, and its light-drenched interior. Close by, vernacular expressions of ethnicity include the patch of corn and tomatoes that flanks the entrance to an inner-city triple-decker, the grapevine that grows over the balcony, and a ceramic appeal to a saint honoured in one small corner of Portugal. A few streets away, entries of wrought iron and marble transform the homes of Italian Canadians in a way that is both reminiscent of the homeland and expressive of material success in Canada.

Because built capital has a lifetime of the order of a century, outlasting a human generation and a single immigrant wave, most city-dwellers occupy landscapes inherited from people of other cultural communities. Relatively subtle marking makes such landscapes comfortable and expressive. In the nineteenth century, Mount Royal Park was the scene of snowshoe exploits of Scottish and French Montrealers. In the 1870s it was "improved" with carriageways and curves in the tradition of romantic English gardens on an Olmsted design. In the century since, new users have appropriated it. Middle-aged couples speak German and Russian as they skate on its lake; small groups of old men chat in Chinese as they walk up the park road; younger Chinese Canadians are doing tai chi exercises; Portuguese Canadians are bowling. At certain hours, young English Canadians walk their golden retrievers and Rhodesian ridgebacks. Spanish-speaking and Afro clubs play soccer round the clock. On Sunday afternoons, Latin American and Caribbean Montreal jives in English, French, Creole, and Spanish, to

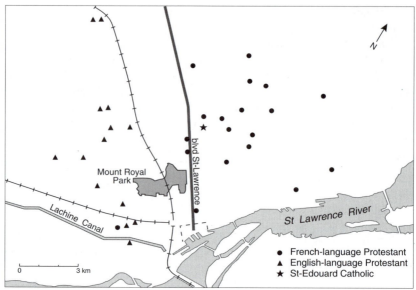

Figure 7.3
Churches of Montreal's black community. The scatter of Protestant churches reflects
residential separation of two linguistic groups in the black community, despite the
willingness of many members to make a fairly long journey to church. A majority of the
French- and Creole-speaking community from Haiti is Catholic, with services centred in the
parish of St-Edouard.

steel drums and rattles, in the shelter of a guardian angel, four stone lions,
and a Father of Confederation, who from his pedestal reminds us, "We are
of different races, not for strife but to work together for the common
good."[28] Those are some of the coexistences.

Another form of appropriation is the adaptation of places of worship. A
former synagogue in rue St-Urbain is now a Russian evangelical church. A
Presbyterian church in nearby Outremont has become the home of a Syrian
congregation, and a Methodist building has been rebuilt, with icon and
dome, as a Russian Orthodox church. A simple map of the hand-me-down
churches of Montreal's black congregations (Figure 7.3) shows a rather
wide scatter and a substantial gap between French- and English-speaking
groups from the Caribbean.[29] In a longer and more complicated history, successive locations of Montreal synagogues map out, in twenty-year jumps
which match the city-building rhythm, the path of Jewish communities up
"the Main" (boulevard St-Laurent), through Outremont to Snowdon, into
suburban Chomedey, Côte Saint-Luc, and Hampstead. The homes that surround the synagogues reflect the climb from confining poverty to expansive
comfort and display.[30]

Another appropriation of landscape is achieved by small business. Greek-Canadian entrepreneurs have expanded from barbershops, bakeries, and corner convenience stores into control of a large proportion of metropolitan restaurants and hotels. Each little island of Greece has its own pool hall and social club. As the community has aged, and become aware of the need to preserve its language and sense of identity, it has assigned the pool halls on Sunday afternoons to rotations of children who practise Greek letters and folk dances. At the Caffé Italia, many segments of the Italian community watch European soccer games transmitted by satellite and, with equal animation, discuss a political statement, a film, or a musical.

Another means of appropriating the landscape is festival, and a metropolitan area is a round of celebrations. The Chinese community spills out of "Chinatown" at the Harvest Moon. A St Patrick's Day parade survives in Montreal despite the raw weather in March. In 1989, under the green and the shamrock, thousands of Irish Canadians twirled their batons in rue Ste-Catherine, while a few blocks away, in rue St-Hubert, forty thousand French Canadians marched with equal enthusiasm, under the blue and white fleur-de-lys, to demonstrate support for the primacy of the French language. At Pentecost, a middle-aged band produces a solemn step for Santa Cruz's communicants, while the June Carifête parade has a jazzier rhythm, and Midsummer Night, celebrated throughout French Canada on 24 June as the feast of St. John the Baptist, has become one of Montreal's most vividly multicultural celebrations.

MULTICULTURAL COMMUNICATION

While people usually describe a landscape from its visual cues, we must also listen to the soundscape, more revealing of ethnicity in its symbolic meanings. Ethnic identity is a sense of family extended to a larger group. We experience it initially in a pre-verbal way, as we are cuddled and combed and cuffed. We situate ourselves in our home culture through the "near" senses of touch and scent and sound. In the larger world of the multicultural kindergarten, children make encounters and assert themselves, sharing their counting-out rhymes and naughty words. From there on, the security of our ethnic identity is threatened and challenged in constant contact with otherness.

Because the transmission of language is renewed for every infant, incentives that influence linguistic choices are perennial sources of family anxiety and public controversy. Transmission of religious traditions and moral standards is often perceived as dependent on the vehicle of language. For parents, a sense of control is eroded by a language shift. Older citizens, who may have sacrificed their own ambitions for the oncoming generation, sometimes feel isolated as they suffer from a deeply personal gap in communications with their grandchildren. The soundscape changes rapidly with each wave of immigration, each new generation, and each technological leap.

Language behaviour is more complex and more variable than a map of residential patterns can convey. In Montreal, for example, ephemeral clienteles of tourists, sports fans, and students vary the linguistic landscape of Ste-Catherine and St-Laurent from year to year, from one season to the next, and even from Friday night to Saturday morning. In workplaces, hospitals, and public spaces of the city, the several groups interact, and immense personal investments are made in maintaining first languages and acquiring second languages. Corporate and public investments are made in language teaching, translation, signs, bilingual publications, and other efforts to ease communication across language barriers. In Canada, the bulk of such investments are made in urban contexts, primarily Ottawa, Montreal, and Quebec. The personal costs are also concentrated in metropolitan areas, with greatest impact on communities of persistent immigration and limited formal education. A ride on the no. 55 bus or the no. 80 bus, up or down Montreal's "immigrant corridor," reveals the magnitude of cultural investments and encounters.

These encounters enrich our experience, but they also make us uncomfortable. Large cities are always places of ambiguity and tension as well as stimulation, and the multicultural nature of Canadian cities possesses an unpredictable quality. In the Montreal soundscape someone is always wondering: "In what language shall I address the policeman?" "In what language will the shop clerk speak to me?" The French and English versions of "O Canada" are capsules of our ambivalence, the exhilarating sense of inclusion and the distressing sense of exclusion that we all experience. The two versions are sung simultaneously at hockey game and convocation. Where do I join in? The land of whose ancestors? Singing the same tune, but using different words and radically different texts, are we sharing the same landscape of imagination?

This emotion-charged landscape is influenced by public policy, and policies formulated at different levels of government interfere like ripples in a pool. In the late 1960s, to accommodate Quebec nationalism and reduce social conflict in the country, the Canadian government espoused policies of *bi*lingualism and *bi*culturalism – French and English – and invested heavily in the partnership of those two "charter groups." A policy of bilingualism and defence of individual linguistic rights was promulgated, consistent with a deep current of individualism.[31]

Quebec's experience of misunderstanding (in 1970) was followed by a renaissance of ethnic communication: rage and laughter poured out in a solidarity of music, poetry, festival, and caricature, and in 1973 and 1976 provincial laws, prescribing the language of negotiation, contract, schooling, and advertising, strengthened the French consonance of Montreal. The legislation provoked a succession of tantrums and then a more solid political mobilization on the part of Italian-, Greek-, and English-Canadian communities.[32]

In a larger arena, concern with problems of identities, rights, and expectations of groups other than the two "charter groups" led to a federal policy of *multi*culturalism and debate over the needs of cultural minorities. The policy of multiculturalism was conceived as an expression of equality and the right of all Canadians to maintain their original ethnocultural affiliations.

Public policies have only limited effect on private behaviour. Because the various policies have not been consistent or coherent, each adjustment has provoked new doubts and old resistances. The degree to which a language is favoured or imposed in various parts of Canada is affected by many other political factors, including the tenacious habits of railway conductors, air controllers, and mail clerks. A sixty-year trend of bipolarization is expressed geographically: the French-speaking population is becoming increasingly localized in Quebec, and the English-speaking population outside Quebec. As Waddell has put it, "Canada seems to be evolving toward a more strict spatial segregation of the two official language communities."[33] Pointing to the gap between policy and behaviour with respect to language, Waddell attributes it to Canadians' symbolic use of language as a marker of the boundaries of ethnic identity. Francophones currently tend to treat French as a marker that distinguishes a "Québécois" from a "Canadian" identity, or "French-Canadian" from "English-Canadian" identity. At the same time, anglophones are adopting French as a marker of a "Canadian" national identity which they wish to distinguish from the "American." Various groups of people are attaching different meanings to language as a symbol.

The policy of multiculturalism is not fully practised, and equality is not a reality in either the social or the political landscape, as we can infer from residual racism, tensions among groups, and a persistent association between class and ethnicity. These problems are often evident in urban schools, since schools are the most elaborately regulated institutions for manipulating ethnicity. As a system for transmitting values, "school" is in its nature totalitarian,[34] and urban schools are early warning systems for the social trends that we have described, notably toward greater ethnic diversity, more visible minorities, and more open expression. In the wake of the new language regulations, the French-language schools of Montreal received new concentrations of visible and audible minorities,[35] and, with a ten-year lag, school boards have recognized the challenge of intercultural communication.

All the cultural landscapes of Canadian cities are expressions of identities, of social histories, and of the solidarities of languages and religions. While people are often preoccupied with rivalries and generation gaps, a larger drama facing every city is a commercial force that undermines all cultural landscapes. Symbolic meanings are eroded as steeples are overtopped by office buildings, and as high-rise apartments are designed without ethnicity, post offices without nationality, corporate logos without language. Cities become stage sets for the plastic identities of international banks and hotel chains. If indeed "money dissolves culture," the determination of all Canadians to preserve their several identities is a defence of values deeper than the dollar.

CHAPTER EIGHT

Work, Labour Markets, and Households in Transition

D. ROSE AND P. VILLENEUVE

One of the defining characteristics of the historical development of industrial cities, especially in North America, is the spatial separation of employment and residence. This separation has been reflected in a somewhat artificial, if often expedient, dichotomy between the "economic" and the "social" in analyses of urban structure.[1] More recently, however, social geographers, notably in Canada,[2] have begun to bridge this gap by exploring the links between the evolution or restructuring of urban economies and the changing social composition of neighbourhoods.

In this chapter we review these developments by interpreting the changing geography of labour and households in Canadian metropolitan areas from the early 1970s to the mid-1980s in terms of three systemic tendencies: the *tertiarization* of metropolitan economies; the *feminization* of the paid workforce; and the increasing *polarization* of occupational structures in certain cities with a large and advanced tertiary sector. Drawing on secondary sources, complemented by empirical analyses based on special tabulations of census material obtained for Montreal and Quebec City, we examine the particular shapes these trends have taken in Canadian cities; much of the more theoretical literature has been informed by the American context or by the dubious concept of a "North American City."[3] We suggest that these three trends may be modifying the texture of social space in Canadian cities in ways that depend largely on the types of households in which the various new segments of the labour force live,[4] and we point to some new and challenging research avenues.

TERTIARIZATION AND LABOUR-FORCE COMPOSITION

Canadian cities have experienced specific forms of tertiarization[5] conditioned by the country's position in the world economy – translated internally into a metropolis-hinterland pattern of unequal development.[6] In this section we explore key aspects of this process – the role of the state in resource and infrastructure development, the trend toward Canadian ownership of key corporations, and the growth of the public and parapublic sectors – and their implications for "centrality" within metropolitan cities.[7]

Canada is both an exporter of raw materials, capital, and producer services (such as engineering services) and a net importer of capital, labour, and finished products. The nature of industrialization and urbanization has influenced the varying success of the state and indigenous capital in capturing within Canada the backward and forward linkages of resource extraction. Yet, and somewhat paradoxically, successful, diversified long-term development requires independence from the original staple resource, as has been achieved in southern Ontario[8] and to a lesser extent in southern Quebec, although its historical staple, cheap labour, has left a much weaker legacy.[9] Thus a diversified space-economy is now managed from the downtowns of Toronto and Montreal in Canada's heartland. In the hinterlands, situations vary greatly, but regional cities have been typically more successful at capturing lower-level tertiary spin-offs from staple production than at retaining higher-level tertiary and manufacturing linkages. Even Vancouver, notwithstanding its increasingly strategic location with respect to the Pacific Rim, remains an "intermediary between world markets and a vast resource environment."[10]

From the 1960s on, governments (both federal and provincial) increasingly took control of the management of natural resources and the profits they generated – which had previously been largely siphoned out of the country[11] – with a view to using these resources to foster domestic economic growth. Thus much of the resource base was nationalized (creating such crown corporations as Petro-Canada and Hydro-Québec), and multinationals faced tougher negotiations with state agencies.

Moreover, the period from the mid-1960s until passage of the free trade agreement with the United States in 1989 also saw increased Canadianization of private corporate ownership structures, especially in the resource sector. Firms under Canadian control tend, much more than those located in Canada but under American control, to buy their services inside the country, especially in the case of management services traded by manufacturing firms.[12] Additionally, data on mergers by ownership[13] point to more Canadian-based mergers in tertiary sectors – here again, then, Canadian ownership seems to reinforce metropolitan growth. As we shall see, ownership also has implications for the relative location of corporate headquarters within metropolitan areas, which in turn affects the social composition of central cities.

Canadianization – both public and private – thus not only reinforced the growth of metropolitan areas but also contributed to the economic vitality of their central business districts. The CBD location of the head offices of Canadian corporations stimulated expansion of adjacent producer-services complexes – the most metropolitan- and CBD-oriented of all economic activities. In some cases this led to very rapid growth, such as that generated by the resource sector in Calgary in the 1970s.[14] Moreover, in Canada, ownership in finance and commercial real estate is heavily concentrated in a few hands; this has helped ensure a reliable and abundant flow of private and public funds for the redevelopment of central cities, often assisted by state agencies.[15]

The long history of state involvement in transportation, as well as the more recent formation of crown corporations in the communications sector, has also significantly influenced the tertiary employment structure of Canadian metropolitan cities, notably Montreal, where the head offices of Canada's largest transportation corporations are still located.[16]

Another key aspect of tertiarization has been employment in provincial governments, which expanded greatly in the 1960s and 1970s, as they took on greater responsibility for health, education, and welfare. This expansion promoted the concentration of tertiary activities in major cities, although effects varied from province to province: for instance, Montreal has a much bigger share of Quebec's universities than Toronto has of Ontario's.[17] The expansion of the public and parapublic sectors was perhaps most dramatic in Montreal and Quebec City, fuelled by the rise of social democracy and economic nationalism in Quebec and linked to the wresting of power, especially in education and social services, away from the Catholic church. Furthermore, expanding the Quebec state also increased the size and the economic and political clout of the francophone middle class, specifically through the expansion of professional and managerial jobs targeted to francophones, traditionally excluded from important positions in anglophone-controlled corporations and institutions. Emergence of a state-centred power base led in central Montreal to attempts to develop a second downtown core, anchored on public- and parapublic-sector complexes specializing in administration, communications, education, and the arts and located northeast of the traditional CBD, bastion of private-sector tertiary employment and power.

These sets of developments also help to explain why the Canadian urban "new middle class" is heavily grounded in the public and parapublic sectors and less associated with corporate "yuppiedom" than stereotypes borrowed from the United States might lead us to believe.[18] Table 8.1 illustrates the importance of direct public-sector tertiary employment in Canada's three largest cities in 1986 (unfortunately these data do not identify crown corporations). The role of public and parapublic services in Montreal (29 per cent) is striking, especially compared to Toronto (24 per cent). Table 8.1 also points to the significance of private-sector producer services, the most rapidly-growing component of the tertiary sector.[19] Not surprising, the 1986 employment data in the table show that producer services are more impor-

Table 8.1
Tertiary-sector labour force 15 years and over by industry divisions (1980 classification) and sex, Montreal, Toronto, and Vancouver CMAs, 1986

Industry divisions	Montreal			Toronto			Vancouver		
	Total	Male	Female	Total	Male	Female	Total	Male	Female
Total, tertiary industries	1,075,260	561,395	513,885	1,388,190	680,040	708,165	577,320	287,985	289,325
Producer services	(29.3)	(33.3)	(24.9)	(33.4)	(36.6)	(30.3)	(30.6)	(36.0)	(25.1)
Transportation and storage	7.2	11.2	2.7	5.3	8.3	2.4	8.0	12.7	3.2
Finance and insurance	6.1	4.0	8.5	8.5	6.4	10.6	5.8	3.9	7.8
Real estate operator and insurance agent industries	2.7	2.7	2.6	3.5	3.6	3.4	3.4	3.7	3.2
Business service industries	8.2	8.8	7.5	10.8	11.5	10.2	8.6	9.6	7.6
Communications and other utilities	5.1	6.5	3.6	5.2	6.8	3.6	4.7	6.1	3.3
Consumer services	(41.9)	(43.8)	(39.8)	(42.7)	(45.2)	(40.3)	(44.8)	(45.9)	(43.7)
Wholesale trade industries	7.6	9.9	5.2	8.8	11.6	6.1	7.2	9.9	4.5
Retail trade industries	17.7	18.3	16.9	17.0	17.1	16.8	16.5	16.4	16.6
Accommodation, food, and beverage industries	7.5	7.4	7.5	7.8	7.9	7.7	10.2	9.7	10.7
Other service industries	9.1	8.2	10.1	9.2	8.7	9.7	10.9	9.9	12.0
Public and parapublic services	(28.8)	(22.9)	(35.3)	(23.9)	(18.2)	(29.4)	(24.6)	(18.1)	(31.2)
Government service industries	7.6	9.3	5.8	6.8	7.7	6.0	6.8	7.6	6.0
Educational service industries	9.0	7.1	11.0	7.8	6.4	9.3	7.3	6.2	8.3
Health and social service industries	12.2	6.5	18.4	9.3	4.2	14.2	10.6	4.3	16.8

Source: Statistics Canada, 1986 Census, Summary Tabulations, all CMAs, Table LF86B05.

tant in Toronto, both in absolute terms and relative to public services, than in Montreal or Vancouver. This concentration of producer services at the top of the urban hierarchy, together with high-level financial services and the head offices of large corporations, reinforces the dominance of Toronto's downtown corporate complex over the Canadian urban system.

Table 8.2 shows how the trends described above affect occupational structures in five Canadian metropolitan areas – Canada's three largest cities and two other cities of different size and position in the urban hierarchy that have also been experiencing distinct forms of tertiarization, each related to the particular forms taken by regional economic development. Halifax has been rapidly developing as a regional control centre for Atlantic Canada, while Sudbury has endeavoured to diversify its economy into a whole range of tertiary industries in order to reduce its dependence on mining and primary processing.

The reinforcement of Toronto's dominant position between 1981 and 1986 can be seen again in Table 8.2, this time in terms of occupational structures: managers and administrators formed a larger share of the work-force than in the other cities, and their numbers grew more rapidly. Vancouver's increasing trade with the Pacific Rim, not only in primary products but also in services, has contributed to a much more rapid growth in numbers of financial directors than has occurred in Montreal. At the same time, the continued and prolonged vulnerability of the BC economy shows up in the decline in construction and processing jobs. Halifax shows substantial growth not only in management but also in scientific occupations. In Sudbury, deindustrialization – the world slump in the nickel market and mechanization at INCO that have cost the community many thousands of jobs since the late 1970s – can be seen in the decline in "occupations not elsewhere classified" (essentially mining) and processing and in sales and administrative jobs, as well as in teaching (as a result of reduced numbers of young families). Yet the period 1981–86 also saw a phenomenal increase in managerial positions, caused presumably by the city's economic development stategy, which has fostered the growth of a diverse range of small businesses.[20]

The full import of the developments discussed in this section for the structure of urban labour forces and labour markets by occupation and gender has yet to be adequately researched. A prior step is to make the link between tertiarization and the processes of feminization and polarization in the occupational structure, as we attempt in the next two sections.

FEMINIZATION OF THE LABOUR FORCE

Together with the rise of tertiary activities, feminization in the labour market is probably the most significant change occurring in Canadian society over

Table 8.2
Experienced labour force, by occupation major groups, and employment change 1981–86, for selected "tertiarizing" CMAs, 1986

Occupations	Toronto 1986 (%)	Toronto Δ1981–86 (%)	Montreal 1986 (%)	Montreal Δ1981–86 (%)	Vancouver 1986 (%)	Vancouver Δ1981–86 (%)	Halifax 1986 (%)	Halifax Δ1981–86 (%)	Sudbury 1986 (%)	Sudbury Δ1981–86 (%)
All occupations	(1,960,170) 100.0	13.0	(1,467,335) 100.0	3.8	(734,290) 100.0	8.6	(160,140) 100.0	12.7	(69,365) 100.0	−8.3
Managerial, administrative, and related	13.4	33.0	12.4	23.2	11.6	29.0	10.8	27.8	8.4	16.8
Natural sciences, engineering, and mathematics	4.4	18.3	4.0	17.9	3.6	4.2	4.0	14.0	3.1	−6.1
Social science and related fields	2.1	33.6	2.0	26.4	2.4	35.7	2.0	17.1	1.8	−3.2
Teaching and related	3.7	16.7	4.4	2.7	3.7	4.0	4.7	9.9	5.3	−8.6
Medicine and health	3.9	18.2	5.3	13.0	5.2	12.0	6.3	11.6	4.7	19.0
Artistic, literary, recreational, religious	2.4	22.1	2.6	10.5	2.2	36.5	2.0	24.4	1.6	6.3
Clerical and related	22.4	5.9	21.1	−0.8	20.0	−0.5	20.7	7.8	19.4	16.7
Sales	9.8	16.5	9.4	4.9	11.1	13.8	9.8	13.0	8.8	−1.6
Service	10.3	14.2	11.3	4.7	14.5	24.8	17.9	13.6	14.3	11.7
Farming, horticulture, and animal husbandry	1.0	4.6	0.7	12.6	1.7	34.6	0.6	15.6	1.1	33.3
Processing, machining, and related	4.7	−2.7	5.2	−10.9	4.3	−14.3	2.2	4.8	4.0	−32.4

Product fabricating, assembly, and repair	8.6	8.1	9.4	−9.7	5.5	−2.9	4.5	1.0	5.8	−13.2
Construction trades	4.6	5.8	4.3	3.0	5.6	−4.8	5.9	23.3	6.6	−8.6
Transport equipment operating	3.0	8.0	3.5	−6.2	3.7	4.6	3.7	9.1	4.1	−2.9
Not elsewhere classified	5.6	9.4	4.4	5.9	5.1	−5.2	4.9	5.5	11.2	−26.5*

Source: Statistics Canada, Census of 1986, Cat. 93–156, Table 16.
* Mainly mining occupations.

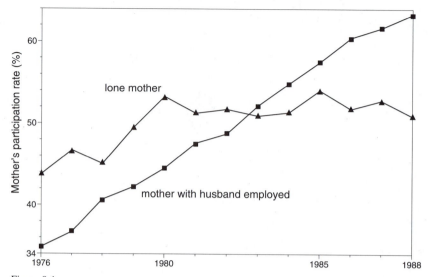

Figure 8.1

Female labour-force participation rates for families with pre-school age children, Canada, 1976–88

the last twenty years. The drastic increase since the mid-1970s in the participation of married women with pre-school children is particularly remarkable (see Figure 8.1 for Canada as a whole; comparable time series data are not available at the metropolitan level). Table 8.3 shows that although female "job ghettoes" persist and have even been reinforced in areas such as health services and clerical work, major inroads have been made in managerial and administrative occupations, especially by young women aged 25–34.[21]

Tertiarization involves restructuring by firms and governments, in response to global changes in productive activities and shifting consumer needs. Yet feminization of the labour market also involves responses from individuals and households to a changing social and economic environment. A whole array of factors combine to account for feminization, most of them seeming more pronounced in metropolitan areas, and – contrary to the US situation – even more so in the inner city than in the suburbs.

First among these factors has been the rise of the two-earner family as the "new norm"[22] as a result of changes in attitudes to paid work, family life, and women's roles. Women have sought out better education and better jobs and have taken on more hours of paid work, often as part of a household strategy to increase living standards or to maintain the household income in the face of men's declining real wages. Depending on local housing-market conditions, access to homeownership or the type of home bought may be contingent on the second income.[23] Similarly, gentrification of inner-city

Table 8.3
Trends in the feminization of the occupational structure, occupation major groups and selected minor groups, Canada, 1976–85

Occupations	Percentage of jobs held by women		Δ employment 1976–85 (%)	
	1976	*1985*	*Women*	*Men*
All occupations	37.1	42.5	36.7	9.1
Managerial and professional	39.6	44.4	71.4	40.7
Managerial and administrative*	20.4	32.3	221.1	72.3
Teaching	56.4	59.9	19.7	3.9
Medicine and health	75.7	77.9	37.9	21.8
Clerical	75.4	79.6	21.2	−4.6
Sales	35.5	43.4	26.4	−9.4
Service	51.0	56.2	49.2	20.6
Processing, machining, and fabricating	18.8	19.1	3.6	1.6
Material handling and other crafts	18.1	19.0	12.0	5.4
Transportation	3.9	6.6	84.0	5.2
Construction	0.7	1.7	99.6	−11.1
Primary occupations	18.4	20.0	14.9	3.4

Source: Dumas (1986).

* Statistics Canada altered this occupational classification over this period. Effective in 1983, a considerable number of jobs were reclassified from "sales" (and, to a lesser extent, from "service") to "managerial and administrative." This has the effect of exaggerating the increase in employment in the latter. The statistical effect is slightly greater for men than for women.

areas in many Canadian cities has rendered single-family homes inaccessible to most families without two incomes.

Second, an increasing awareness has developed, brought about by many factors, including the women's movement, that economic autonomy is an almost necessary condition of more equal gender relations. This seems to be most pronounced in certain centrally located neighbourhoods conducive to social networking[24] among certain groups of women and among households of professional couples, including those who could be termed "marginal gentrifiers."[25]

Third, households and families have become increasingly mutable. Couples marry later, and marriage rates have dropped. Common-law unions now form about one-quarter of all couples whose main breadwinner is between 25 and 44;[26] fewer than one in four of these unions last for more than five years, leading either to marriage or separation. Divorce rates are now around 30 per cent for couples married during the early 1970s; these rates are higher among marriages following common-law unions, perhaps because of the type of people most likely to form such unions (see chapter 4).[27]

Higher rates of separation and divorce have contributed to large relative and absolute increases in female-headed lone-parent families; in 1986 al-

Table 8.4

Labour-force participation (%) of mothers according to ages
of children, selected CMAs, 1986

CMA	All children under 6 years	All children 6 years and over
Vancouver	62.6	65.0
Edmonton	61.5	71.1
Calgary	65.4	70.8
Regina	68.4	71.8
Winnipeg	61.7	67.0
Toronto	66.4	69.0
Ottawa-Hull	64.9	65.7
Montreal	60.2	55.3
Quebec	67.9	47.7
Saint John	60.4	54.3
Halifax	64.5	60.7
St John's	70.3	56.4

Source: Statistics Canada, 1986 Census, Summary Tabulations,
CMAs, Table LF86B08.

most one in six families with minor children was headed by a lone parent –
female in well over 80 per cent of cases.[28] The resulting need for women to
be "breadwinners" has undoubtedly helped feminize the labour market –
although, as Figure 8.1 shows, labour-force participation rates of lone mothers fell behind those of mothers in two-earner families after the 1982–83 recession. The gap is most pronounced for lone mothers of very young
children, largely no doubt because of the high cost and difficulty of finding
good-quality childcare (see chapter 18).[29] At the same time, the increased
chances of finding employment may contribute to women's decisions to live
independent of a partner.

Local variations in the above processes and in those factors motivating
women, especially mothers, to be in the paid labour force may explain the
considerable differences in labour-force participation among CMAs (Table
8.4). In Quebec City, mothers with no children under six years old (probably an older group of women) have the lowest participation rates among
CMAs whereas mothers whose children are all pre-schoolers rank among the
highest; this finding highlights the rapidity of socio-cultural change affecting
younger francophone women in the province of Quebec. The high rate observed for St John's may be related in part to spousal unemployment or low
wages or to migration from rural areas or small towns to a metropolitan milieu where employment prospects are better for women and child care is
more available. Despite these differences among CMAs, time-series data
show that variations among provinces as a whole are decreasing over time.[30]

We have been careful thus far not to overemphasize "top-down" explanations of labour-market feminization – that is, the growth and particular characteristics of tertiary-sector employment.[31] Yet there are connections between factors encouraging women to be in the labour market and developments that influence the types of employment opportunities available. For instance, affirmative action in the public sector probably contributed to the higher-than-average aggregate occupational mobility of women in public administration during the 1970s in the Montreal CMA.[32]

Preliminary data point to interesting links between feminization in different cities, the types of tertiary jobs available, and a city's position in the urban hierarchy. For instance, from 1975 to 1987, women comprised 84 per cents of new entrants to the labour force in the Montreal CMA, compared to 61 per cent in Toronto.[33] Yet 1986 census data show that in most branches of tertiary employment sectors, Toronto's labour force remains more feminized than Montreal's. This is perhaps because of the former's massive growth and concomitant low unemployment levels and the need for new and young workers, who tend more often than not to be female,[34] combined with the pressures put on married women to take on paid work because of the high cost of living. Yet there is a notable exception: the finance and insurance sector is more feminized in Montreal than in Toronto.[35] Montreal's secondary status in the urban hierarchy as a financial centre means that this sector contains a lower proportion of "top jobs" (still primarily occupied by men) and is thus more open to women.[36]

Table 8.1 shows how public services contribute to employment opportunities for women, especially in Montreal, where 35 per cent of all women with tertiary-sector jobs were in public and parapublic services, as compared with 29 per cent and 31 per cent in Toronto and Vancouver, respectively. Furthermore, data for 1971 and 1981, cross-tabulating sectors by occupations, show that in Montreal expansion in the public and parapublic sectors has enabled more women to obtain professional and managerial jobs.[37]

Governments can also influence the type and gender composition of tertiary jobs in different locations. For example, the proportions of federal-government jobs held by women are highest in smaller metropolitan cities where administrative support positions dominate.[38] In Sudbury, this resulted largely from the relocation of income-tax processing to that city.[39] Provincial governments frequently adopt similar policies of decentralization: Sudbury was designated a regional health care centre (500 new jobs were created from 1981 to 1986, the majority being held by women[40]) as a result of community lobbying in the wake of deindustrialization.[41] Such strategies are often influenced more by the political goal of reducing regional inequalities than by considerations of efficiency or labour costs; thus new employment patterns (including reasonably paying jobs for women in small cities) are not necessarily the same as those that would emerge if private-sector ter-

Table 8.5

Women's share of employment, 1986, and employment change by sex, selected occupations, for five "tertiarizing" CMAs, 1981–86

Occupations	Toronto % jobs held by women 1986	Toronto Employment change 1981–86 % men	Toronto % women	Montreal % jobs held by women 1986	Montreal Employment change 1981–86 % men	Montreal % women	Vancouver % jobs held by women 1986	Vancouver Employment change 1981–86 % men	Vancouver % women	Halifax % jobs held by women 1986	Halifax Employment change 1981–86 % men	Halifax % women	Sudbury % jobs held by women 1986	Sudbury Employment change 1981–86 % men	Sudbury % women
All occupations	45	10	17	43	0	9	44	5	13	44	10	17	43	–8	11
1130 Directors-general, senior managers	15	59	164	16	52	176	14	63	94	14	51	160	19	102	92
1135 Financial directors	35	20	79	31	–2	41	34	7	42	29	5	80	35	–20	125
1171 Accountants and auditors	28	5	47	41	10	70	40	20	96	40	2	81	45	–16	35
2183 Systems analysts, programmers, etc.	30	57	91	34	47	82	28	73	62	31	104	95	29	33	116
2343 Lawyers and notaries	22	20	87	28	23	67	23	28	87	29	7	100	7	17	–50
3131 Graduate nurses	96	112	32	94	3	26	93	56	28	97	64	18	98	67	37
4131 Bookkeepers	84	1	2	78	–21	–14	85	–16	–7	85	–5	17	84	46	28
4133 Cashiers	88	64	20	87	33	13	88	60	14	90	65	15	95	–27	–17
4143 Electronic data-processing operators	79	39	49	77	35	53	83	50	18	84	26	60	85	89	94
5135 Salespeople, sales reps, n.e.c.*	54	32	19	49	10	17	51	40	20	54	35	17	59	40	–1
5172 Real estate agents	41	25	31	43	16	19	35	14	30	43	43	81	45	–7	41

6125 Waiters and waitresses	67	18	7	72	13	4	71	36	27	75	21	2	86	67	1
6191 Janitors, domestics	38	−1	7	30	3	7	38	13	−11	44	07	5	53	3	7
Processing, machining and related	20	−4	2	17	−13	−1	16	−17	0	17	2	22	8	−32	31
Product fabricating, assembly, and repair	32	9	7	34	−8	−14	18	−3	−2	11	−2	36	7	−13	34

Source: Statistics Canada, Census of 1986, Cat. 93-156, Table 16.

* Not elsewhere classified.

tiary employment were more predominant. At the same time, and somewhat levelling out these differences in the aggregate, there is little proportionate variation among cities in consumer-oriented private and parapublic services (other than health care), which are very highly feminized.

Table 8.5 illustrates the links between different processes of tertiarization and feminization of selected occupations. Here, although absolute numbers remain small, we see major gains for women in senior management positions and business service occupations, especially in the three largest cities. The higher proportion of women in senior management and accounting in Sudbury is probably linked to the growing small-business orientation of the city's private sector. The decline of manufacturing has hit men more severely than women, except in Montreal, while women's jobs in manufacturing rose significantly (from a small base) in Halifax and Sudbury. Low-level service occupations are more feminized in these two cities than in Toronto, Montreal, or Vancouver, as are clerical jobs – indicating perhaps a greater range of occupational options for women in the largest metropolises.

The social implications of feminization on a community can be illustrated for Sudbury. As Table 8.5 shows, the female labour force increased by 11 per cent in five years, while the male labour force shrank by 8 per cent. This growth in the presence of women in the work-force, in an economy traditionally dominated by mining and resource-processing, must surely be affecting cultural values, gender relations, and service provision. As more small and medium-sized places in Canada seek to diversify from dependence on traditional staples, the feminization of employment and its implications for social geography are becoming crucial issues.[42]

More broadly, and with more ambiguous results for women's work, greater flexibility in some branches of the labour market[43] has increased job precariousness and part-time work.[44] These trends are particularly marked in certain white-collar and service sectors, as well as in some branches of manufacturing, where women were already heavily concentrated.[45] Discussions of increasing labour-market segmentation, subcontracting, deunionization, and so on now fill the literature.[46] Technological change is also evoked as a crucial factor mediating these changes.[47] For reasons often originating in the household sphere, women's occupational and career paths are more often interrupted than those of men,[48] making them good subjects for a whole array of strategies aimed at greater labour flexibility.[49] Of course, such flexibility can sometimes be advantageous to the work-force, as in voluntary job-sharing.

POLARIZATION OF OCCUPATIONAL STRUCTURES

The rise of tertiary activities and the feminization of the work-force have been happening in a turbulent period that has seen systemic occupational restructuring in all economic sectors. There has recently been considerable Canadian research on the distribution of occupational skills[50] and wage levels, enabling empirical exploration of the phenomenon of a "vanishing middle class." In the 1970s continued expansion of the welfare state and maintenance of relatively strong unions muted these tendencies,[51] yet employment income became more polarized from the mid-1960s to the mid-1980s, especially in Ontario.[52] Young workers, in particular, are increasingly concentrated in minimum-wage tertiary-sector jobs – again, especially in Ontario[53] – and their earnings have fallen further behind those of older workers.[54]

Occupational polarization is most rapid in large cities with a strong "advanced tertiary" sector (which requires many managers and professionals and labour-intensive support services for buildings and for high-level staff); a rapidly growing consumer-services sector (fast-food restaurants and services and goods associated with "conspicuous consumption"); and a sizeable residual and highly feminized labour-intensive manufacturing sector (most important, the garment industry, which tends to employ mainly immigrant women).[55] While contrasts in Canadian cities are not as stark as in the "global cities" of New York or Los Angeles,[56] Toronto and Montreal would seem likely candidates for increased occupational polarization in the future.[57] Ontario's greater income polarization is probably a product of Toronto's economic boom in the 1980s;[58] unemployment among young people is lower there than in Montreal, for instance,[59] because so many young workers have been absorbed into "flexible" and minimum-wage jobs in consumer services. Research on Montreal has found increased polarization in the aggregate employment structure within such sectors as finance and labour-intensive manufacturing.[60] The trend was stronger for the female work-force than for the work-force as a whole, because of the simultaneous feminization of managerial and professional positions and the continued relegation of a great many women to low-level tertiary and manufacturing jobs. Here, then, we see the complex interweaving of the processes of tertiarization, feminization, and polarization.

IMPLICATIONS FOR THE CHANGING GEOGRAPHIES OF WORKPLACE AND RESIDENCE

The three relatively macro-scale tendencies discussed so far in this chapter may also be creating new geographical patterns of socioeconomic and life-course differentiation within major Canadian cities, caused by the intersection of changes in employment structure with shifts in household composition and practices.

Tertiarization in major Canadian metropolitan economies seems to be simultaneously creating two different tendencies: a new centrality within cities, as high-level employment, especially in head offices and producer services, maintains the pivotal role of CBDs, and the emergence of suburban subcentres of office activity, sometimes combined with lower- and middle-level consumer-oriented employment. Interpreting these tendencies in the light of feminization and polarization, and in relation to the household characteristics of that labour force, produces some interesting and under-researched issues.

As Gad and Holdsworth[61] have shown, corporate capitalism has used the high-rise office building to enhance its centrality since the end of the nineteenth century. Yet head offices of large firms have continuously shared downtown high-rise facilities with many smaller business service firms. More recently, and associated with the standardization and automation of information processing, many semi-skilled and predominantly female clerical workers of large firms have relocated to suburban back offices, while management and its support staff stay downtown.[62] In Canada this trend is less established than in the United States; even in Toronto it does not take the orthodox form noted in US studies,[63] perhaps in part because of urban planning practices. Other Canadian metropolitan areas have experienced more-or-less isolated relocations of routine office work from central areas to the suburbs, such as an on-line batch processing division of Revenu Québec in the Quebec City region,[64] and the head offices of B.C. Tel in Vancouver.[65] Ownership may also influence headquarters location. Gad[66] reports that foreign firms account for a large proportion of those locating their Canadian head offices in the suburbs of Toronto, while Canadian-owned firms seem to be staying in the centre. The former are, of course, less involved in higher-order decision making and less attached to the local market and local institutions.

Some authors argue persuasively that centrality in location is indeed the norm for high-level decision-making activities and that in Canada – unlike the United States, where urban decay repels corporate headquarters in all but a few major cities – a downtown site continues to be favoured for head offices.[67] A recent and exhaustive study of trends in office location in Montreal found that for a firm's key decision-makers, especially in financial and business services, personal contacts are still crucial, notwithstanding advances in communications technology.[68] A central location of offices, and the whole social and cultural infrastructure of downtown services, facilitate such contacts.

The link between centrality in this sense and the reshaping of the social fabric of downtowns and adjacent residential areas is relatively unexplored in Canada. The above-mentioned types of consumption practices of key personnel are an integral part of their "productive work" (as decision-makers, communicators, and analysers of information) and also part of their own personal "reproductive" activity (eating and socializing with colleagues and competitors, for example). To what extent do these practices extend outside normal business hours, and how do they influence residential location?[69] This could depend on whether "executive" housing is available downtown or close by (as in Toronto, Vancouver, and, increasingly, Montreal) or mainly in suburban areas (as in Edmonton).[70] Such decisions and practices are also influenced by household composition and gender relations, family situation, and partners' employment locations.

The relationship between centrality and feminization, both of the work-force and in terms of residential location, is also important. Huang and Gad[71] point out that the feminization of Toronto's labour force is so pronounced that an increasing proportion of the central-city work-force is female, notwithstanding the proliferation of suburban jobs for women. Whether or not this translates into inner-city residence by women workers depends on a complex interaction of occupation, household situation, and life-course position – sometimes mediated by ethnocultural identification and structured by local housing-market conditions and by considerations such as personal safety. In some of Montreal's inner-city areas, occupational polarization translates into a large social gulf between different groups of women residents who work in the central city and live in the same neighbourhood. Marginal and established gentrifiers, typically Canadian-born professionals and managers, may live alone or be full- or part-time lone parents or members of two-career couples.[72] In contrast, immigrant and ethnic-minority females may work in low-level service positions or in the garment industry (where jobs are still located close to the traditional sources of cheap labour);[73] they may live with their parents or in extended families, as lone parents or with partners who work a different shift so as to cope with childcare needs.[74] Variations on this theme, with emphasis on the social networks developed by women and on inter-class relations, have been studied for Quebec City neighbourhoods.[75] Women professionals who form part of a two-earner family (their spouse usually also being a professional) are more likely to live in traditionally elite and more homogeneous neighbourhoods, still close to the centre, where single-family housing is more readily available.[76]

Stage in the life-cycle has always been a crucial demarcator of social space in Canadian cities, but increased feminization of higher-level jobs has allowed growing numbers of women to make their own decisions about residential location, alone or with partners (or even with ex-partners with whom they share childcare).[77] These decisions enable them to reduce the spatial separation between home and employment and thus often to simplify the day-to-day carrying out of family- and employment-related tasks.[78]

At the same time – and although all the literature for Canadian and US cities shows that women's journeys to work are still shorter than those of men[79] – continued downtown concentration of office jobs, combined with residential suburbanization and the rising cost of inner-city housing, has resulted in longer trips to work for many women. Lone parents must often work centrally to obtain decent salaries[80] but may have to reside in the suburbs, trying to hold on to the marital single-family home,[81] or live in walk-up apartment districts undergoing déclassement.[82] In such situations, good public transit becomes crucial. Data on changes in commuting patterns by gender, occupation, and economic sector in Montreal do indeed show a positive relationship between occupational and spatial mobility, in which the presence of rapid transit no doubt plays a part.[83]

For many women, especially suburban dwellers, short journeys to work may "trap" them in part-time and low-wage service employment if they are to reconcile a job and domestic responsibilities. For other groups of suburban women, the situation may be more analogous to that of urban professionals – a short work trip taking them to a subcentre with middle- or upper-level jobs, mostly in consumer-oriented public services.[84] Feminist proposals to reduce the separation between home and work and to create local employment opportunities may also result in captive labour markets.[85]

Such considerations open up new questions concerning a firm's choice of central or suburban headquarters. Ley[86] has compared two Canadian-owned utility firms, B.C. Hydro, which has kept its head office in downtown Vancouver, and B.C. Tel, which has relocated to suburban Burnaby. Such moves do not affect all head office personnel in the same way. A large proportion of managers at B.C. Tel noted a loss of personal contact with other businesses downtown, while clerical workers – most of whom live in the suburbs – like the shorter time spent commuting. Yet Ley also found that at B.C. Tel female workers were more satisfied than male workers with the suburban location, whereas at B.C. Hydro they were more satisfied with the downtown location.[87] The reason could be that the female clerical workers at B.C. Hydro do not have the same life-cycle characteristics as those at B.C. Tel, particularly since the former are less concentrated in the suburbs than are the latter. It may well be useful to study links between the household sphere and the sphere of employment, an argument supported by Ley's further analysis of the relocation of B.C. Tel[88] and by Simpson's analysis of commuting distances in Toronto, where many employee decisions about where to work were made from a predetermined residential location rather than vice versa.[89]

Feminization of the labour force may have other intriguing yet little-explored effects in suburban areas. While child-oriented two-parent families still live predominantly in the suburbs, these areas are becoming socio-economically more heterogeneous. The "second income" has enabled many families in which the male spouse has a modest blue-collar job to become homeowners in areas traditionally the preserve of the white-collar middle class or in new suburban municipalities.[90] Also, because men and women generally still have different kinds of jobs, often located in different parts of the metropolitan area,[91] they may have very different types of social networks and may take home widely different experiences of work culture, social values, and political priorities.[92] The implications of these changes for neighbourhood-level organizations and local and even provincial or national voting patterns remain a matter of speculation (see chapter 14).

GENDER AND METROPOLITAN LABOUR MARKETS

The interaction of these socioeconomic and demographic factors thus profoundly affects the functioning of metropolitan labour markets in Canada. A familiar assumption is that job-specific labour costs are more or less uniform within the commuting field used, for instance, by Statistics Canada to define CMAs. Each CMA is taken as forming a single local labour market. Yet this assumption may not always hold for large metropolitan areas, which may have a number of intricately interwoven but spatially distinct labour markets, as Scott suggests.[93] The traditional view particularly ignores the significance of gender in labour-market segmentation and in work-trip length.

Studies of commuting patterns reveal certain key dimensions of local labour markets, especially with respect to labour-force segmentation by gender. Research is needed to integrate approaches typical of journey-to-work studies and those typical of spatial-divisions-of-labour studies.[94] Here, we illustrate the potential of such an approach by bringing together various sets of changes in aggregate patterns summarized with the aid of simplified journey-to-employment matrices compiled by gender and economic sector for the Montreal metropolitan area (Table 8.6). The Montreal labour force has been categorized into a number of economic sectors, and each has been cross-tabulated by sex and according to place of residence by place of employment, using only two broad geographic zones, "city" and "suburb." These categories have been made identical for both the 1971 and the 1981 censuses.[95]

These data serve to document aspects of the changing division of labour between places of residence and places of employment. First, we observe a more pronounced centrality among the total female labour force than among the total male labour force. In 1971, 43 per cent of female workers both lived and worked in the city, compared to 29 per cent of male workers. By 1981, these percentages had dropped sharply for both females and males, but the gender differences were maintained. Second, the shorter average trip length of women is evident from the lower proportion commuting from suburb to city or vice versa, compared to men: 71 per cent of women were employed in the zone of residence against 63 per cent of men in 1971, and 67 per cent against 65 per cent in 1981. Clearly however, the gap in

Table 8.6

Percentage distribution of labour force at place of employment by place of residence, metropolitan Montreal, 1971, 1981

	Place of employment											
	Female workers						Male workers					
	1971			1981			1971			1981		
Place of residence	City	Sub.	Tot.	City	Sub.	Tot.	City	Sub.	Tot.	City	Sub.	Tot.
Total labour force												
City	43	8 =	51	27	8 =	35	29	13 =	42	19	12 =	31
Suburb	21	28 =	49	25	40 =	65	24	34 =	58	23	46 =	69
Total	64	36	100	52	48	100	53	47	100	42	58	100
Labour-intensive manufacturing												
City	49	10 =	59	33	10 =	43	29	14 =	43	19	14 =	33
Suburb	20	21 =	41	25	32 =	57	24	33 =	57	23	44 =	67
Total	69	31	100	58	42	100	53	47	100	42	58	100
Transportation, utilities, wholesale, and retail trade												
City	37	9 =	46	20	9 =	29	28	13 =	41	16	12 =	28
Suburb	21	33 =	54	20	51 =	71	23	36 =	59	21	51 =	72
Total	58	42	100	40	60	100	51	49	100	37	63	100
Finance, insurance, and real estate												
City	41	6 =	47	26	5 =	31	32	6 =	38	24	6 =	30
Suburb	35	18 =	53	38	31 =	69	40	22 =	62	40	30 =	70
Total	76	24	100	64	36	100	72	28	100	64	36	100

Source: Statistics Canada, special tabulations obtained by the authors.

work-trip length between the sexes is closing, although our data do not reveal whether this trend is affecting women in two-earner families, lone parents, and single women equally.[96]

Third, interesting differences can be observed between economic sectors. We have selected here three sectors that vary widely in locational patterns and feminization. Traditionally, labour-intensive manufacturing in Montreal has been centrally located and has employed a fairly high proportion of women (39 per cent in 1971 and 44 per cent in 1981), mainly because of the dominance of the clothing industry. A major shift during the 1970s decentralized workplaces and residences of female workers faster than those of male workers, while levels of reverse commuting remained fairly steady. However, by 1981, women in this sector were still more centrally located, in terms of both employment and residence, compared to women in other sectors as well as to men in the same sector.

Transportation, utilities, and wholesale and retail trade correspond to what may be called distributive services. In 1971, only one-quarter of workers in this sector were women, but their share of jobs jumped to 34 per cent in 1981. The role of the retail trade within this sector gives it a fairly marked "neighbourhood orientation" (that is, a higher-than-average proportion of jobs are located in the suburbs) for women's and men's jobs. Note the large increase, from one-third to 51 per cent, in the proportion of female workers both residents of and employed in the suburbs. In contrast, finance, insurance, and real estate form a much more centrally located sector, where the proportion of female workers, already at 47 per cent in 1971, jumped to 59 per cent in 1981. By 1981, this sector had the least pronounced gender differences with respect to location and commuting between city and suburb, with many women commuting long distances from suburban residence to downtown office.

In sum, these data suggest that the degree of convergence between female and male workers in terms of job and home location and commuting patterns differs between economic sectors, being strongest in rapidly growing tertiary subsectors such as finance, insurance, and real estate. Studies at a less aggregated level are, however, necessary to shed light on the relationships between feminization of the work-force and reduced household size, the persistent failure of men to participate adequately in household tasks, and the partial transfer from the domestic sphere to the market of certain basic activities such as eating.[97]

SOME FUTURE RESEARCH DIRECTIONS

Of the three major tendencies discussed here – a particularly Canadian form of tertiarization of employment structures and feminization and polarization of the work-force between occupational categories – it is the third whose significance for the social geography of Canadian cities is the most difficult to interpret. Information about occupational structure or incomes or both, available from published statistics for cities, is given for individual men and women without any indication of the type of households in which they live. Although patriarchal relations mean that resources are not always fairly distributed within households, and although kinship networks beyond the household may also affect resource-pooling and mutual aid,[98] we suggest that the household should be a key unit of analysis, since households are gentrifiers, maintainers of stable inner-city immigrant neighbourhoods, and tenants in deteriorating suburban apartment buildings. Household composition, number of earners in a household – an increasingly important line of cleavage in living standards in Canadian society – and the types of job held by members can either mitigate or intensify the effects of occupational polarization on the redistribution of per-capita income. Thus, even though in some inner-city areas more than half the resident population lives in one-person households, analyses of occupational and employment-income polarization alone cannot form an adequate basis for making inferences about increasing social heterogeneity within, or differentiation between, city neighbourhoods.[99] There ought to be more systematic and geographically disaggregated research relating trends in the occupational composition of the work-force to trends in household type and to data on household income.

Our goal in this chapter has been to illustrate and interpret certain aspects of the interactions between the changing geographies of employment, household structures, and residence in Canadian cities. Recent research points strongly to gender relations as a key explanatory factor. Gender appears as one of the most pervasive dimensions of labour-market segmentation and a crucial element in current economic restructuring; while the interlocking trends of feminization and tertiarization are structural processes, they take different forms and have different social effects, depending on the previous sectoral and occupational mix of the cities in question. Changes in the size and composition of households, and also in the number of earners per household, are profoundly modifying the spatial structuring of urban labour forces. Centrality, whether in older inner cities or in new suburban subcentres, may acquire renewed value for small "outwardly oriented" households just as it has in industries with high ratios of external to internal transactions of firms. Yet this type of work is still in relative infancy; major gaps still exist in our understanding of these linkages and provide us with exciting research challenges.

CHAPTER NINE

Housing Markets, Community Development, and Neighbourhood Change

L.S. BOURNE AND T. BUNTING

The social geography of Canadian cities is intimately linked with the provision of housing.[1] Housing is not only shelter and a major capital good, it is the largest component in the average household's expenditures and wealth, as well as a visible status symbol. The overall quality of life for most households is directly tied to satisfaction with and attachment to the home and to its immediate surroundings.[2] Individual dwelling units provide the physical setting for most activities associated with family nurturing and social reproduction, but, when clustered together, these units also constitute the basic building blocks of urban neighbourhoods and communities.

This chapter, building on the discussion in chapters 4, 6, and 8 in particular, demonstrates the varied and complex ways in which housing and neighbourhoods shape the social geography of Canadian cities. It begins by illustrating the institutional context within which housing is produced in Canada and the immense variability over time and from region to region in both the production and consumption of housing. Attention then turns to the processes by which housing is produced and occupied in particular places by particular kinds of people – that is, the matching of dwellings and households in a specific housing market and within varied neighbourhood settings. We then examine the important roles played by governments and public agencies, or more broadly the state, in the provision of housing and residential land and in regulating neighbourhood change. We conclude with a discussion of future housing-market trends and an assessment of their implications for urban social structure, the quality of living environments, and the opportunities that Canadian cities provide.

HOUSING IN URBAN CANADA: TEMPORAL AND STRUCTURAL TRENDS

The housing stock occupied by Canadians, as chapter 4 demonstrated, has undergone a profound transformation since the Second World War. Most obvious, the stock has grown immensely in size, from 2.6 million dwelling units in 1941, to over 10.1 million in 1991. Housing quality has also improved dramatically. In 1941, 44 per cent of all dwellings in Canada lacked one or more basic facilities, such as indoor plumbing or running water. In the census of 1986 that figure was less than 3 per cent.[3] Statistics on housing in need of major repair show similar decline. At the same time, average dwelling size, whether measured in floor area or rooms per person, has increased significantly. Today very few Canadian households share their dwellings with other households.

Notwithstanding that many in our population still lack access to adequate or affordable housing, Canadians as a whole have become much better housed since 1945. This is reflected not only in the number, size, and quality of new housing starts but in the growing number of improvements and alterations being made within the existing stock.[4] As physical needs in housing have declined in importance, however, other factors (notably affordability) have taken their place on the national policy agenda.

Equally dramatic has been the transformation of the stock in terms of building type and tenure. In 1941, 40 per cent of urban dwelling units were owner-occupied; by 1986, rates of owner-occupancy had risen to 63 per cent. The built form of housing also changed. The 1950s and early 1960s were dominated by the construction of single-family suburban housing, often in the classic form of the one-story bungalow sited on a relatively large lot. The 1960s and early 1970s witnessed much larger and more diversified production: town-houses, split levels, and especially high-rise apartments. The public sector also invested heavily in the building of subsidized dwelling units – most visibly, high-rise public housing projects. The 1970s and 1980s experienced an even wider variety of supply – a mix of housing forms, densities, and tenure types, including in-fill housing (various types of residential intensification), condominiums, non-profit or third-sector housing, and the conversion of older, non-residential buildings to residential use. Although the supply of new housing has always fluctuated over time, from the late 1970s on, except for the mid-1980s, the average number of housing starts has gradually fallen, in part because of the recessions of the early 1980s and early 1990s, and specifically because of decreased construction of private rental apartment units.[5]

Three trends in housing supply stand out in the changing geography of neighbourhoods in Canadian cities. One is the shift in policy emphasis and in the relative flow of private investment from new construction to modification and renovation of the established stock, reflecting both the needs of an ageing stock and the rising expectations and values attached to older housing (Figure 9.1). A second notable trend is the introduction since the late 1960s of new forms of tenure, notably condominiums and co-operatives. After a slow start, condominium sales expanded in the mid-1980s and came to account for as much as 30 per cent of all new home sales in cer-

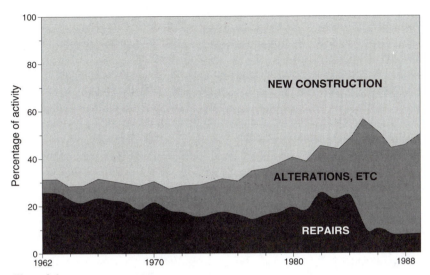

Figure 9.1
Distribution of residential construction activity by type, Canada, 1962–89

tain local markets. In some cities, luxury condominiums have created not only a new combination of built form and housing tenure but new kinds of life-styles and communities. Widespread conversion of existing units from private rental to condominium status (some as questionable short-term speculative "flips") has further increased tightness within rental housing markets, especially in larger and rapidly growing places.[6]

The third trend involves a continual redefinition of the state's role in housing, especially the position of public subsidized housing. Traditional forms of public housing, homogeneous in residents and owned and managed by the state, have given way to a much broader concept of social housing, as a blending of types and levels of subsidy, including non-profit co-operatives, and to the emergence of entirely new suppliers of non-market housing (e.g. private and municipal co-ops, non-profit organizations, churches, and unions). This trend reflects a desire to reduce the social stigma and costs attached to traditional public housing and to discourage the formation of ghettoes of the disadvantaged, through greater income mix within the socially assisted sector.

REGIONAL DIVERSITY IN URBAN HOUSING MARKETS

In Canada, as in most other countries, housing markets are intensely localized. The attributes of the stock and the demands of households differ widely from one region and urban area to another. The geographically uneven and historically variable nature of regional growth in Canada, however, adds to these differentials. In fact, fluctuations in regional economic performance tend to become telescoped on particular metropolitan housing markets within growing regions. As the west boomed in the 1970s, for example, pressures on the local housing markets of Calgary, Edmonton, and Vancouver became intense. In the period 1982–88, the pendulum of economic growth shifted to southern Ontario, leading to massive real price increases, especially in the greater Toronto region. Most recently, the Vancouver market has undergone another real estate boom, and average house prices there are now similar to Toronto's. In cases of rapid economic growth such as these, hyper-inflation in house prices and rents results because the supply of housing (and local government services) are simply unable to respond to dramatic and localized growth pressures in the short run and because fears of further inflation in the future increase the prices that households are willing to pay today.

To illustrate the degree of regional diversity in housing markets across Canada, Table 9.1 presents a selection of indices of housing conditions for the twenty-five CMAs in Canada in 1986.[7] Note, first, the immense differences in scale: from 42,000 units in Saint John to 1.2 million and 1.1 million units in Toronto and Montreal, respectively. Rates of production, as measured by new housing starts, also vary greatly, in relation to the size and demographic composition of a city's base population and most notably to variations in regional growth and migration rates.

Differences in dwelling unit type, size, and tenure from place to place are also influenced by these same factors, as well as by historical inheritance and regional buildings styles, though there is less variation around the national figure. On average, 57 per cent of Canadian households live in single, detached dwelling units.[8] Among metropolitan areas, this figure ranges from a low of 28 per cent in Montreal, where cultural, economic, and historical factors interact to produce a unique dominance of higher-density housing types,[9] to a high of 70 per cent in smaller and slow-growth cities such as Windsor, Thunder Bay, and Regina, which also have relatively unlimited access to low-cost, developable land. Regional differences in levels of owner-occupancy tend to mirror the proportionate share of single, detached units. Throughout eastern and central Canada, rate of ownership are significantly higher than corresponding rate of single, detached units (particularly in Montreal). This discrepancy attests to the importance of duplex and row housing within the older, pre-1939 housing stock and to the more recent popularity of condominium tenure.

The cyclical pattern of regional economic growth, shifting in the 1982–88 period from the west to central Canada, is clearly evident in real estate values.[10] Price increases have been greatest in cities within commuting distance of Metropolitan Toronto: Hamilton, Oshawa, Kitchener, and St Catharines-Niagara.[11] Rental costs exhibit similar trends, although the range in rent levels is much narrower than for the purchase cost of housing (see chapters 4 and 15). On the whole, the incomes of home-purchasing households (as indexed by NHA mortgage approvals) increase with the average price of housing, but almost always at a lower rate.

Although place-to-place differentials in housing and neighbourhood price appreciation may not be as severe or visible in urban Canada as in the United States, they are considerable. Moreover, it can be shown that housing tenure and location influence households' social and economic status beyond the effects of income.[12] Those Canadians lucky enough to have purchased or sold housing in certain markets (e.g. Vancouver, Toronto, Calgary, Halifax) at particular times, and in selected neighbourhoods within those cities, have become substantially wealthier. Those who rent (except for some in rent-controlled units) or who had the misfortune to purchase housing at the wrong time or in declining urban areas or neighbourhoods, have usually lost money, especially if circumstances require them to relocate to an area undergoing a real estate boom. Daily newspapers provide continuing evidence of how the "surplus value" attached to housing is being redistributed among cities and urban neighbourhoods, as well as among social groups and between generations, through the changing fortunes of our housing system.

HOUSING IN PLACE: SOCIAL PROCESSES OF HOUSING PROVISION

Most housing in Canada is produced for profit by the private sector, as a market product, yet that production takes place under unique circumstances and within a given social and political context. As such, it can only be partially examined as an economic good through conventional economic analysis.[13] Moreover, housing is both a capital good and a consumption good. Usually, most housing researchers focus on the suppliers of new housing, a capital good, as produced by the home-building industry. But if we employ the broader concept of housing as a flow of services – as a consumption good – then landlords produce housing services in the rental sector and self-builders and owner-occupiers produce in the ownership sector.[14] Renovators and converters, whether professional or lay, are also housing suppliers in both sectors.

Adding further complexity, it is now widely recognized that housing is produced within a highly volatile institutional framework and in response to a host of macroeconomic and political factors. The housing sector is very responsive to fluctuations in the growth of the national economy, particularly in unemployment and income levels, and to demographic change and rates of household formation. The flow of new housing supply is also dependent on fiscal conditions and policies, especially with respect to interest rates. It

Table 9.1
Regional variations in housing, CMAs, 1986

CMA	Number occupied dwelling units (000s)	Number new housing starts 1981–86 (average annual rate)	% stock single, detached	% owner occupancy	% housing units built before 1946	Average size, new homes, single, detached (m²)	Owner-occupied house values ($000s)	Average monthly rentals ($)	Average family income of borrowers of NHA-approved mortgages ($)
St John's	47.9	1,103	55.8	68.3	16.8	84.7	78.9	461	42,302
Halifax	103.8	2,568	51.7	58.3	18.2	117.9	94.0	481	45,858
Saint John	41.7	517	53.3	61.6	29.0	108.8	59.7	367	41,318
Chicoutimi–Jonquière	44.9	808	49.5	61.5	18.1	93.0	61.0	396	43,242
Quebec City	214.4	4,838	41.4	53.0	18.6	89.3	66.6	419	44,073
Sherbrooke	48.5	1,103	–	–	17.4	–	61.0	384	–
Trois-Rivières	47.5	1,035	–	–	22.1	101.2	56.5	361	45,015
Montreal	1,115.4	19,159	28.0	44.7	18.8	103.2	83.8	421	48,354
Ottawa	302.7	8,380	41.9	52.4	13.9	126.1	124.0	492	54,743
Oshawa	68.0	1,367	61.8	70.2	16.2	130.8	109.8	471	57,020
Toronto	1,199.8	25,099	43.2	58.3	17.3	177.4	156.4	501	69,619
Hamilton	201.3	2,649	59.7	64.6	22.2	127.5	110.8	414	59,700
St Catharines–Niagara	124.6	1,233	71.6	72.0	27.0	117.1	78.3	388	51,497
Kitchener	110.2	2,383	55.5	61.9	17.9	120.9	100.1	410	52,883
London	129.4	1,937	56.3	57.8	23.0	137.4	87.8	411	58,377
Windsor	91.6	567	68.5	67.2	29.4	115.2	76.3	397	58,623
Sudbury	51.6	419	61.0	64.5	16.7	112.8	59.6	328	51,454
Thunder Bay	43.7	444	69.5	69.2	26.4	125.2	80.3	413	53,087
Winnipeg	236.3	3,827	59.7	60.8	22.4	121.9	79.6	412	48,085

Regina	67.6	1,503	69.6	65.7	13.9	102.2	73.6	445	42,701
Saskatoon	74.0	2,252	62.9	60.0	12.8	107.7	78.6	421	43,586
Calgary	248.6	6,075	56.0	57.9	6.6	128.8	105.8	489	53,605
Edmonton	283.4	5,944	57.7	57.1	5.7	119.9	91.2	465	47,744
Vancouver	532.2	12,037	53.3	56.5	15.5	141.7	132.1	494	55,367
Victoria	105.5	1,810	56.5	59.8	19.5	137.8	107.1	459	58,300

Sources: Census of Canada; CMHC.

is therefore susceptible to fluctuations in capital markets, as manifest in the availability and cost of credit for building loans and mortgages. What researchers analyse as statistical undulations in aggregate demand and supply curves, and what we observe directly in the changing nature of urban neighbourhoods, are the combined outcomes of these variable capital flows.

Unlike most other forms of commodity production, however, housing is actually produced in place – that is, it is assembled in specific locations and on individual sites. At this scale, then, housing production becomes physically tied to land, to the local planning process, and to the availability and costs of services. The latter include hard services, such as water, roads, and sewerage (the physical infrastructure), as well as soft services, such as educational, recreational, and community facilities.[15]

To produce housing in place requires bringing together a large number of local actors and agencies. First is the building industry itself – the firms, unions, individuals, landlords, renovators, and market intermediaries which design, build, or modify and renovate housing. The second is the institutional network, consisting of the local agents of capital markets, such as investors, brokers, and mortgage lenders, which provide financing for construction and ultimately for home purchase. They may, and often do, discriminate in terms of where and whom they lend to, among both builders and households (the "redlining" issue). The third sector is made up of consumers, the occupants of the physical stock produced by the industry, and the communities they collectively occupy.

Finally, there is the local state as expressed in the agencies, policies, and regulatory instruments of municipal and provincial governments. Governments initially control the release of land for development and subdivision, provide basic infrastructure, directly or through special taxes or levies, and set servicing standards. They subsequently define zoning and building by-laws and set occupancy regulations (e.g. density of buildings and persons per dwelling). As the media consistently tell us, however, governments do not always make neutral decisions, for example, with respect to the various actors in the supply process and to the distribution of price levels for new housing. They are of course subjected to strong pressures from influential special-interest groups, including local community organizations and the development industry itself. Whether these actions can be said to constitute a "conspiracy" against certain groups, and specifically regarding restrictions on the supply of low-cost housing, remains highly debatable.

What is clear is that the production of housing and living space in cities involves provision not only of capital, bricks, and mortar but of an entire bundle of services. The latter include serviced land and those related services (such as sewers, parks, schools, and community centres) that are provided through the initiatives of local government and largely with public

Table 9.2
Residential house-building in Canada: number of firms
and revenue, by size of firm, 1984

Output ($ value)	Number of firms	% of total	Revenue ($mil.)	% of total
< 500,000	7,449	86	1,005	27
500,000–1 million	680	8	467	12
1–10 million	505	6	1,210	32
> 10 million	44	1	1,098	29
Total	8,678		3,780	

Source: CMHC (1987b).

financing. All urban housing then is at least in part "collectively" produced, in effect as variations of social housing.

The actual system of housing provision in Canada has changed significantly since 1945, but in an evolutionary rather than a revolutionary fashion.[16] To begin with, the housing and residential development industries have undergone numerous corporate and financial reorganizations. Firms have tended to become larger and more vertically integrated, in part through government encouragement and incentives, while producing a more locally homogeneous product. Building codes and servicing standards have also been raised, because of strong demand from consumers and higher servicing requirements set by provincial and local governments.[17] These, of course, have been among the major factors driving inflationary trends in urban real estate. The institutional context has also changed, particularly the organization and behaviour of financial and mortgage lending agencies and their intermediaries[18] and through the uncertainty created by the frequent rewriting of federal and provincial housing policies.[19]

Despite these changes, a surprising amount of house-building in Canada, especially single-family, duplexed, and row housing, is still produced by small firms (Table 9.2). The relative ease of entry into the industry attracts entrepreneurs and workers in boom times and rapidly discards many of them in times of recession.[20] This ease of entry provides opportunities for small-scale operators, many of them new immigrants, to obtain work or establish businesses in the construction and real estate trades. It also, however, introduces a further element of uncertainty and inefficiency into the industry and adds to the difficulties of setting and then monitoring standards of building and of doing business in the residential sector.

The building industry also tends to be geographically fragmented and decentralized. The majority of firms usually focus on a single region and a few urban areas, because of the need to understand the peculiarities of a complex local regulatory framework (e.g. zoning and building by-laws) and in order to be more responsive to local politics, tastes, and fluctuations in demand. In some regions, notably Atlantic Canada, a sizeable proportion of the housing stock is still the result of self-building.[21] In most metropolitan areas, however, a large and integrated residential development industry has evolved over the post-war years. Some of the larger firms emerging in the 1970s, such as Cadillac-Fairview, Tridel, Daon, and Bramalea, incorporated a full range of functions from initial land development through marketing and project management. Massive post-war residential developments such as the comprehensively planned and exclusive "corporate" suburbs (e.g. Don Mills and Erin Mills) that surround our cities, or high-rise inner-city redevelopment projects (e.g. St James Town), would not have been possible without this corporatization of property development.

One specific consequence of this continual corporate reorganization of the housing industry was a national debate in the 1970s on land-hoarding, speculation, and the apparent exercise of oligopolistic power. General concern over the rapid rise of housing prices, and specific concern that some large land-owners were intentionally restricting the flow of developable land in order to raise prices, led to the establishment of a federal task force on the supply and price of residential land.[22] Although some observers claim to be able to document such a conspiracy, and the literature is replete with examples of excessive profits and unethical practices in land development, the results of the task force were widely viewed as inconclusive. It was difficult to define the geographical area over which a concentration of land ownership was to be measured. Is it the entire metropolitan housing market area, or just part of it? Further, wide regional variations in urban land prices undermined the need and rationale for national solutions. To some the problem was largely confined to a few metropolitan centres, notably Toronto and Vancouver. It was also almost impossible to separate the price effects of oligopolies in the private sector from the effects of government restrictions on land supply and increased demand pressures attributable to rising real incomes and rapid household and population growth. In the late 1980s, during yet another round of housing price rises, the debate on the cost of land for housing continued.[23]

Also in the 1980s, some corporate financial interests, including developers and mortgage lending agencies, began to shift their investment portfolios toward non-residential uses and to pre-construction phases of development – notably rural land acquisition and development approval. The reasons for this shift are numerous and complex, including the effects of increased real interest rates, a reduction in federal tax shelters, and additions to an already cumbersome and costly municipal approval process. Whatever the reason, much of urban Canada, especially in growing regions, has seen more speculation in land, rising standards of servicing, considerable conflicts over land use, and, consequently, higher prices.[24] In most urban areas, land prices have increased much faster than housing prices over this decade.[25] This factor – combined with the greater financial pressures facing local governments, which further restrict the provision of infrastructure, as well as higher lot levies (or development charges) and servicing standards – has resulted in a widespread shortage of developable land for residential purposes and thus of affordable housing relative to demand.

THE MARKET: MATCHING HOUSEHOLDS AND DWELLING UNITS

The local housing market matches households and dwellings within more or less distinct spatial submarkets.[26] Individual blocks of land, or entire subdivisions, tend to be built at the same time and with roughly similar kinds of housing in terms of price and tenure, producing an elaborate patchwork of social areas segregated by socioeconomic status, life-cycle, and ethnicity (see chapter 3).

The local housing market is an exchange process in which most of the participants at any given time are traders, moving within part of the stock or urban area to accommodate changing preferences, living needs, and/or employment conditions.[27] Traditionally, changes in a household's life-cycle or family status (e.g. household formation, birth of children, "empty nesting") generate most of the consumer turnover in housing (see chapter 6). These factors may, however, decline in importance in the future as the population ages and as the nuclear family becomes less of the norm.

The operation of these local housing markets is crucial for this discussion. First, the inherent dynamism and complexity of those markets distinguish housing from other durable consumer goods. For example, the magnitude of units and the immense variation in product type and location make it necessary for most consumers to use agents and other information sources to get access to the market. Second, and related, the urban housing market, more than most other markets, brings together in one place – the neighbourhood – changes in external factors, which act on both demand and supply, and local conditions and policies, which continually redefine the bundle of housing services (see Figure 9.2). Third, the market involves many actors; each set of actors brings to the market a somewhat different set of resources, information, and goals. Some are undoubtedly more influential than others, and their relative importance will vary from place and over time with changing economic conditions, politics, and social attitudes. Our urban social landscape is equally diverse in its origins and motivations, as is the choreography of community organization and neighbourhood change.

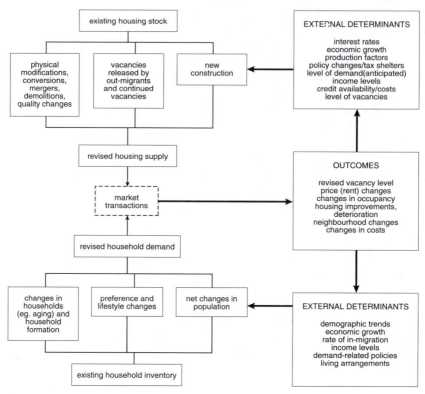

Figure 9.2
Components of an urban housing market

THE NEIGHBOURHOOD CONTEXT AND DYNAMICS OF RESIDENTIAL CHANGE

Neighbourhoods, despite their origin in the residential supply process, are considerably more than local collections of housing units.[28] For their residents they provide a familiar place, a focus for daily activities, and a set of informal social support networks. They offer, in varying degrees, a source of identity, of security, and shared living experiences, as well as the benefits (and costs) of membership in a common social space. Homes and neighbourhoods together take on non-economic values of profound personal and social significance. Membership in a neighbourhood also offers access to a specific basket of local public and private goods and services and, of course, the opportunity to exclude others from those goods. Where those goods and services differ widely in quality or availability (and cost) from one part of an urban area to another, location increases in importance as a housing variable accordingly. The price of a similar housing package can be inflated considerably in "good" locations or ones that have a fashionable cachet. The reverse holds true where externalities are perceived to be negative.

Two well-known inner-city neighbourhoods in Toronto provide excellent examples of this contrasting process. The east-end "Beaches" district, with its cottage-like housing and tree-lined streets fronting on a sandy, boardwalked stretch of Lake Ontario, has experienced relatively rapid real estate price inflation. If the media are to be believed, the area has become one of the more "trendy" places to live. In contrast, real estate values in "South Parkdale," once considered one of Toronto's finer neighbourhoods, have been in relative decline, even though the area contains attractive housing adjacent to one of the city's largest parks. South Parkdale has come to be perceived as socially undesirable; former single-family houses have been converted to rooming houses and group homes, many occupied by ex-psychiatric patients (see chapter 16).

When households move they also shift their demand, accumulated wealth, and social influence to growing areas, while withdrawing these same assets from other areas. In many large US cities, this process has often led to a ring of resource-rich suburban communities surrounding an increasingly isolated and impoverished central city.[29] This contrast has been much more muted in Canada, and in recent years some have argued that the trend has been reversed, at least at this spatial scale.[30]

Individual neighbourhoods may change for many reasons. New neighbourhoods are added at the suburban margin or adjacent to small towns in the rural surroundings of urban areas.[31] These new communities mirror the social conditions and planning/architectural styles of the day. Older neighbourhoods change, as noted, subject to a combination of external pressures and internal conditions. The former, for example, may include revisions in federal immigration policies, the provision of new public infrastructure (e.g. hospitals, roads, transit lines), and redevelopment of sites for commercial purposes or the in-migration of a new class of households (leading, in the two extremes, to either gentrification or ghettoization). The internal pressures may include ageing of the resident population, changes in the social attachment to a particular neighbourhood, physical subdivision and tenure conversion within the housing stock, as well as modernization, refurbishing, and other forms of upgrading (or downgrading) of existing housing by in-migrants or long-time residents.[32]

To illustrate this diversity we draw on a simple typology of neighbourhood types in Canada.[33] While inevitably generalized, the following classification of four neighbourhood types at least captures the principal attributes and directions of change in urban housing environments: stable, upgrading, downgrading, and redeveloped (Table 9.3). These four types can

Table 9.3
A typology of neighbourhood types and housing characteristics, by income

Direction of change (neighbourhood types)	Income level/social status characteristics		
	High	Medium	Low
1. Stable	All housing types	All housing types	All housing types
2 Upgrading			
Incumbent	Elite upgrading	Middle-class upgrading	Lower-income and older immigrant areas
Revitalized	Elite/prestige gentrification, renovation, mega-housing and in-fill	Marginal gentrification, renovation, and in-fill	Publicly sponsored rehab.
		Third-sector housing	Social-housing initiatives
3 Downgrading			
Incumbent	–	Declining income/ housing quality	Low-quality stock
Residualized	–	Declining public goods	Skid row, transients, the poor
		Speculative activity	
4 Redeveloped			
Public	–	Co-ops	Social housing
Private	Luxury apts, condominiums	Modest rental/ price apts	–

be further disaggregated according to the principal source of pressures for change (external-internal; public-private) and to the average income or social status of existing and in-coming inhabitants. The largest proportion of neighbourhoods in contemporary urban Canada would appear to fit into the first category – they have remained more or less stable since 1945 in their relative social attributes and position within the matrix of residential sub-markets of an urban area. Others have displayed modest upgrading or, in a few cases, relative downgrading.[34] These shifts are the net result of income and occupational changes, gradual alteration in the composition of in-migrants and/or out-migrants, and the removal, alteration, or addition of housing units.

Relatively few neighbourhoods have undergone extreme transitions, through revitalization (upgrading or gentrification), residualization (concentration of socially dependent populations), or large-scale renewal and redevelopment. Those that do change rapidly attract most attention in the media and in the academic literature, precisely because they are concentrated, particularly in the inner city, and because of the rapidity and visibility of the transitions involved. They are also the venues of most urban-based community movements whose primary purpose is to resist change.

Clearly, there is no single model of neighbourhood transition, nor of the life-cycle of a neighbourhood, that is applicable across the country or throughout any urban area. To illustrate the point, we can refer to a model developed by Smith and McCann in a study of Edmonton.[35] They demonstrated that while most neighbourhoods do undergo systematic changes over time, through shifts in the composition and density of land uses as housing and residents age, they do not display a fixed life-cycle. Parts of the stock are continually renewed (or replaced), and new households take up residence, but the social status of the area may or may not alter as a result. Whether the latter change occurs or not depends on other factors.

Systems of housing provision, local markets, and neighbourhood dynamics are clearly interdependent. Changes may be led by factors that are explicitly housing variables and/or by social, economic, or political factors. Once set in motion, however, they usually become cumulative and self-reinforcing. As Logan and Molotch argue, neighbourhood futures are primarily "determined by the ways in which entrepreneurial pressures from outside intersect with internal material stakes and sentimental attachments" to the community.[36]

Perhaps the most significant change in the location of housing opportunities available to Canadians has been the result of the massive geographical restructuring of local housing markets. This process, most prominent in large and growing cities with service-based economies, has combined extensive suburban and sprawling low-density exurban development with widespread gentrification, upgrading, in-fill, and condominium construction, particularly in the inner city. The former has dispersed housing opportunities over an immense geographical area, with all of the associated transportation and environmental costs, and social difficulties. The latter has reversed for many cities the traditional "filtering down" of older housing and neighbourhoods.[37] Aggressive public policy has facilitated this rejuvenation. As suggested earlier, rejuvenation of the older stock and the boom in new condominiums, in both suburbs and inner city, seem increasingly to differentiate Canadian cities from many larger American cities.

THE ROLE OF THE STATE

Governments, as stressed above, participate in every stage of housing provision and neighbourhood construction, from supplying parcels of serviced residential land to developing and distributing housing. Federal, provincial, and municipal governments act directly in the market as suppliers, as regulators of private-sector development, and as providers of public goods and services. The federal government, for example, controls capital financing, mortgage insurance practices, levels of subsidies for social housing, relevant tax rates, and overall economic conditions influencing the supply of credit, materials, and labour. Provincial and local authorities dictate the rate at which the potential supply of housing in aggregate is realized on the ground, as well as the physical and social environment in which it is produced. Municipalities then become the final gatekeepers for residential development. Together with provincial governments they designate raw land for development and set standards for the provision of collective services. They also administer a complex set of legislation that defines permitted land uses, specifies restrictions on building design and density, and manages the development approval process.

The form of regulatory framework varies widely among provinces and municipalities. Nevertheless, at least for the major urban centres, there are numerous commonalities.[38] Almost everywhere the regulatory environment for residential development is becoming more complex, demanding, and cumbersome, just as are the tasks set before it. Servicing standards in new areas are now such that almost all residential services must be in place, and largely paid for (in part through increased use of lot levies), before the first dwellings are built. This arrangement restricts the housing choices available to many households through high standards and because of spending limits on new infrastructure for sewers, roads, transit, parks, and schools. The costs of these services raise the initial or entry price to the consumer and effectively reduce the proportion of households, particularly first-time buyers, that can afford new suburban housing. These costs also act to redistribute the locations at which housing demand is expressed, as, for example, in directing some potential urban homeowners to search further and further outward into rural areas and small towns for lower land prices.[39]

The geographical dispersal of housing markets has also created "accessibility ghettoes" in some new and more remote residential tracts. Housing there is certainly relatively more affordable, but their location adds to discrepancies with older neighbourhoods in the quality of local services and overall transportation costs. One counter-intuitive effect has been further inflation in the inner-city housing market, augmented by gentrification and the preference shifts that it represents. Factors external to the housing market, such as the expansion of commercial uses into residential areas near the CBD, have also increased inflation in land. In some cities, the land-rent gradient has begun to take on the pre-1939 pattern of large differentials between the high cost of real estate in central areas and lower costs in those suburbs furthest from the urban core.

Although local governments do not determine who lives in what housing or in which areas, except in some sectors of socially assisted housing, they clearly do influence the probability that certain households will live in some neighbourhoods rather than others. For example, by assigning public facilities with perceived negative externalities to particular communities (e.g. a new expressway, group homes, or parking structures), governments can quickly alter the social composition of a neighbourhood. Alternatively, by changing the zoning, for example, either to prevent or to permit redevelopment, basement apartment conversions, or the partitioning of lots, they can redirect social change in very different ways. When these policy directions coincide with the movement of private residential capital in or out of the local housing market, or with the in-migration of numbers of atypical households, be they disadvantaged "minority" groups or upper-income gentrifiers, transition can be rapid, as has been found in many American cities.[40]

The most explicit form of government involvement – provision and management of public housing – is limited in Canadian cities. Overall, less than 5 per cent of the stock is owned and managed by governments, and little new urban public housing is being built, except for senior citizens (and for Native Canadians in rural areas). Currently, efforts to house the poorer members of society have shifted back to the private market and to the third sector (co-ops, non-profits, and so on). The latter have created some of the more innovative and mixed-income social-housing projects in the country, particularly as in-fill within older inner-city neighbourhoods or on cleared redevelopment sites (e.g. False Creek in Vancouver, St Lawrence in Toronto).[41]

Although new units of social housing are not large when compared to total housing starts, these initiatives have been significant, but in ways different from public housing projects in the 1950s and 1960s. The large renewal projects that blotted the landscapes of many inner cities, and tended over time to create ghettoes of the disadvantaged, will not be repeated.[42] Unfortunately, the subsequent tendency has been to concentrate socially assisted and high-rise housing projects in the suburbs, such as in the Jane-Finch corridor of suburban Toronto. In contrast to the American experience, this approach has offered new housing for low-income households and recent immigrants in outlying suburban areas, but all too often on an excessively large scale and before adequate social and community services were in place. In any case, the prevailing image of a homogeneous social geography in the suburbs of Canada will never be the same (as illustrated in chapter 12). The future effect of third-sector housing initiatives, however, will be more muted, precisely because these projects will be more geographically dispersed, smaller, and of mixed income levels. They will also tend to be better organized, and to date at least they have been better managed than older public housing.

HOUSING ISSUES,
NEIGHBOURHOODS, AND
FUTURE SOCIAL GEOGRAPHIES

In this chapter we have attempted to identify some of the major factors that have influenced the production and distribution of Canada's post-1945 housing stock and to outline their consequences for urban communities and households. By way of conclusion, we adopt a broader perspective which anticipates some of the major structural parameters of housing and neighbourhood-related change as our urban society moves into the twenty-first century.

First, and perhaps most obvious, are the external economic changes that will follow as Canada becomes further integrated into the North American and global economies. Ley, and Rose and Villeneuve, for example (see chapters 8 and 11), have demonstrated how occupational restructuring has changed patterns of housing consumption and the socioeconomic composition of neighbourhoods. Bunting and Filion have argued that the traditional gap between housing in resource-rich suburban communities and older housing in the inner city, where substantial disinvestment took place in the past, has been closing.[43] Despite considerable reinvestment in gentrified neighbourhoods, however, the inner city still houses the most needy in our society, including the homeless, the growing ranks of minimum-wage service workers, the un- and under-employed, and other socially dependent populations.[44]

Table 9.4 illustrates this increasing polarization with respect to income distributions in the three largest metropolitan areas. Not only have household incomes in central-city neighbourhoods decreased with respect to their respective metropolitan averages between 1971 and 1986 (except for Toronto), but incomes within those cities have become more variable. The skewness index, measuring the widening difference between average and median incomes, suggests a growing number of low-income households co-resident with a relatively small but increasing affluent population.

A second major determinant of change in housing consumption relates not so much to economic factors of production as to social factors of reproduction. In chapter 8, Rose and Villeneuve emphasized the link between productive and reproductive spheres – how transformations of the economy reshape employment patterns and family composition, and vice-versa. In chapter 4, Miron demonstrated how demographic change, notably the baby boom and bust, and increasing post-war affluence have reshaped living arrangements. Smaller households and fewer children may reduce the traditionally high demand for housing units with large amounts of interior and exterior space. Increased female participation in the paid labour force in turn will continue to focus attention on access to the workplace, which will be-

Table 9.4
Changing income distributions within selected urban areas, 1971–86

	Toronto		Montreal		Vancouver	
	1971	1986	1971	1986	1971	1986
City/CMA ratio						
Average income*	90.8	90.9	86.1	79.7	93.8	89.7
Median income*	82.8	79.1	85.8	73.9	89.5	80.9
Skewness index†						
CMA	11.9	16.6	14.7	17.1	12.6	18.4
City	22.8	34.1	15.2	26.3	18.0	31.3

* Household income.

† Index calculated as (average-median income/median) *100.

come a more complex factor. Two-earner households will further polarize household incomes and thus the money available for housing and related services.

Further conjunctions between social and economic trends will be expressed in tastes and preferences. Swings in taste and fashion affect local housing markets, particularly for those trend-setting groups that are relatively unconstrained by problems of access or affordability. They tend to drive residential real estate markets by expressing their wealth through home purchase and renovation.

We might also anticipate continued difficulties for the building and land development industries in meeting the housing needs of all Canadians. The immense financial and organizational problems facing residential construction will probably persist. After the current recession, and the collapse of many development firms, construction and land costs may again increase faster than the general rate of inflation. Labour costs will also rise as shortages of skilled trades appear. Zoning and environmental restrictions will no doubt continue to restrict the flow of raw land available for new development.

In contrast, overall federal government expenditures on housing in general, and support for the social housing and co-op sectors in particular, are unlikely to grow, given current constitutional trends and fiscal conditions.[45] As a result, the existing stock will be crucial to meeting future housing requirements. Here the trend toward social, ethnic (racial), and geographical polarization of housing conditions noted earlier may well continue. Pressures for conversion to condominiums, intensification of land use, the intrusion of "mega-homes," and new construction will continue to force prices up in desirable neighbourhoods. In contrast, housing located in declining re-

Table 9.5
Changes in home ownership rates by income quintiles, various years

Income group (quintile)	% of households owning unit				% Change 1967–81
	1967	1973	1977	1981	
Lowest	62.0	50.0	47.4	43.0	−19
Second	55.5	53.6	53.3	52.4	−3
Middle	58.6	57.5	63.2	62.7	+4
Fourth	64.2	69.8	73.2	75.0	+11
Highest	73.4	81.2	82.3	83.5	+10

Source: Statistics Canada, *Household Facilities by Income and Other Characteristics*, Cat. 13–567. Ottawa: Supply and Services, 1983. From Hulchanski (1989).

gions and marginal neighbourhoods will go up little in value. In both growing and declining communities, as a result, attention will focus increasingly on the maintenance of the existing housing stock, but for quite different reasons and residents. Maintenance problems will be most widespread within the older, privately owned rental stock subject to rent controls, especially lower-income, high-rise apartment towers.[46]

Uneven housing appreciation may also lead to increased differentials in wealth resulting from housing consumption. These differentials are already apparent with respect to homeowners and renters, in part because of the well-known tax advantages that the former receive. Table 9.5 shows that rates of homeownership for higher-income households increased over the period 1967–81 while rentership rates rose for low-income households. These differentials are also related to changes in household size and composition (e.g. number of workers) and particularly to age. Intergenerational inequalities will probably increase, as inheritance becomes a more important factor in housing consumption.

How will these housing trends affect living conditions in Canadian cities? Given the circumstances outlined above, we can, for example, anticipate that accessibility and affordability will decrease in tandem.[47] In prosperous areas of the country, in the Toronto and Vancouver regions in particular, rising housing costs have reduced overall housing consumption, most notably between 1986 and 1990. Apparently more families have had to double up, and some grown children remain longer in the family home, thus reducing the formation of new households. Under these conditions the extended family could re-emerge. Inflated prices may also redirect migrants and development toward areas where real estate is less costly or create inflated wage demands to offset cost-of-living differentials. Although real land costs and housing prices have declined in many cities since the onset of the latest recession (in 1990), rising unemployment has meant that fewer households can take advantage of this situation. Other households are now trapped in relatively overpriced housing, with their real incomes declining. The boom-bust cycle in residential prices has come full circle.

Restricted access to housing may thus become more, not less, a feature of life in Canadian cities. This is the result of the intersection of several factors, most notably the limited supply of affordable housing (by price and tenure) in accessible locations, the continuing effects of rent controls (in Ontario) on rental supply,[48] high interest rates, exclusionary municipal zoning and lot fees, and budgetary limitations on new social housing. These problems will continue to be particularly acute in the larger and growing urban areas and for low-income consumers, especially single parents, female-led households, new immigrants and visible minorities, those on fixed incomes, elderly renters, and the homeless.[49]

On the whole, it is questionable whether future generations will enjoy the increasing quantity and quality of housing that most Canadians have experienced in the decades since the Second World War. Either way, innovation will be needed on all sides – not just among private and public agents of housing production, but among consumers and community groups as well – if the housing needs of Canadians are to be met into the twenty-first century. The design of dwellings and neighbourhoods, for instance, will have to be more flexible and more sensitive to the needs of a wider variety of household types, an increasingly multicultural population, and a different mix of incomes and choices of living arrangements. Further, we must find a new balance between, on the one hand, the desire for capital appreciation through the ownership of housing and urban land and the dependence of local governments on the property tax and, on the other hand, the need for affordable prices and fair rents. The future social geography and quality of life of our neighbourhoods and cities will mirror how well we collectively meet these challenges.

PART THREE

Places: Selected Locales in Urban Canada

Integrating Production and Consumption: Industry, Class, Ethnicity, and the Jews of Toronto

D. HIEBERT

We tend to think of our workplaces and neighbourhoods as separate components of our lives, connected only by the tedious daily commute. This conceptual separation reflects the reality of a disjuncture between home and work locations in the modern city: in Toronto, for example, fewer than 7 per cent of the population live and work in the same census tract.[1] This pattern is reinforced by an array of private and public institutions, such as mortgage lending agencies and the urban planning profession. Planners in particular play a crucial role in defining urban spatial patterns and normally work with an assumption that separate commercial/industrial and residential land uses are desirable.[2] But planners do not operate in isolation from the wider social and economic system; it appears that most people who live in cities accept the notion of designated commercial and residential areas.

This separation is equally entrenched in geographical discourse on the city, where discrete subdisciplines focus on industrial, commercial, or residential patterns.[3] However, there appears to be growing recognition that the relations between work and home are worthy of far more attention, that the spheres of home and work are not discrete in terms of lived experience and should be seen as mutually dependent rather than separate. Recent research has demonstrated that the character of the work environment is reflected in the residential landscape which, in turn, affects social relations in the workplace. In broader terms, the social and spatial structures of society are inextricably connected.

This chapter outlines the major theoretical issues surrounding links between home, community, and work. It reviews briefly the various ways in which class, gender, and ethnicity have been conceptualized as the principal dimensions underlying the socio-spatial structure of the capitalist city. It examines two traditional theories of urban spatial structure, based on human ecology and neo-classical land-use theory, and offers a brief exegesis of recent attempts to explain the causes and consequences of urban residential differentiation. I try to demonstrate that our understanding of urban socio-spatial structure has expanded as our conception of class, ethnic, and gender divisions in society has become more active – as social categories they are constitutive, as well as the outcome, of social relations. We should now move toward an *inter*-active view of these three dimensions.

This theoretical discussion will be followed by a detailed description of how class and ethnicity, workplace and homeplace, were intertwined in a specific time and place: the inner-city district of Spadina in Toronto during the early twentieth century. I shall focus on how the area came to be simultaneously a working-class neighbourhood, an immigrant reception area, and a garment production complex. In this context, I hope to illuminate the various recursive relationships between work and community, class and ethnicity, and society and landscape.

THEORETICAL ISSUES: WORK, HOME, AND RESIDENTIAL STRUCTURE

One of the earliest general theories of urban land use in North America was formulated by a group of sociologists at the University of Chicago early in this century. Combining a Social Darwinist perspective with a keen ability to observe the world around them, Burgess and Park derived the now familiar concentric zonal model of the city, with a central business district surrounded by a sequence of progressively more wealthy residential zones. Ultimately, the model is founded on the premise that urban spatial structure is the logical outcome of the innately competitive "biotic" forces of human nature. The urban land market therefore reflects competition between individuals which, if unchecked, could lead to social disorganization.[4]

In the city, this competition results in a socio-spatial sifting whereby individuals with similar cultural and economic characteristics gravitate to proximate locations. According to Park, this process fragments urban space into "natural areas," each becoming a nexus of internal co-operation.[5] Natural areas therefore develop their own moral orders, and this phenomenon reduces the level of conflict and moral breakdown in the wider society. While the Chicago model assumes that employment sources and immigrant neighbourhoods are both situated in or near the city centre, it generally ignores the links among employment, income, and residential structure. Thus it plays down the need for those with low incomes to reduce their housing and journey-to-work costs in favour of a cultural-moral explanation of inner-city immigrant areas.[6]

This interpretation of urban spatial structure receded in importance following continuing development of a neo-classical theory of urban land use. The latter stripped away all sense of the "social," as land economists focused on the activities of abstract, rational individuals with universally held preferences.[7] Urban spatial patterns represent the aggregate decisions of individuals who wish to maximize their utility with respect to land and access to the centre of the city (where, it is assumed, all jobs are located). Given that the poor cannot afford commuting costs, they settle near the city centre, while the affluent prefer large tracts of land on the periphery. Although this model recognizes the connection between workplaces and the location of blue-collar neighbourhoods, it is silent on such related issues as whether preferences are indeed universal; non–income-based residential segregation (e.g. racial ghettos); the relationship between residential differentiation and community formation; and the role of welfare and other actions of the state.

Thus human ecologists, while concentrating on underlying biotic and cultural explanations for residential differentiation, look at income variation primarily insofar as it sorts individuals into "natural areas." Conversely, neo-classical approaches to residential behaviour underplay cultural and social characteristics of a population in favour of income differentials. Yet, as the proponents of factorial ecology have shown, these perspectives have not proved irreconcilable. Inspired by the work of Shevky and Bell,[8] social geographers found that they could use factor analysis to examine simultaneously the influence of income variations, ethnicity, and familial status on residential structure.[9] Ultimately, however, the type of data and the theory of stratification employed limit these studies. Differences in income, ethnicity, and domestic cycle are assumed to be relevant, but sub-groups (e.g. middle-income; Italian) are seen unproblematically. Thus the assumption of social stratification is not grounded in a clearly articulated theory of social structure and conflict.[10]

Since the 1970s, social geography has been newly invigorated by reengagement with social theory. In particular, class, ethnicity, and gender (conspicuously absent from earlier theories) are increasingly seen as systems of stratification which must be explained rather than assumed. This new emphasis requires a more direct link to theories of the sources and consequences of social conflict, such as neo-Weberian theory, Marxism, and structuration theory. According to these perspectives, class, gender, and ethnicity arise from the action of groups engaged in a struggle for social dominance. The contours of this struggle are established in the workplace, community, and home, although emphasis varies on these three sites of conflict. All would agree, however, that class, gender, and ethnic differences are embedded in the labour market, provide axes for formation of communities of interest, and act to sift populations in residential space. As geographers and sociologists have begun to deal with these issues, their collective findings present considerable challenges to existing theories of urban spatial structure.

Geographers have long been interested in the land market and, as we have seen, in the spatial manifestation of economic differences. However, their focus has expanded from differences in "socio-economic status" across the city to the more active concept of class.[11] This has been a priority especially among Marxist geographers, who argue that residential differentiation is an important ingredient in the reproduction of capitalist social relations. Marxist researchers have therefore sought to explain the causes of residential differentiation and to understand its possible consequences.[12] Richard Harris, for example, suggests that class-based residential segregation serves to mute conflict between classes on a day-to-day level, since separation promotes mutual ignorance and quiescence. In times of social ferment, says Harris, the proximity of individuals in the same material situation favours formation of class consciousness.[13] However, empirical research has rarely uncovered enough class-based segregation to justify the assumed significance of class.[14] Clearly, other factors also allocate individuals in urban space.[15]

Gender differences and their relation to socio-spatial divisions in capitalist society have also stimulated recent scholarship in social geography that charts the reciprocal relations between spheres of production (work) and reproduction (home) in cities.[16] Mackenzie, for example, has shown that both the household economy and gender relations within households are structured by labour-market characteristics, especially availability and quality of jobs for men and women.[17] Similarly, she has demonstrated that issues in the household (such as availability of day care) eventually spill over into the work environment. Gender relations are therefore closely associated – through household and work – with the wider labour/capital relation in society; that is, gender and class are intertwined. The most obvious connection is, of course, the different positions of women and men in the labour force: women are more likely to be employed part time and in occupations with lower pay.[18]

Geographers have also sought to understand the socio-spatial attributes of racial and ethnic groups.[19] Initially, these groups were seen in passive terms as collections of individuals with fixed biological or ancestral attributes whose "active" characteristics were limited to resisting assimilation. An emerging, more transactional view of ethnicity assumes that ethnic affiliations are not given but are rather an outcome of social and economic forces[20] – ethnic groups do not simply exist on their own terms; they coexist. Ethnic identity "cannot be explained in terms of some all-embracing, superorganic notion of culture" but should be seen as a product negotiated both within and between groups.[21] This process usually involves a struggle for social dominance which, in a particular urban context, may be achieved through ownership and control of residential space,[22] and it will frequently lead to ghettoization of certain groups.[23] Yet segregation is a double-edged sword: while it reinforces the dominance of the "charter" population, it also presents economic opportunities to individuals in the ghettoized group, such as provision of in-group goods and services.[24] Also, such spatial concentration may foster community consciousness and may assist in the collective struggle for social and political equality.

Emergence of ethnic communities in specific neighbourhood settings has long been a theme within ethnic research. This general orientation could provide a useful corrective to studies of class and gender (at least within social geography), which seldom examine the way in which these abstract social cleavages promote formation of spatially oriented communities of affiliation. However, the literature on ethnic groups has yet to deal adequately with the role of labour-market segmentation in reducing or sharpening boundaries between ethnic groups – i.e. the relationship between ethnicity and social class.[25] We know, for example, that educational attainment, occupation, and income all vary across different ethnic groups just as they do between men and women.[26] It is important that geographers begin to establish the links among class, ethnicity, and gender, as it is in combination that they most profoundly affect the socio-spatial ordering of society.

What is needed, therefore, is a series of case studies that delve into situations that reveal class, ethnic, and gender inter-relations, which in turn requires careful examination of the intersection between work and community life and the home.[27] The scope of this chapter is not quite so ambitious, but I hope to bring together several crucial pieces of the puzzle via an examination of Jewish immigrant garment workers in Toronto during the early twentieth century. More specifically, I shall concentrate on the way in which class and ethnic affiliations among Jews evolved in relation to the intersecting domains of workplace, community life, and neighbourhood.

THE SPADINA AREA: CASE STUDY OF AN INDUSTRIAL DISTRICT

The Spadina area of Toronto, bounded by Simcoe, Bathurst, and Bloor streets and the shore of Lake Ontario (Figure 10.1), encompasses just under 4 km². Two major cycles of development left their mark on this area during the latter half of the nineteenth century. The southern portion (south of Queen Street) was first settled by relatively wealthy families who sought to be near the various public institutions located in the district, such as the Ontario legislative buildings, Government House, and Upper Canada College, as they were then located. The area north of Queen Street was settled in more piecemeal fashion, juxtaposing large lots and homes east of Spadina Avenue with more modest dwellings to the west. However, non-residential land uses began to "invade" the Spadina district during the 1870s and 1880s, and the area's upper-class residents gravitated to more impressive properties elsewhere in the city.[28] By the turn of the century, Spadina was

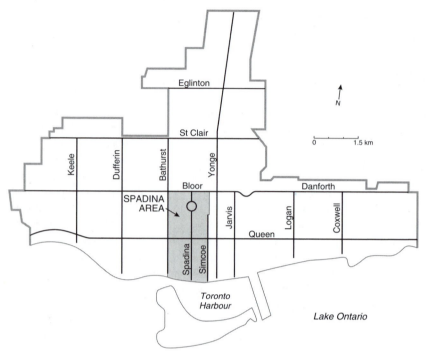

Figure 10.1
City of Toronto, showing the Spadina area

a densely populated area of intermingled industry and retail activities as well as working- and middle-class neighbourhoods. In one major respect, however, the Spadina district had changed little. Even in 1901, 92.3 per cent of the residents of ward 4 (the census division which encompassed the Spadina area) were of United Kingdom origin, while most of the remaining 7.7 per cent were from western Europe.[29] Although the animosity between these groups – especially between Irish Catholics and Protestants – should not be underemphasized, the ethnic composition of the area's population became far more varied during the next several decades.[30]

The Spadina landscape went through a third far-reaching transformation between 1901 and the Depression. Toronto's retail, wholesale, and financial sectors all expanded rapidly during this period, and the city's former immigrant reception area – bounded by Simcoe, Yonge, Queen, and College streets – gave way to these more profitable land uses. Meanwhile, immigration to Toronto reached unprecedented proportions, and newcomers were forced to find alternate areas of affordable shelter. There were few opportunities available to them, but their needs for accommodation coincided with the continuing cycle of downward filtering in the Spadina area. Thousands of eastern-European Jews, Italians, Greeks, and Slavs entered the area between 1901 and 1931, rapidly shifting the ethnic balance and forming distinct neighbourhoods in the process. By 1931, the overall proportion of English-speaking ethnic groups in ward 4 had dropped to slightly less than 40 per cent. Eastern-European Jews were, by far, the largest of the new immigrant groups, accounting for 31 per cent of the total population by 1931.[31]

Meanwhile, the handful of factories in the area were joined by dozens more between 1901 and 1931. The twin movements of immigrants and factories toward Spadina did not merely coincide; they were closely interrelated. Initially, the few industrial establishments in the area attracted immigrants, especially those who sought employment in labour-intensive manufacturing (e.g. eastern-European Jews who arrived with artisanal skills). Thus a close relationship between home and work evolved within the area.

Gradually, however, this connection between factories and immigrants began to take on new meanings. Between the turn of the century and 1931, hundreds of immigrants themselves became entrepreneurs, and many chose to channel their resources into manufacturing. Moreover, most chose to locate their firms within areas with which they were familiar: around their homes and property holdings in the Spadina district. The majority achieved economic success through participation in the garment industry. The following section explores the significance of this constellation of events with specific reference to the reciprocal relationship between Jewish immigrants and the structure of Toronto's garment industry. First, I describe the process whereby Jews came to work in and, later, control much of Toronto's clothing industry. This includes discussion of how Jewish entrepreneurs helped to restructure the industry and, in turn, how their participation in the clothing industry affected social relations within the Jewish community. Second, I shall turn to the "home" side of the same equation and examine the ongoing evolution of the Spadina neighbourhood, a landscape that juxtaposed home, work, and community relations all within walking distance.

WORK, HOME, AND COMMUNITY LIFE AMONG JEWISH IMMIGRANTS IN TORONTO

State-sanctioned persecution of Jews in eastern Europe led to one of the largest exoduses of all time; between 1880 and the First World War, nearly one-third of all eastern-European Jews emigrated. Jews around the world responded to the plight of their co-religionists by organizing a network of philanthropic agencies to promote orderly resettlement of Jewish emigrants, and tens of thousands were directed toward Canada. Approximately one-fourth of these immigrants settled in Toronto. Most arriving eastern-

European Jews possessed few skills pertinent to an industrializing society, and few employment opportunities were available to them. Those without artisanal experience generally found unskilled work or took up peddling, while former crafts persons sought manufacturing jobs.[32]

Within Toronto's turn-of-the century economy, the garment industry best realized the potential of this rapidly growing source of inexpensive labour. The level of clothing production was expanding rapidly in order to meet steady growth in consumer demand; the garment labour force was relatively unorganized, and so established (Anglo-Saxon) workers could not prevent their employers from hiring cheaper, immigrant labour; clothing production was one of the few economic sectors that Jews were allowed to enter in eastern Europe, and many had been trained as tailors; the structure of clothing production in Toronto (i.e. piece work) allowed a fairly high degree of flexibility, and Jewish workers were able to observe the Sabbath by extending the hours they worked on other days; and, finally, the manufacture of clothing had remained virtually unmechanized and therefore still required artisanal skills.[33] By 1901, 22 per cent of Jewish heads of household in central Toronto were garment workers, although they represented only a small fraction of the total garment labour force in Toronto.[34] The connection between the Jewish community and the clothing trades strengthened during the next three decades. As the garment factory work-force expanded by approximately 7,000, eastern-European Jews captured most of the newly created jobs and began to replace indigenous workers. By 1931, nearly 40 per cent of all Jewish household heads were clothing workers, and Jews accounted for over one-third of workers in the industry.[35]

A similar transformation occurred in the patterns of ownership within the industry. At the turn of the century, Toronto's clothing industry was still oriented largely to custom production, although ready-made clothing, produced in factories, was rapidly overtaking the custom sector.[36] The trend toward capital concentration and factory production gained momentum between 1901 and the First World War, and the Toronto clothing industry came to be dominated by a few very large firms. The largest of these, the T. Eaton Co., adopted an aggressive policy of vertical integration and was simultaneously engaged in production, distribution, and retailing. Its influence extended well beyond the Toronto market; the company realized much of its success via its catalogue and mail-order sales throughout Canada.[37] By 1914, Eaton's had established factories in Winnipeg and Montreal, but most of the clothing it sold was produced in a complex of several factory buildings located in an area of Toronto known as The Ward (2 km east of the Spadina area).[38] Its market share in the Toronto clothing industry peaked in the late 1910s and 1920s, when the company employed over 5,000 garment factory workers in Toronto alone, thereby accounting for approximately half of the city's clothing production.

By the 1920s, however, the dominance of large establishments was challenged by a growing group of small clothing workshops and factories owned largely by immigrant entrepreneurs, primarily eastern-European Jews. In contrast to the assembly-line style of production at Eaton's, where as many as 150 individual workers would each perform an extremely narrow operation on a single garment, smaller firms had a more flexible production run.[39] It these shops, individual workers completed relatively large segments of the assembly process. In fact, Jewish entrepreneurs frequently organized their businesses around as few as five separate stages of production: designing, cutting, sewing, finishing, and pressing.[40] Sometimes all these operations could be performed by various family members, and the home itself doubled as a workshop.[41] More typically, however, a few would-be entrepreneurs (often with designing, cutting, and sewing-machine operating experience) formed a partnership and hired three or four other workers to fill in the remaining functions. Since their workers were acquainted with a wider range of operations and skills, these firms were best suited to compete in the more volatile sectors of the industry, such as women's outerwear. Some Jewish entrepreneurs pursued flexibility even further and switched between different sectors of the industry (e.g. from menswear to womenswear production, or vice-versa) month by month, as opportunities arose.[42] Competition between standardized and flexible styles of clothing production (and, indeed, in other industrial sectors as well) continues even today as North American and European manufacturers attempt to remain profitable in the face of more cheaply produced offshore garments.[43]

Jewish-owned garment firms proved highly competitive, and their number jumped from a handful in 1901 to over 400 in 1931.[44] By 1931, Jewish entrepreneurs owned well over half of all the garment factories in Toronto.[45] In pursuing innovative styles of production, Jewish factory owners helped reverse the trajectory of change in the entire industry. Between 1915 and 1931, the size of the average garment firm in Toronto dropped by more than 50 per cent, the number of sub-contracting relations between firms increased, and labour/capital relations became strained in new ways.[46] These changes are evidence of a "crossover," whereby a specific cultural group (Jewish entrepreneurs and workers) altered Toronto's economy (the garment industry) by introducing practices that they had developed in other settings. Jewish factory owners in Toronto emulated the success of similar firms in other North American cities, such as New York, Chicago, and Montreal.[47] Jewish workers also helped change the structure of garment production in Toronto by transferring union organizational skills that they had learned in other North American cities.

In addition, Jewish entrepreneurs relocated a large segment of Toronto's clothing industry. At the turn of the century, virtually all clothing manufacture in the city took place in the downtown area or just to the north of it, par-

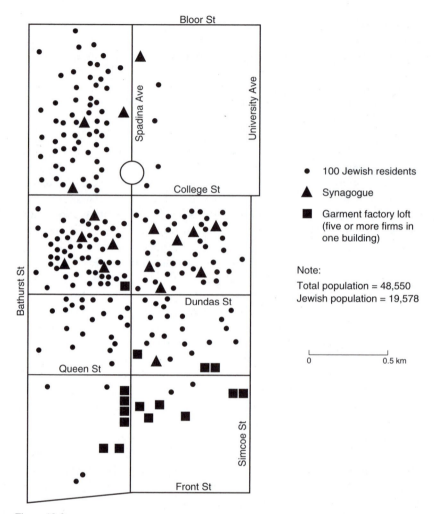

Figure 10.2
Spadina area in 1931, showing distribution of Jewish residents, synagogues, and garment factory lofts

ticularly at the Eaton's factory complex. By the late 1920s, the Spadina area held well over half of all Toronto's garment factories, and the combined mass of these factories began to approximate the scale of the Eaton's production complex.[48] This relocation away from the central business district was in large measure motivated by a search for cheaper land and rents in an industry characterized by intense competition. However, the choice of Spadina was linked to its proximity to Jewish entrepreneurs, investors, and workers (Figure 10.2). For example, all but one of the garment lofts constructed between 1920 and 1929 in this area were built on land owned by Jews, designed by Jewish architects, and owned by Jews.[49]

Although the number of garment factory lofts had multiplied in the Spadina district, most of the area retained a residential character, which was dominated by immigrants.[50] Even Jewish factory owners chose to remain in Jewish neighbourhoods, even though they could easily have afforded to reside in "better" areas of the city.[51] Motivation for this choice hinged on two factors: Jews were not particularly welcome in many other areas of the city,[52] and participation in community/religious activities frequently required residential proximity, especially for Orthodox Jews, who for religious reasons wanted to be within walking distance of a synagogue on the Sabbath.

The concentration of Jewish homes and community institutions also provided sufficient consumer demand for a range of commercial activities which helped establish Spadina as a distinctive commercial landscape. By the 1910s, Jewish immigrants began to introduce a "new" form of retailing on Kensington Avenue and a few adjacent streets near the intersection of Spadina Avenue and Dundas Street. Here, Jewish small proprietors created a kind of country market, selling fruit, vegetables, poultry, and other goods which were displayed in their front gardens.[53] The success of Kensington market illustrates several ongoing processes in the Spadina area. First, Kensington Avenue itself reveals the ethnic micro-segregation that occurred in the Spadina area: 78 out of 82 homes and flats on the street in 1931 were occupied by Jewish families (as opposed to 1 out of 82 in 1901). Second, the market flourished because it was situated among thousands of recently arrived Jewish households, all of which were familiar with this type of "shtetl" retail environment. Third, the existence of Kensington market reveals selective recreation of Old World traditions in a New World context: Kensington was a hybrid landscape, with New World architecture juxtaposed with an Old World market atmosphere.

By turning to examine the relationship between Jews and the garment industry from the opposite vantage point – the spheres of home and community among Jews – we shall find that a second "crossover" had occurred: the social structure of the Jewish community increasingly came to reflect the cleavage between workers and capitalists. Differential participation in the garment industry led to widening disparity between wealthy and working-class Jews. Despite the economic vulnerability of small manufacturers, there were striking differences in the level of material well-being between those Jews who owned factories and those who merely worked in them. For example, Jewish factory owners enjoyed higher rates of homeownership and lived in homes that were worth, on average, 50 per cent more than those of blue-collar garment workers.[54]

The economic distance between Jewish workers and entrepreneurs was also reproduced in community institutions and cultural practices. Wealthy eastern European Jews began to assert their status within the community in several ways: by attempting to gain control over philanthropic organizations, establishing an elite social club, and, increasingly, by converting from Orthodox to Reform or Conservative variants of Judaism.[55] Conversion was at least partly motivated by economic considerations, since Orthodoxy demanded of its adherents a work-free Sabbath and specific styles of dress, both of which were seen as "unbusinesslike." By the 1920s, members of the "*Yahudim*-bourgeoisie"[56] were also beginning to launch careers in the political arena, as either Liberals or Conservatives.

Jewish workers established a different set of institutions embedded within their own cultural practices. Most of these appear to have evolved around one of two strategies for material advancement. Workers who adhered to traditional values normally participated in a network of institutions centred on the Orthodox synagogue and the Landsmanschaft group (i.e. individuals who emigrated from the same village or region).[57] Judging by the number of clothing union members during these years, few Orthodox Jews were active unionists. Instead, most sought to ensure their welfare through mutual-aid organizations, and collective medical plans, death benefits, and unemployment insurance were all introduced into the structure of Landsmanschaft groups.[58] Radical Jewish workers, in contrast, were among the most militant unionists in Toronto, and they also built their own cultural and benefit institutions, such as the Jewish Labour Temple, which housed a radical press, a reading room, and a wide array of educational activities. The Labour Temple was also used as the headquarters of the largest garment union in Toronto.

Over time, the institutions of wealthy Jews and the two working-class sub-groups came into conflict. One of the earliest examples of this clash of ideologies occurred in 1912, when Jacob Gordon, the most prominent rabbi among Russian Jews, attempted (without success) to ban a Sabbath-day picnic by the Jewish National Radical School.[59] Ongoing divisions within the community were revealed even more clearly in 1923, when an attempt by a group of influential rabbis to establish a Kehillah (an organization to co-ordinate a common approach to Jewish rituals, such as preparation of kosher meat)[60] was opposed by another group supported by the major Jewish clothing union.[61] Thus, through the 1920s and into the 1930s, class distinctions and their accompanying ideologies affected issues which, initially at least, appear to have been non-political.

Class relations among Jews were also becoming more direct as well as more pervasive. At the turn of the century, Jews were already stratified by class but rarely occupied directly conflicting class positions.[62] In the garment industry, for example, given the small number of Jewish factory owners, few Jews were in a position to hire other Jews, and the potential for conflict between Jewish workers and employers was decidedly weak. By the 1920s, however, Jews were directly employed in large numbers by their co-religionists. This polarizing of individual fortunes among Jews led to new sources of friction in the community. Previously, Jews differed over the nature of religious observance, desirability of assimilation, and other concerns that were only partially economic in nature.

By the late 1920s and the 1930s, the most divisive community issues involved the struggle between Jewish employers and their workers. While wages and working conditions improved between 1923 and 1929, they quickly deteriorated as the economic viability of the clothing industry grew more precarious.[63] Clothing unions, led by eastern-European Jews, were instrumental both in precipitating wage increases prior to 1929 and in framing a more militant stance toward the downward pressure on wages after 1929.[64] The growing labour movement involved numerous struggles between Jews, since "the boss was usually someone's uncle."[65]

The most bitter conflict took place in 1934, when a strike was organized at the Superior Cloak Co., owned by Sam and Louis Poslun. The incident galvanized the entire Jewish community, and it was widely discussed and written about in local Jewish newspapers. The Poslun brothers initially responded to the strike by hiring scab labour. A series of violent confrontations ensued, and, a week later, the Posluns relocated their company to Guelph (about 100 km northwest of Toronto) and attempted to hire a new work-force. Rather than defusing the conflict, this tactic incited workers, and the union sent daily convoys of picketers to the new location. Authorities in Guelph, arguing that they were not prepared to deal with a large-scale, violent strike, requested that Superior leave the city. The workers' victory was short-lived, however; the Posluns declared bankruptcy a few weeks later.[66]

The residential cohesion of Jews with widely varied – and at times antagonistic – class backgrounds meant that workplace confrontations were quickly carried over into the neighbourhood as a whole. Thus spatial proximity facilitated the two "crossovers" identified in previous paragraphs. Kensington market, while only a few steps away from an Orthodox synagogue, was also within steps of a complex of over a dozen garment factory lofts, which were owned by Jews and held Jewish garment firms and a Jewish labour force. In other words, for the eastern-European Jewish community, the Spadina landscape was "institutionally complete": it came to hold a full range of economic and social land uses, many of which represented contradictory interests, all within walking distance. It was also, therefore, a place where home and work relations were inextricably linked. As we have seen, this was illustrated most graphically during labour disputes, when even the most minute points of various conflicts were debated in Jewish newspapers and individuals in the community were forced to take sides. Many of the barriers with which we are so familiar today, especially the divide between our personal spheres of homes and work, were virtually nonexistent in the Spadina context. Day-to-day life took place in a rather narrow compass of only a few blocks, and the lives of the residents of the area were enriched by the very density, heterogeneity, and vibrancy of the activities that it enclosed.

CONCLUSIONS

What can you eat on Spadina?
What is this dish rice-fried?
When you are looking for a bagel,
You get a cookie with a message inside.[67]

Spadina continued to be an area of rather direct home, community, and work relations after the 1930s. Thousands of "displaced persons," especially from eastern and southern Europe, moved into the area in the years immediately after the Second World War. Still, Jews accounted for nearly one-third of the population even in 1951. Between 1951 and the present, however, most Jews left Spadina in favour of the Bathurst Street corridor, approximately 2 km north.[68] This relocation coincided with a change in the labour-market participation of Jews: they gradually abandoned the garment industry for other pursuits. By 1986, Jewish men were three times – and Jewish women four times – less likely to be hired in the manufacturing sector than the average worker in Toronto.[69]

Meanwhile, newer immigrants have "invaded" Spadina in yet another cycle of ethnic change. Since the mid-1960s, an increasingly large Asian-origin community has migrated to the area; by 1981, fully one-third of residents were Chinese, Vietnamese, or from some other part of the Far East. Like the Jews, these groups have also established their own retail, residential, and cultural landscapes, complete with visible juxtapositions of old-country and new-country attitudes. Residents of the area still tend to work nearby, if not at home. In 1981, 6 per cent of all heads of household worked in their homes (twice the Toronto average), while 14 per cent were employed in the census tract in which they resided (the Toronto average was 6 per cent).[70]

I earlier emphasized the importance of charting the intersections and interdependencies between the principal systems of social stratification. Approaches are needed that highlight the ways in which class, gender, and ethnicity shape the socio-spatial structure of Canadian cities. I investigated class and ethnicity here in the context of Toronto's Jewish community. What is often referred to as "cultural baggage" (e.g. Landsmanschaft ties) was used as a resource by both Jewish workers and capitalists to enhance their material and social well-being. Jewish workers added a new dimension of organized militancy to the garment work-force. Jewish entrepreneurs discovered latent opportunities for more flexible styles of garment manufacture. In exploiting these possibilities, they recast Toronto's clothing industry into a more dualist mould, where small and large firms coexisted by specializing in different aspects of garment production.[71]

Meanwhile, a simultaneous "crossover" occurred in the opposite direction as the social dynamics of the Jewish community came to reflect the struggle between owners and workers within the garment industry. This is not to argue that all previous conflicts between Jews were non-economic; rather, the participation of Jews as both workers and entrepreneurs in the same economic sector transformed class relations among Jews from indirect forms to direct ones. Tensions within the community took on new meanings as Jewish workers learned that their co-religionists could be just as unrelenting in their desire to reduce wages and restrict union activities as non-Jewish capitalists.

This double crossover sheds light on at least three theoretical issues within social geography. First, ethnic and class relations can be inter-penetrating. Radical Jewish workers created a sub-culture that was at the same time Jewish and aggressively pro–working class; ethnic and class forms of exploitation were not seen as entirely distinct. Further, neither of these forms of stratification was seen as the more central. Put more simply, Jewish workers were non-reductionist. Second, the spatial proximity of home, community, and work among Jews contributed to rapid transfers of conflicts and struggles between these spheres. Thus the landscape created by Jews became both a container of relations among Jews and a catalyst for intra-community discord. Finally, to echo a point persuasively argued by Michael Mann: the real world of socio-spatial relations is far "messier" than our theories allow.[72] In the Toronto case study, ethnicity (often theorized as a cultural dimension) came to incorporate economic tension, just as class (often theorized as an economic dimension) cannot be understood without reference to cultural practices.

CHAPTER ELEVEN

Past Elites and Present Gentry: Neighbourhoods of Privilege in the Inner City

D . F . L E Y

In November 1986 the *Globe and Mail* began a series of investigative articles under the banner headline "Toronto the Rich." The reporter had in mind in particular the wealth of the central city: "From the sleek luncheon haunts of King Street West, to the 100 per cent cashmere crush in the carriage-trade shops of Yonge Street and St. Clair Avenue on Saturdays, to the graceful Jaguars gliding into the parking garage at First Canadian Place, the wealth of Toronto seems pervasive," he gushed.[1] But evidence of conspicuous consumption is, of course, distributed more broadly over the residential areas of Metropolitan Toronto, and, standardizing against household size, there is not much difference between the city of Toronto and the metropolitan area in mean household incomes.[2] Of the four districts of high income, two (Rosedale and Forest Hill) are established neighbourhoods in the city and two are in the suburbs (the Bayview Avenue corridor in North York, and central Etobicoke).

In the intra-metropolitan distribution of wealth, city residents are certainly holding their own. Between 1970 and 1980 their real per-capita income grew by 40 per cent, faster than the metropolitan area as a whole, and the city has borne the brunt of escalating property prices in the past decade. While house prices in the market area of the Toronto Real Estate Board increased some 290 per cent, 1977–87, in the eleven districts within the city of Toronto inflation ranged between 339 and 433 per cent.[3] This trend is in part a product of what the *Globe*'s correspondent described as "galloping gentrification," the social upgrading of many inner-city neighbourhoods as middle- and upper-middle-class households move into renovated or redeveloped dwellings in previously more affordable districts. As a result, a current review of high-status districts in Toronto and other cities identifies three major types: old elite areas in the inner city and inner suburbs which have maintained a high rank; new inner-city neighbourhoods, further subdivided by the prevalence of either renovation or luxury redevelopment (often of waterfront condominiums), both processes creating new enclaves of wealth;

and new suburban districts, often associated with exclusive recreational amenities.[4] This chapter is concerned with the first two members of this typology.

The geography of wealth in Toronto is repeated with minor variations in other large Canadian cities. In Montreal eight of the ten census tracts reporting the highest mean household incomes in 1980 were in the horseshoe of old inner suburbs including Westmount, Outremont, and Mount Royal. In Vancouver the wealthiest tract was in the traditional inner-city elite district of Shaughnessy; overall five of the top ten tracts comprised inner neighbourhoods, with the remaining five in the newer suburbs on the North Shore, primarily in West Vancouver. The spatial distribution of economically privileged households in Canadian metropolitan areas presents one of the consistent distinctions between Canadian and American cities.[5] In the United States, with few exceptions, wealth fled the central cities, but in Canada a more stable spatial structure endured: old elite neighbourhoods survived and sometimes expanded with gentrification, while new wealth appeared in the suburbs. Overall, metropolitan areas with an above-average concentration of high-status residents in their inner cities are national or regional service centres, usually with significant government employment (Table 11.1).[6]

Thus the social geography of Canadian cities reveals a select pattern of privileged inner and inner-suburban communities which have persisted, in eastern Canada for up to a century and in western Canada since their construction in the early twentieth century. From the Old South End in Halifax to the Uplands in Victoria are a series of elite districts whose landscape and social practices lend a distinctive impress to Canadian cities. To them are being added other inner-city districts, such as Toronto's Don Vale (Cabbagetown) or Vancouver's Kitsilano, which are currently experiencing marked social upgrading. Neighbourhood studies have comprised an important component of social geography.[7] In this vein, the present chapter will examine in particular older elite neighbourhoods, with a lesser emphasis on gentrifying districts, and ask a set of questions of them. What are their characteristic landscapes? What are the distinctive social practices and patterns of social interaction of their residents? What is their longevity, and what are the politics of neighbourhood preservation? How may gentrification be understood within the framework of urban elites? And, finally, recognizing the tight bonding between society and space, how can we account for regions of privilege within Canadian cities in terms of the broader structure of Canadian society?

RESIDENTIAL LANDSCAPES
OF THE URBAN ELITES

As in the United States, elites in Canada have traditionally shown deference to European styles and precedents in the staging of their residential land-

Table 11.1
Index of inner city social status among Canadian metropolitan areas, 1971–86

	Social status score, 1981	Social status score, 1986*	Increase in social status, 1971–81	Increase in social status, 1981–86*
St John's	21.1		8.4	
Halifax	31.1	35.1	13.0	4.0
Saint John	14.4		2.5	
Quebec City	20.6		7.4	
Montreal	19.0	28.1	6.4	9.1
Ottawa-Hull	26.1	33.4	10.0	7.3
Oshawa	10.4		1.9	
Toronto	23.4	28.8	8.7	5.4
Hamilton	15.8		5.7	
St Catharines–Niagara	14.3		3.8	
Kitchener	16.0		6.8	
London	17.3		6.2	
Windsor	15.0		4.6	
Sudbury	14.4		3.6	
Thunder Bay	15.9		4.2	
Winnipeg	14.4		4.0	
Regina	18.9		5.1	
Saskatoon	27.7		8.0	
Calgary	26.0		9.0	
Edmonton	23.4	27.1	6.5	3.7
Vancouver	25.1	31.5	9.3	6.4
Victoria	22.3		9.5	
All CMAs	21.1		7.5	

Sources: Ley (1985; 1992).
* Based on a study of six CMAs.

scapes.[8] In the presentation of their houses they have been particularly receptive to the picturesque and the romantic pastoralism of the English country estate. In the late nineteenth century they favoured an eclecticism of classical styles relayed primarily from Britain, especially Gothic and Italianate forms, while in the first half of the twentieth century the Tudor mansion occupied a privileged position. One of the principal reasons for the primacy of British precedents in Canada was the strong personal sentiment of elite members for their own native or ancestral land. In Westmount the 1901 census showed that 88 per cent of the population identified with a British ancestry, including 27 per cent Scottish and 45 per cent English; by 1941 this figure had fallen only to 74 per cent.[9] Among the new urban centres of the west, it was not much different. In the young city of Vancouver the elite in the 1886–1915 generation had strong ties to Britain; of the foreign-born, almost three out of four were from the United Kingdom, and of the native-born perhaps half were the children of British immigrants.[10]

Strong British and imperial consciousness provided the cultural matrix of these neighbourhoods until well into the twentieth century. This argument has been amply documented for Montreal's Square Mile in the nineteenth and early twentieth centuries; this district, west of downtown and immediately east of Westmount, was populated by the successful British and particularly Scottish merchants of Canada's then primate city. "Few neighbourhoods in the Empire were more British than the Square Mile. Inspired by the thumping strains of 'Land of Hope and Glory' and the old Queen's Diamond Jubilee, Lady Drummond spoke for her Montreal-born friends as well as herself when she was heard to remark: 'The Empire is my country. Canada is my home.'"[11] Indeed still in 1919, "The Drummonds, Gaults, Meighens, Molsons and the rest set the fashions and the fashions were British."[12] British royalty were frequent visitors to the Square Mile, and the daughters of Square Mile families were presented as debutantes at Buckingham Palace well into this century. Lord Strathcona, head of the Hudson's Bay Co. and founder of the Canadian Pacific Railway, owned stately homes in England and Scotland, as well as his three mansions in Canada, and his cousin and a number of other neighbours retired to the "Old Country." Even at play the British connection survived with foxhounds for the Montreal Hunt imported from England.[13] At the outbreak of war in 1914 the call of empire was irresistible. The Molson family sent to war thirty fathers and sons, five of whom were killed in action, and a further thirteen wounded. A Square Mile diarist attended a commissioning service in 1915 and noted the words of the bishop's "fine stirring sermon – be Christian, be British."[14]

The residential landscape that emerged from such sentiment was irrepressibly European in its inspiration. Sherbrooke Street had much in common with Knightsbridge in London or Princes Street in Edinburgh. The elite's mansions bore names like Ravenscrag and Craiguie, and their stylistic sources and architects were steeped in British precedents. From the 1860s to 1914 a succession of picturesque styles – Italianate, Gothic, classical, medieval and renaissance eclecticism (sometimes integrated in a Scottish romantic genre), Second Empire, and Château style – graced the leafy, elm-arched avenues, all of them a refraction of the fashions of the British elite (Figure 11.1).[15] Some buildings were copied directly from imperial prototypes: the Bank of Montreal on St James Street was modelled after the Commercial Bank of Scotland in Edinburgh.[16] Similar landscapes occurred in other major cities; indeed with the construction of the Parliament Buildings in Ottawa in the 1860s, Gowans considers that picturesque eclecticism became Canada's first national style.[17] In Toronto an eclectic classicism "gone picturesquely wild" embraced residential and institutional building in the last

Figure 11.1
Ravenscrag (built 1863) in 1902, the home of Sir Hugh Allan, founder of a family shipping
line, among many other business ventures. Designed in the Anglo-Italianate style, it
resembles a fifteenth-century Tuscan villa and also Queen Victoria's summer retreat,
Osborne House. Since 1943 it has been the Allan Memorial Institute, a psychiatric centre
for the Royal Victoria Hospital. *Sources*: McCord Museum, McGill University, and
MacKay 1987.

decade of the Victorian era, to be followed by a more ordered Edwardian
classicism in Rosedale and elsewhere which provided "solid versions of English gentry mansions."[18] In Vancouver's West End, Victorian eclecticism
was frequently refracted as it diffused via Toronto, Montreal, or California.
Nonetheless the exuberance and individuality of styles fitted well with the
optimism of the imperial frontier, while the layering of historical references,
variously Roman, Greek, medieval, and renaissance, offered both comfortable and romantic nostalgia and instant status to a new aristocracy emerging
in a resource-rich province.[19] In this landscape of personalized ostentation
gaudiness seemed next to godliness.[20]

   Much of this landscape of the early elites is now residual or much changed
before the pressures of downtown expansion and residential redevelopment.
A later generation of privileged neighbourhoods, including Westmount
(Montreal), Forest Hill (Toronto), and Shaughnessy (Vancouver), has succeeded them. From the 1890s and the beginnings of Westmount to the 1920s
and the opening up of Forest Hill, the new architectural vocabulary of the
Tudor revival graced the streetscapes of the elite (Figure 11.2). Its popular-

Figure 11.2
Walter C. Nichol house (built 1912–13) in 1992. A Tudor revival mansion built for the
newspaper publisher and later lieutenant-governor on The Crescent in Shaughnessy (minor
repairs under way on the porch roof).

ity, however, showed the same symbolic appropriateness for elite sentiment
as earlier styles: "Its English ancestry and picturesque appearance made the
half-timbered Tudor Revival the favourite mode in Shaughnessy Heights."[21]
The landscape of Shaughnessy showed too that neighbourhood included
more than the home itself: the subdivision developed by the Canadian Pacific Railway (memorializing in its street names CPR executives from Montreal's Square Mile) provided for a pastoral estate of large lots, curving
streets, generous set-backs – a leafy parkland with architectural controls to
shape an exclusive inner suburb, to which Vancouver's elite flooded from
the urbanizing West End after 1910.

While British precedents were less significant in the construction of hum-
bler homes, they continued to provide a model for the upper middle class
and the elite well into the inter-war period. Thus in 1927 local architects in
Vancouver described a "Canadian Home of Charming Character" in purely
English terms as a "splendid spacious home in Elizabethan half-timbered design which possesses in full measure the comfortable homelike atmosphere
of this style."[22] Nor was this icon to pass quickly from the scene. A neighbourhood plan for Shaughnessy approved by the city in 1981 set out to preserve the romantic landscape of the Tudor country house. Its design controls

forbade new structures that did not adhere to the standards of "the English picturesque landscape tradition."[23]

This tradition has become a publicly sustained myth, expressive of a world of shared elite values. The conformity inherent in neighbourhood-sponsored design controls is indicative of a tightly bonded social world where landscape offers a non-verbal communication of inclusion and exclusion. This is true equally of exterior and interior design. Interviews in Shaughnessy homes in 1979 showed that the traditional Tudor of their external appearance was matched by a similar conservatism of interior decoration, where styles have been learned and accepted from parents, and where commonly shared taste is an expression of a shared subculture.[24] In marked contrast, a newer elite has gathered in the postwar suburb of West Vancouver, a landscape of more modern design conventions, including ranch-style homes and post-and-beam structures in the International Style. A large number of West Vancouver households are immigrants, almost all have moved from elsewhere, and upward social mobility is a common trait. Among this entrepreneurial group (over one-quarter of the sample were connected with the property industry), the interior design of homes is a study in practised individuality. Women stove to be different in their taste, not only from their friends, but also from their own preferences of a few years earlier. The house was to be expressive of individual artistry and creativity. Yet was this much different from the parvenus of the West End three generations earlier, who had practised the same landscape convention of expressive individuality? With the passage of time, in contrast, the impress of socialization among an established Shaughnessy elite has mediated between self and landscape and dampened the oscillations of personalized taste into the more stratified layering of subcultural consensus.

ELITE COHESION AND THE POWER OF PLACE

The label of the urban village has been applied primarily to working-class ethnic communities where common interests, generational continuity, and a local array of shops, churches, and voluntary associations sustain a supportive provincialism. Yet established elite areas may be no less provincial, and no less self-contained; indeed, as in London's Highgate Village, Boston's Beacon Hill, or the Village of Forest Hill (incorporated as an independent municipality in 1924), the spirit of localism is an important strategy of both self-identity and self-preservation. For the elite, neighbourhood as a place contributes to its status repertoire and social cohesion.

In their study of Forest Hill, Seeley and his colleagues emphasized the public function of the home for "the staging of 'productions,'" notably acts of hospitality incorporating friends, neighbours, and esteemed outsiders.[25]

Exterior and interior design created a setting intended for public display and sometimes elaborate public performances. The diaries and social histories of the upper class are replete with the sumptuousness of frequent social events. "We had parties, we had parties, endlessly we had parties," enthused the son of Senator Ballantyne, a resident of Montreal's Square Mile, of neighbourhood life in the 1920s.[26] Diaries commonly record entertaining or being entertained. Not unusual was a 1907 gathering of neighbours-cum–business associates with an esteemed foreigner: "Lunched at W.C. Van Horne's to meet Mr. Elihu Root (u.s. Secretary of State). In the evening dined at Lord Strathcona's where the Root family was stopping. Sat next to Madame Forget, Sir Thomas Shaughnessy, both very pleasant, Lord Strathcona, a little deaf, the only sign, with his grey hairs, of the passing of time."[27] Shaughnessy's grandson was later to remember the coming-out party for his older sisters and its sit-down supper for several hundred in their home beside McGill College.[28]

Besides the round of entertainment and intermittent society spectacles there was also a more steady rhythm to neighbourhood interaction. A regular feature of Victorian and Edwardian society was the at-home tea party held by women on fixed days of the week or month which cemented the contours of social worlds. The elaborate protocol to home visits maintained the niceties of social distance even within the elite stratum.[29] It was deemed in bad taste to initiate a social visit to a more status-worthy household. However, an at-home card could be left, and a visit acting on the invitation might later be returned by a reciprocal invitation. A similar form of scrutiny might occur through attendance at a garden party, a less privileged occasion because it did not involve access to the sanctity of the home. An acceptable performance as a garden party guest might, however, be followed by the more esteemed invitation to the home itself. This screening and selective incorporation constructed the configuration of a status hierarchy with definable edges. Within the bounds of the accepted elite, social interaction was often dense and even spatially orchestrated. In Vancouver's West End, the 1908 *Elite Directory* indicated five separate groups of at-home meetings on every day of the week. They formed individual territories with little geographical overlap, indicating the degree of spatial management exercised by the elite in order to ensure social cohesion. The existence of at-home regions where the event fell on the same day thus closed access to neighbours, who could anyway interact more readily at dinner parties or on other occasions, and threw open interaction with geographically removed elite members by ensuring that at-home days with them did not conflict.

Beyond the home was an ensemble of neighbourhood land uses with symbolic and functional significance to the sustenance of the elite, including private schools, clubs, churches, and fashionable stores and hotels. Schools and clubs helped to mould and exert closure around a world of status and privilege. Institutions such as McGill and Upper Canada College were neighbourhood assets, sustained by the fees and sometimes spectacular philanthropy of those who lived around them and who recognized their capacity to reproduce family dynasties. In Forest Hill "it is the massive centrality of the schools that makes the most immediate physical impact on any outside observer," and three of the nation's most prominent private schools are adjacent to the area.[30] Private schools continue to consolidate socialization into privileged social networks. In 1970 a sample survey in Shaughnessy revealed that the parents of 80 per cent of respondents had attended private school themselves, while almost 90 per cent paid school fees for their own children. The role of private schooling in their own socialization was marked; almost four out of five adults stated that over half of the people with whom they exchanged home visits were friends made at school.[31] Nationwide studies have underscored that the right school and university also define the shortest path to career success. Among Canada's economic elite in 1972, 40 per cent had attended private schools and 43 per cent of those attending universities were educated at Toronto or McGill.[32]

A second abiding feature channelling the biographies of (male) members of the elite is the institution of the private club. Clubs date from the earliest years of elite districts and originally were neighbourhood institutions. Merchants in the Square Mile successively formed three private clubs during the nineteenth century within the borders of their neighbourhood, two of which in 1972 remained among the top three in Canada in terms of membership among the national elite.[33] In the West End, too, elite clubs, like private schools, were within the district or on its perimeter; in 1908 almost three of every four members of the Vancouver Club were West End residents.[34] Affiliation was retained by members of the elite as they moved out of the West End after 1910 and was sustained into the 1970s, when two surveys in Shaughnessy found that 67 per cent (in 1970) and 30 per cent (in 1979) of sample households in single-family dwellings continued to be members of the Vancouver Club.[35] The club provided both a basis for integration for local elites and a point of incorporation for national and international people of prominence. Homes, too, or at least certain homes, were staging points or bridges for the convergence of local and national networks. I have already referred to the presence of an American secretry of state at Lord Strathcona's house. As a dominant figure in the Canadian Pacific Railway and with extensive US business interests, Strathcona was a substantial power broker, and the guest book at his home included the names of a small army of European royalty and North American politicians, judges, clerics, and generals.[36] In the West End, one of the key homes for visiting dignitaries was owned by the CPR superintendent, H.B. Abbott, brother of Sir John Abbott, resident of the Square Mile, CPR lawyer, dean of law at McGill, mayor of Montreal, and prime minister of Canada 1891–92.[37] In this manner local and national spheres of influence and personal and business interests were tightly interwoven.

Beyond home, school, and business club, a set of voluntary organizations with a primarily female cast further bound the elite. In Vancouver's West End, a semi-autobiographical novel showed the wife of a well-to-do family, newly arrived from England, quickly joining the Ladies Aid, the Women's Auxiliary, the Women's Council of the YWCA, the Victorian Order of Nurses, the Anti-Tuberculosis Society, and the Ladies Minerva Club.[38] The rites of noblesse oblige drew the elite into extensive philanthropic and charitable activities. In 1970, over half of a sample of Shaughnessy households belonged to charitable associations, mainly local hospitals.[39] In Montreal, too, the elites gave generously of time and money to such neighbourhood institutions as the Royal Victoria Hospital, McGill University, and its affiliated women's institute, Royal Victoria College. McGill was a favoured beneficiary, receiving at least three gifts of over $10 million from rich Square Mile philanthropists.[40]

Leisure activities similarly showed overlapping networks around a set of sports and cultural associations. Patronage of the arts was a favoured activity, and municipal galleries and museums benefited from the donations of private collections from their executive members. Private sporting and recreational clubs repeated the social bias of other elite institutions and provided yet another way-station on the social round of prominent citizens, as they shuffled between, for example, membership of the prestigious Royal Vancouver Yacht Club and executive positions in the major business clubs and the mayor's office (Figure 11.3). "My, aren't we a bunch of joiners," observed an upper-middle-class informant from Forest Hill, and her assessment has been borne out by every study of the institutionally complete world of the urban elite.[41] The families of Shaughnessy women in 1979 commonly belonged to at least one, and often two or three, of the Vancouver Club, the Vancouver Lawn Tennis Club, and the Junior League. These clubs provided the nexus of social life: 75 per cent of the women observed that their friends were also fellow club members.[42]

The selective channelling of biographies through private schools and clubs among those who share a common residential space has an inevitable consequence. Intermarriage is a feature of any village society, not least in the provincial world of traditional urban elites. In Forest Hill in the 1950s "the marriages that do occur are not notably different from those that might have been arranged in a caste system based on race, creed, color, and – above all – money."[43] As one informant put it: "I know that my own daughter is going to go to private school later, not because I think it is better for her, but because it is socially necessary."[44] And so the social geography of an older elite district reveals a remarkable degree of intermarriage. In Montreal's Square Mile the examples were legion: the industrialist John Redpath

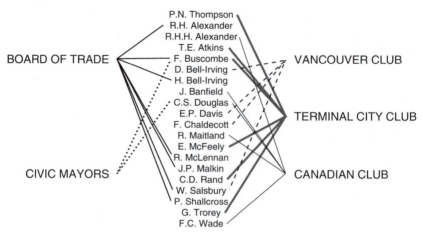

Figure 11.3
Institutional linkages among Vancouver's elite: membership of the Royal Vancouver Yacht Club, business clubs, and the mayor's office, 1906. *Source*: Robertson (1977).

enjoyed the statistical improbability of being both father-in-law and brother-in-law to George Drummond, later president of the Bank of Montreal; Sir Hugh Allan, shipping magnate, industrialist, and financier, and his brother married two Smith sisters; Lord Mount Stephen, the first CPR railway baron, married into the Meighen family, the Van Hornes into the Molsons (the families lived across Sherbrooke Street from each other), and so it went on.[45] One is reminded of the power of place in James Beshers's argument concerning the convergence of society and space.[46] Emergence of segregated social areas is the result of initial social distancing between groups, and in an ongoing iteration, space then consolidates the social structure by limiting contact with out-groups. Selective friendship and marriage partners are regulated by the friction of distance and reinforced by a complete set of social institutions. The common conspiracy of society and space serves to sustain a dynasty and its landscape of privilege.

THE LONGEVITY OF ELITE DISTRICTS

And yet this conspiracy may be broken. A century after the river baron John Torrance built his forty-two–room mansion amid acres of fields and gardens in Montreal's Square Mile, the house had become a tenement,[47] a fate that befell many of the fine mansions of the nineteenth-century elite before the Depression quickly extinguished its remaining stability and pretensions. In Vancouver's West End the cycle was much faster; by 1908, apartments were invading the district, and by 1914, one-third of the 1908 elite had departed, barely twenty-five years after the West End had made its spectacular appearance. In the absence of zoning, there were no institutional controls to protect the district, and its stability stood or fell according to the individualistic ethos of its landholders. The ideology of individual gain, commercial expansion, and urban boosterism was perceived to be as much a part of nature as the waves that broke (far more gently) against English Bay. The neighbourhood was more dispensable than the ideology, and by the time a zoning by-law was introduced in the 1920s, two-thirds of the affected owners signed a petition rejecting the city's proposal to preserve part of the West End as a landscape of single-family dwellings.[48]

This land-use progression followed precisely the filtering cycle prescribed by Homer Hoyt in 1939 from his review of the evolution of high-rent districts in 142 American cities.[49] The laws of nature were, of course, in reality the rather more historically contingent laws of the market, and, in a later paper, Hoyt acknowledged that his sectoral model of the evolution of urban social areas was based entirely on urban land economics and included no consideration of actions for a broader public interest (such as zoning), nor the non-market values of households that might wish to preserve a district for essentially symbolic reasons.[50] These ideologies are neither geographically nor historically constant; just as the pressures moulding urban neighbourhoods have become increasingly diverse since the inter-war period, so too their impact is geographically variable, with the market mechanism and rapid change more constrained in some places than in others.

Indeed, countries such as Canada and Australia do seem to offer a rather different urban model to that of the United States. In Melbourne, a detailed study of elite addresses over the period 1913–62 showed considerable stability; rather remarkably, the same square mile of the inner city was the single most favoured address throughout the half century.[51] Around this secure core, a secondary sector trailed some kilometres to the south along the coastline. Systematic studies of Canadian cities point to a similar evolution. Up to 1914, elite districts were commonly displaced with little resistance, as commercial and industrial growth revalued residential land. But the displacement was usually incremental, so that the Square Mile slowly edged west to Westmount, and in Toronto continuity has been even greater. Maps showing the highest-status residential districts in the 1890s bear "remarkable similarities" with the map of 1860; at the same time, looking forward, "the consistency in the patterning of these regions between 1899 and 1950 or 1960 is striking."[52] By 1899, South Rosedale was already a substantial upper-class community; however, if in 1899 it was on the advancing edge of upper-class penetration, by the 1960s it was the front line of defence of high-status north Toronto before the advancing edge of the central city and its pressures for commercial and apartment redevelopment. Nonetheless, the line did hold, buttressed by favourable zoning, itself confirmed in a ring of districts by the Central Area Plan of 1976, which arrested commercial and high-density residential pressures from impinging further into residential areas. Consequently, a comparison of Toronto with Buffalo or Cleveland shows a far stronger elite presence in Toronto's inner city and much less downfiltering, with its associated outward movement of wealthy households to newer properties.[53]

The Toronto pattern is repeated in smaller metropolitan areas. In Calgary the inner district of Mount Royal, developed by the CPR as an exclusive neighbourhood prior to 1920, maintained its elite associations sixty years later, though by then its northern portion had been pinched by apartment redevelopment at the expanding southern margin of the downtown area.[54] In Edmonton the geography of elite addresses changed very little between 1937 and 1972: "Much the same area has been occupied by Edmonton's high status citizenry for the last thirty-five years."[55] Though somewhat eroded on its downtown margins, the same core area was clearly identifiable and, like Melbourne, had developed a secondary sector toward the edge of the city. In light of the rapid growth of Edmonton over the same period, the elite is now (as in Melbourne) far from the suburban vanguard. Elsewhere one finds similar patterns. The configuration for the present location of high-status neighbourhoods on the Halifax peninsula existed by 1890, with subsequent slow diffusion over space, including in-fill of nineteenth-century country estates by new development (Figure 11.4).[56] At the other end of the nation, the elite district of The Uplands in Victoria, where development began in 1912, has shown sufficient staying power to encourage thoughts of its "immortality."[57]

In short, the evolution of high-status neighbourhoods in Canada bears many affinities with the Melbourne model. Since 1920, elite districts have expanded but have rarely been displaced; their outward expansion has created a high-status sector which by and large has been leap-frogged by more modest development, so that it no longer represents the advancing edge of the built-up area. The locational stability of elite districts is remarkable, considering the growth and redevelopment pressures exerted within Canadian cities. The Toronto metropolitan area, for example, grew six-fold from 1921 to 1981, with all of the pressures for downtown expansion and redevelopment that this implies, yet the innermost tracts of Rosedale experienced remarkably little diminution of their status. How has this stability been accomplished?

Two factors identified in Melbourne have helped to preserve elite districts in Canadian cities. First, unlike the unprotected landscape of the Square Mile or the West End, high-status subdivisions laid out early in this century incorporated careful land-use controls. In Westmount and, from 1924 to 1967, in Forest Hill, municipal autonomy added political muscle to land-use protection and the preservation of a high-status landscape. Shaughnessy, thwarted in its efforts at incorporation as a separate municipality in the early

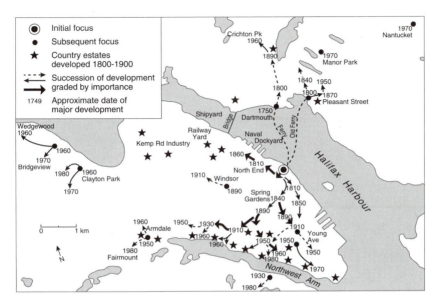

Figure 11.4
The diffusion of elite neighbourhoods in Halifax, 1749–1981. *Source*: Millward (1981).

1910s, nonetheless, in a political compromise, secured protective zoning through a special act of the provincial legislature; later, The Uplands in Victoria obtained a similar privilege which again, through meticulous land-use controls, institutionalized its high status. In other cities, including Halifax, Calgary, and Winnipeg, elite subdivisions laid out before 1920 were protected by restrictive covenants; in Calgary's Mount Royal, for example, the terms included prohibition of commercial land uses, minimum lot sizes and house prices, and single-family dwellings as the only housing type.[58]

Second, these restrictions were enforced by vigilant citizens' groups. The first ratepayers' association in Toronto was formed in Rosedale at the turn of the century, and the Shaughnessy Heights Property Owners' Association in Vancouver has had an active political existence for over fifty years.[59] These groups have effectively negotiated protection of their neighbourhoods, and, if anything, over the past twenty years their lobbying has extended protective by-laws and enhanced social exclusion. A new by-law for The Uplands in 1967 provided sweeping powers of development control, including design approval for every new building.[60] The 1981 Shaughnessy Plan, carefully shepherded through the public-planning process by the Property Owners' Association, sets as its first objective "to preserve and enhance the architectural, landscape, and historic elements of First Shaughnessy"; a second objective is to discourage demolition and promote restoration of older houses, and a third is to retain the predominantly single-family nature of the district.[61]

In Melbourne, elite status was also sustained by continuous replenishment, not only of new households but also of new units. The estate-like quality of such districts does permit modest in-fill without diminishing their status. The Shaughnessy Plan, for example, allows some in-fill on large lots of over 23,000 square feet, "provided that the buildings are subordinate and complementary to the existing principal building." So, too, managed conversion of large, old houses is permitted with careful zoning controls, to a maximum of four units per house, provided that the average unit size exceeds 1,800 square feet.[62] The first such conversions in the mid-1980s produced units priced at over $400,000.

OLD ELITES AND NEW GENTRY

In his discussion of the evolution of high-rent districts, Hoyt described a process of ageing, obsolescence, conversion, and downfiltering, which would pass wave-like through prestigious districts, beginning on their downtown margins, as wealth, in turn, moved out toward the edge of the city. In some Canadian inner cities over the past fifteen years precisely the opposite process has been set in motion.

Between 1970 and 1980 average household incomes rose by more than 20 per cent in real terms in 41 census tracts in the city of Toronto and in only 17 tracts in the rest of the metropolitan area; declines of real income occurred in 32 tracts in the city and 50 tracts outside it.[63] These trends have been sustained and may well have been accentuated since 1980, as the inflation of central-city housing costs has far exceeded that in the suburbs. Indeed, rather than conversion and downfiltering, deconversion and upfiltering are a major force in Toronto's inner-city housing market. Between 1976 and 1985 a net loss of over 10,000 housing units occurred in the city through deconversion.[64] As most of these units were offering rooms and apartments at affordable rents, their loss has accentuated a crisis of affordable housing, acute at present in Toronto and Vancouver and endemic in most major cities since 1970.

A principal cause of deconversion is gentrification, the movement of wealthier households into districts formerly occupied by less affluent residents.[65] Gentrification invokes a dramatic reversal of Hoyt's thinking, but one in which existing elite areas remain centrally implicated at both the macro- and the micro-scales. First, an increase in social status is tightly bound to inner cities in metropolitan areas such as Toronto or Ottawa which already exhibit higher concentrations of elite residence (Table 11.1). Second, at the scale of the census tract, the best predictor of rising social status in the 1970s was a tract's proximity to an existing elite area in 1971. In Toronto, six of the seven sites of earliest renovation and middle-class settlement in the 1960s wrapped around the downtown fringes of upper-middle-class Rosedale (Figure 11.5) and the professional households of the

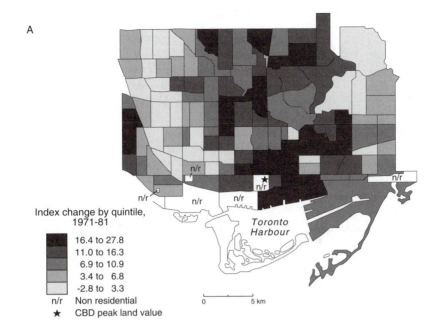

A

Index change by quintile,
1971-81

- 16.4 to 27.8
- 11.0 to 16.3
- 6.9 to 10.9
- 3.4 to 6.8
- -2.8 to 3.3
- n/r   Non residential
- ★   CBD peak land value

Toronto
Harbour

0   5 km

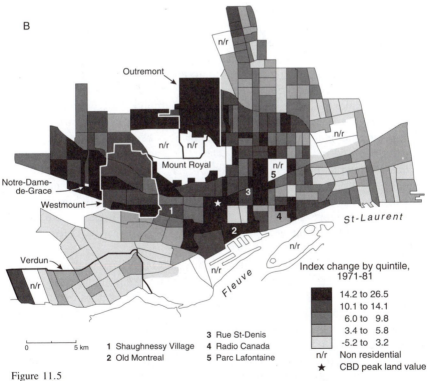

B

Outremont

Mount Royal

Notre-Dame-
de-Grace

Westmount

Verdun

St-Laurent

Fleuve

Index change by quintile,
1971-81

- 14.2 to 26.5
- 10.1 to 14.1
- 6.0 to 9.8
- 3.4 to 5.8
- -5.2 to 3.2
- n/r   Non residential
- ★   CBD peak land value

0   5 km

1 Shaughnessy Village
2 Old Montreal
3 Rue St-Denis
4 Radio Canada
5 Parc Lafontaine

Figure 11.5
Patterns of social upgrading 1971–81 in (A) Toronto and (B) Montreal. *Source*: Ley (1988).

Annex. In Montreal, gentrification has advanced outward from the old elite districts around Mount Royal, notably Westmount and Outremont (Figure 11.5); in Shaughnessy Village, recolonization by the middle class is occurring in the best-preserved corner of the old Square Mile, while along the St-Denis zone, francophone professionals are repossessing the 1880–1920 residences of the French elite.[66]

Gentrification is most likely to have begun in a census tract on the downtown side of an elite district, which in 1971 retained some social status, where rents and dwelling values were still appreciable, but where there was fading grandeur. Incomes were mixed, and households were small, incorporating large numbers of single people, including students and the elderly; rooming-houses and some apartments were present. These districts, in the "zone of discard" of the retreating elite edge or in lower-middle-class areas near an elite district, were precisely the districts where further conversion and subsequent apartment redevelopment should have occurred, according to models of neighbourhood change.[67] But a new valuation of these districts was made by a young professional cohort, attracted by the opportunities of living downtown, close to work and cultural events, attracted also by environmental amenities such as parks and the waterfront, by the distinctive architecture and leafy streets of an older neighbourhood, and, for early arrivals, by social diversity and affordable prices.[68] Like the old elite, members of this group were well educated and politically articulate, and their effect on city politics was appreciable. They were frequently among the leaders of the reform councils in Toronto and Vancouver in the 1970s; in Montreal in the 1980s, and in other cities as well, including Ottawa, Halifax, and Edmonton, they pressed for neighbourhood preservation and residential downzoning in the inner city. In newly created programs of neighbourhood planning, their land-use agendas were, in varying degrees, institutionalized into planning and zoning by-laws. A new inner-city landscape emerged in these neighbourhoods, representing an ideology of preservation and renovation, rather than the 1960s model of redevelopment and intensification. Where redevelopment did occur it was usually low rise and sensitive to context, unlike the massive 1960s high-rise redevelopment of Vancouver's West End or Toronto's St James Town that had preceded it (Figure 11.6).

But behind the apparent stability of preserved landscapes, substantial social changes were taking place. Some 116,000 new residents with high-status "quaternary" occupations were added to the inner cities of Toronto, Montreal, Vancouver, Ottawa, Edmonton, and Halifax between 1971 and 1986; simultaneously, over 215,000 residents with less prestigious occupational status moved out. Despite a national recession, between 1981 and 1986 the pace of middle-class settlement quickened.[69] Though far from complete, the embourgeoisement of parts of the inner city is under way

Figure 11.6
St James Town from Don Vale; contrasting landscapes of inner-city change in Toronto.
Alpha Avenue (foreground) was one of the first streets in Don Vale to gentrify, and many
dwellings are now protected by heritage designation.

along a range of dimensions. In certain districts, local retailing has been
transformed. Along Fourth Avenue, in Vancouver's Kitsilano neighbourhood, fewer than one-quarter of the shops existing in 1966 survived at the
same location in 1976.[70] The street has been "making a comeback as a recreation and shopping area for the upwardly mobile condominium set.
Trendy restaurants, clothes shops and other specialty stores are springing up
like mushrooms after a spring rain."[71] The same could be said of Yorkville,
Parliament Street, or Queen Street West in Toronto, St-Denis in Montreal,
or Spring Garden Road in Halifax. The institutional life of neighbourhoods
is also undergoing transition; school enrolments are sharply down (by 40 per
cent in Kitsilano, 1967–75), and church attendance has plummeted. Social
interaction is far more likely to revolve around the pub or the exercise class
than the church or the school hall.[72]

Deindustrialization of the inner city has led to a marked decline of blue-
collar jobs as manufacturing firms have either shut down or relocated to the
suburbs. However, aside from their own professional and managerial occupations in central city, the new middle class in the inner city are creating
new market niches. By the mid-1980s, home renovation was a bigger business in Ontario than new housing construction, and additional employment

opportunities came from auxiliary home services, including landscaping, security, and various forms of furniture and equipment supply and repair. New jobs include a revival of quasi-domestic service, with the return of the nanny and home cleaners for busy professional households.

While one must be cautious not to overgeneralize, gentrification is creating new landscapes of privilege in the inner city. This was not the intention of the protective zoning of the 1970s, nor of the pioneer households, but it has frequently been the consequence. The enhanced environmental amenity of gentrifying districts which has followed local area planning (and, sometimes, federal designation as neighbourhood improvement areas) has increased market demand and inflated property prices. So, too, the social diversity prized by early gentrifiers has turned toward increasing conformity, which begins to evoke the protected exclusion practised by the old elite. The counter-cultural flair and idiosyncrasy that often characterized these neighbourhoods in the early 1970s is significantly (though not wholly) extinguished. The transition is now moving full circle, and, like the exclusive neighbourhood labels of the old elite, in gentrifying areas the mutual reinforcement of landscape and identity is revealed in a new form of "residential credentialism."[73] In the built environment, this credentialism includes promotion (sometimes creation) of neighbourhood names (for example, the invented titles of Toronto's Cabbagetown and Montreal's Shaughnessy Village), an approved tasteful style for home renovation, and, in redeveloped units, a sophisticated (commonly post-modern) design vocabulary.[74] In their eclectic use of classical elements and studied picturesqueness, some streams of post-modernism bear more than a passing similarity to the eclectic picturesque styles of the Victorian elite.

CONCLUSION:
ELITE NEIGHBOURHOODS
IN SOCIETAL CONTEXT

A broader discussion of privileged neighbourhoods than is possible here would relate more fully the residential landscapes of the elite to the workplaces in which their wealth and power are generated. Clearly these phases of life are not demarcated as distinctly as is sometimes suggested. The homes of the Van Hornes or the Shaughnessys were not simply a haven for serene consumption. We can be sure that when Elihu Root was a guest at Lord Strathcona's home, more than a little about railway development in the American Midwest was discussed over port and cigars; the gentlemen's club represented a yet fuller synthesis of domestic culture and political economy. In gentrifying neighbourhoods today, the computer-facilitated office in the childless professional home defies too categorical a distinction between home and work.[75]

Elites derive their power and wealth from their control of scarce resources, whether material or human. This empowerment in turn is expressed in their occupancy of residential space and their manipulation of it to advertise and to reproduce positions of privilege. The wealth of the old elite in Canada was predicated upon commodities, land, transportation, and the trading functions associated with conveyance of staple products to world markets. Those sources of wealth, of course, remain, and to them have been added the human resources, skills, and specialized knowledge of senior members of the so-called new middle class.

There is, as sociologists have shown, considerable stability to a national pattern of social stratification, but the strongholds of a national elite are not impenetrable. For much of Canada's history, the presence of British icons, Victorian eclectic or Tudor revival, in elite residential landscapes pointed to the stability of a powerful Anglo-Canadian stratum. The infiltration of the ideas of Frank Lloyd Wright and the International Style from the 1930s on was a harbinger of the loosening of that structure and its cultural symbols. In the late 1980s, in the aspiring "world cities" of Toronto and Vancouver, in particular, we see a new departure. The entry of the Pacific Rim as a major player in the global economy has been accompanied by a flood of Asian capital and wealthy immigrants, especially from Hong Kong, into elite districts.[76] A new cultural presence announces this new economic stature. The landscape symbolism of the old elite is not the vocabulary of the parvenus, and demolition of old mansions and their replacement by new expensive structures of a style as yet unnamed declares a new layering of privilege, a new synthesis of identity and landscape, of society and space.

CHAPTER TWELVE

From Periphery to Centre: The Changing Geography of the Suburbs

L.J. EVENDEN AND G.E. WALKER

By definition, the suburb is a thing apart, a human habitat wholly dependent on the city's prior and adjacent existence. It represents the most important element of growth in modern expanding cities, being distinguished in the landscape by its location at the urban periphery and re-creating there in modified ways the intricate forms of the city's built environment. It provides the dominant setting for the rearing of families, the learning of property and political relations, and, through life-cycle stages, the constant recasting of roles in an urbanizing society and polity. It has also been described as a phase in the circulation of capital.[1] To the extent that the suburb is in time drawn into the greater mass of the urban area, its physical form and socio-economic and political structures contribute to the character of the emergent city. Whatever the emphasis, the city itself is incompletely studied without consideration of suburban development.

The attention accorded specifically to suburbs in Canada has arisen largely from the practical need to provide sufficient housing for a changing and growing urban population and from the motivation to study new and emerging land and urban development processes. Further, a considerable body of literature on the rural-urban fringe has accumulated, spurred originally by concerns for the welfare of agriculture in the face of urban expansion and for the planning difficulties occasioned by urban sprawl. But collections of houses alone do not constitute suburbs, although their designs, settings, and prices will condition suburban character; development processes are not specifically suburban issues, although they may occasion distinctive forms of debate where suburbs are concerned; and the rural-urban fringe is conceptually distinct from the suburban landscape. But the suburb, distinguished by its housing, development, and landscapes, is also more. It includes community facilities of all sorts, such as places of employment, education, religion, recreation, and health care. Distinctive attitudes and activity patterns also appear, and an identity in the political sphere might also take form, at any scale from community or neighbourhood to incorporated local or regional district. In this way a suburb becomes a place, at a scale suitable for its individual identity to be realized and for social identity to be comprehended within the larger urban context.

GROWTH AND CHANGE

By the end of the Second World War, more than half the Canadian population lived in urban centres, and by 1971 this proportion had swelled to over 75 per cent, including a major component living in metropolitan areas. During this quarter-century the outer edges of all major cities experienced astonishing rates of subdivision and development, much of it taking the form of low-density "sprawl." Urban sprawl refers to spatially unco-ordinated processes of development, an anarchic cultural landscape made possible by the increasingly widespread ownership and use of automobiles. For planning agencies, to wrestle with the problems of sprawl was also to confront the growth of urban and suburban areas. Gradually, through development of institutional controls, a degree of spatial coherence was imposed on the landscapes of the urban fringes.

During the 1960s, expansion of urban "core" areas by census boundary redefinition reallocated much of the fringe populations to the cores. This acknowledged by implication the consolidation of older suburban communities into the main urban areas. In certain cases, notably in the prairies, consolidation of fragmented local governments into single cities further obscured the differential growth of suburban and central areas. But the general pattern, by which non-central area expansion outpaced that of the central area, is clear (Table 12.1). Causes of growth included natural increase; immigration from overseas, especially to central-city areas; intra-urban mobility, more and more at the expense of central cities, and rural or small-town migration to the fringes of the cities. These were among the first causes to be noted, and they remain key variables.[2] But their relative strengths, interactions, and locational dynamics are constantly in flux.

The Roots of Our Understanding

Although suburban development and life have emerged as dominant experiences in Canada, they have not been reported comprehensively in the geographical literature. In considering the entire urban system of the country, however, Bourne found that despite inner-city redevelopment and new trends in suburban growth, such as localized patterns of ethnic change, social contrasts between inner city and suburban areas remained, in 1981, consistent with earlier experience.[3] Such nation-wide consistency in the social patterning of cities is impressive, but local variation is notable. Hence our

Table 12.1
Population of CMAs, central cities and non-central cities, selected years

Place	1951 CMA (000s)	1951 central city (000s)	1951 Non-central city (000s)	1971 CMA (000s)	1971 central city (000s)	1971 Non-central city (000s)	1981 CMA (000s)	1981 central city (000s)	1981 Non-central city (000s)	1986 CMA (000s)	1986 central city (000s)	1986 Non-central city (000s)
Calgary	142	129	13	403	403	0	593	593	0	671	636	35
Chicoutimi–Jonquière	91	45*	46	134	62*	72	135	120*	15	159	120*	39
Edmonton	194	160	34	496	438	58	657	532	125	786	574	212
Halifax	138	86	52	223	122	101	278	115	163	296	114	182
Hamilton	282	208	74	499	309	190	542	306	236	557	307	250
Kitchener	108	45	63	227	149*	78	288	189*	99	311	210*	101
London	168	95	73	286	223	63	284	254	30	342	304	38
Montreal	1,539	1,022	517	2,743	1,214	1,529	2,828	980	1,848	2,921	1,015	1,906
Ottawa–Hull	312	246*	66	603	366*	237	718	351*	367	819	360*	459
Quebec	289	164	125	481	186	295	576	167	409	603	165	438
Regina	73	71	2	141	140	1	164	163	1	187	175	12
St Catharines–Niagara	189	61*	128	303	177*	126	304	195*	109	343	196*	147
Saint John	81	51	30	107	89	18	114	81	33	121	76	45
St John's	81	53	28	132	88	44	155	84	71	162	76	86
Saskatoon	56	53	3	127	127	0	154	154	0	201	178	23
Sudbury	81	42	39	155	91	64	150	90	60	149	89	60
Thunder Bay	74	66*	8	112	108	4	121	113	8	122	112	10
Toronto	1,262	676	586	2,628	713	1,915	2,999	599	2,400	3,427	612	2,815
Vancouver	586	345	241	1,082	426	656	1,268	414	854	1,381	431	950
Victoria	115	51	64	196	62	134	234	64	170	256	66	190

Windsor	183	120	63	259	203	56	246	192	54	254	193	61
Winnipeg	357	236	121	540	246	294	585	565	20	625	595	30
Total	6,401	4,025	2,376	11,877	5,942	5,935	13,393	6,321	7,072	14,693	6,604	8,089
Non-central/central		0.59			1.00			1.12			1.23	
% of CMA		63	37		50	50		47	53		45	55

Sources: Ross (1984); Statistics Canada, Census Publications.

* Combined total of two central-city populations.

Note: Annexations, amalgamations, and changing methods of CMA and CA (or "greater city") delineations since 1941 mean that populations are not easily compared between censuses. Data given for 1951 are based on 1971 boundaries, as reported in Ross (1984). Data for 1981 and 1986 are based on then current definitions. Thus only 1951 and 1971 figures are directly comparable in all cases. Places are included if they were CMAs by 1971.

understanding must be pursued from several perspectives, some of which are discussed below.

Issues of development and design have been emphasized by planners, while publications such as *City Magazine* have, with a sense of urgency, analysed large-scale construction.[4] Early post-war sociological work included Lacoste's comprehensive study of greater Montreal[5] and detailed studies in and around Toronto. *Crestwood Heights*, which had considerable impact in the mid-1950s, summarized a major project based on concern for the mental health of children and young people in the schools of Forest Hill Village, an upper-middle-class community.[6] But it presented an extensive analysis of the family and of relations between men and women in a long-established inner suburban setting during a time of economic prosperity. This understanding underlies the more recent emphasis on gender roles in inquiries set in today's larger suburbs. In his introduction to the book, David Reisman compared "Crestwood Heights" to Park Forest, Ill., and, in prescient anticipation, remarked that "there is almost nothing in the book which strikes me as peculiarly Canadian," although he does note somewhat dated and "provincial" behaviour.[7] Some three decades were to elapse before a major counter-argument, from urban geography, challenged this sense of continental urban character.[8]

If the study of the Anglo-Gentile-Jewish society of Forest Hill did not strike the Harvard sociologist as "peculiarly Canadian," S.D. Clark's empirical inquiries, set within Canada, could be considered nothing else.[9] Clark's studies yielded insights not only into the growth patterns of suburbs, especially those in the process of initial establishment and set at a considerable remove from the city, but also into their changing ethnic character. He reported that on the fringes of towns in northern Ontario, as well as in adjacent areas of Quebec, a poverty-stricken and predominantly francophone population of rural background inhabited the loosely coalesced suburbs. In contrast, poor suburbs north of Toronto initially were populated almost exclusively by long-established Canadians of British ancestry.[10] The origins of the latter within Canada were diverse, some having come from the city itself in an attempt to escape the pressures of urban living and expensive housing. But others had come from small towns and the countryside, some expelled by agriculture's reduced need for labour, others unable to scratch a living from marginal lands. The migration of people from the prairies to British Columbia was also notable. Some migrants settled in the countryside, within about thirty miles of Vancouver, in areas which, in the 1950s, were to become dominated by urban sprawl. These areas have now evolved into high- and medium-density suburbs. Thus, although suburbs contained dislocated urban populations invading a disintegrating countryside,[11] rural people were also approaching the city without immediately becoming part of it. Suburbs became the staging grounds for the initiation of rural migrants into urban

ways. In this, the experiences of anglophone and francophone Canadians
were similar.

But the suburbanizing experience did more than bring together two types
of migrants. Ethnic and class relations were also implicated. Some suburbs,
as reported for Toronto, became receiving areas of British immigrants.[12] Because such people did not have to learn a new language, and because in this
century most of the British immigrants were already skilled in urban ways,
they could and did enter suburbs directly. For this reason, and because the
census has been inconsistent in how it records the origins of populations,
their ethnicity has perhaps been overlooked.[13] Thus the typical suburb was
more "Anglo-Saxon" than the inner city during the years of rapid urban expansion following the end of the Second World War. But the contrasting anglophone and francophone suburban developments in and around Montreal
point to the complexity of the overall pattern. In fact, ethnic diversity has
increasingly characterized suburban growth across the country.[14]

For some people, the move to the suburbs was a move up the social scale;
others stagnated socially whether they moved to the suburbs or remained
fixed.[15] Foggin and Polèse concluded that in greater Montreal "Déménager
hors du centre-ville ne signifie pas automatiquement s'intégrer à la classe
moyenne."[16] Indeed, by 1980 large numbers of "réfugiés" – low-income
blue-collar workers, often living in mobile homes on poor land – were reported in the exurbanite populations in the townships south and east of Montreal, centred on Sherbrooke.[17] This situation paralleled Clark's earlier
findings that migrants to outlying suburban or sprawl settlements often faced
bleak prospects. Lack of education and training, limited employment prospects, victimization suffered by some in property relations in ill-developed
subdivisions, remoteness from the city and its opportunities, and isolation
from family and kin networks all came together to prompt the planner's
reaction that "Hell is a Suburb."[18]

While suburban developments have thus displayed a wide spectrum of social conditions and are not associated exclusively with an emergent middle
class, geographical study and planning have emphasized forms of landscape
development as an aspect of the study of social characteristics.[19] This approach, directly relevant to the question of poor living conditions, focused
attention on the forms of emerging settlement on the urban fringe. The
Lower Mainland Regional Planning Board of British Columbia, set up in
1949, was among the leaders in conceptualizing the nature and problems of
urban sprawl.[20] Partly in response to the impact of sprawl, local jurisdictional authority was gradually transformed throughout the country.[21] Metropolitan or regional forms of government were created, urban annexations
approved, and zoning powers strengthened in an attempt to cope with the apparent disorder of urban fringe developments. These initiatives have resulted
in tidier suburban landscapes, in which subdivisions tend to be clustered,

thereby promoting the formation and maintenance of communities. Some have been quite experimental, as in the attempt to create a French-Canadian form of the garden city[22] or in various co-operative attempts.[23] And it has been common for new suburbs to blend with pre-existing rural communities or small towns. For example, districts of post-war single-family housing, punctuated on the horizon by apartment blocks, now extend northward from Quebec to surround Charlesbourg, one of the oldest settlements in the country and distinguished to this day by the radial plan at its core.[24] In this way the most recent forms of urban expansion reach back in time to link with earlier forms of settlement, providing a certain continuity in the Canadian experience while demonstrating different phases of historical development.

Planned Suburban Environments

Public regulation has been complemented in the private sphere by the increasing prominence of large developers.[25] Such national and multinational companies have the necessary organizational and technical skills and can capitalize developments at such scale that they can, and do, create new suburbs all of a piece. They can construct the suburb and target the mix of populations, employment, and local facilities to create a planned, integrated whole.[26] In such a scenario, community development itself is not a matter of change or voluntary congregation.

The case of Don Mills is widely taken to be precedent-setting. The Don Mills Development Corp. was formed in 1952 by E.P. Taylor to build a new town on the northeastern edges of Toronto, in the township of North York. Planning principles emphasized design controls over density, a focus on neighbourhoods (four were constructed), green spaces, and an overall main focus in the town centre. A variety of housing to meet the needs of a planned mixture of population and employment was to be provided locally to promote self-containment. Traffic was to be partly controlled by a hierarchy of collector roads, and pedestrian walkways were separated from streets carrying vehicular traffic. Lots were wider and larger than was common in cities, allowing for more private green space and creating lower densities. Construction was managed through contracts to builders, and commercial land by leasing arrangements. But mixed housing and local jobs did not materialize to the degree envisaged, and this large-scale corporate suburb became a dormitory community of above-average housing. It did, however, set a precedent: " The corporate plan superseded the planners' plan,"[27] and elements of the plan, in process and landscape outcome, became common. "Today, the plan is so ordinary that it elicits almost no response. But in 1952, it was breaking new ground."[28]

Although corporate or developer suburbs were by no means new to Canada[29] the claim for the special significance of Don Mills is not without foundation. Its sheer scale, and the involvement of one of the country's leading financiers, link it to issues of the flow of capital. In addition, its importance to the planning profession, and its location in Toronto, made it the leading example of new ways to cope with the suburban "explosion" through application of the planned neighbourhood unit. But Don Mills was not alone. The typical residential subdivision built in Canada after 1945 appeared to be based on the neighbourhood model.[30] In Alberta, its adoption was perhaps facilitated by rapid suburban growth, extensive municipal ownership of land, integration of municipal government by annexation, and planning advocacy.[31] Calgary's Fairview neighbourhood, for example, was developed at the same time as Don Mills, and its location in the Macleod Trail sector, extending south from the city, indicates its accessibility to the central city and proximity to the transitional ribbon developments of an earlier era (Figure 12.1).[32] It has been argued that the neighbourhood unit, which Fairview exemplifies, represents a stage in the evolution of suburban developments in Calgary, from grid to modified grid to neighbourhood unit and "sector," or town centre.[33] Perhaps these stages have application elsewhere, as shown in the modified grid neighbourhood in the suburban municipality of Laval in metropolitan Montreal.[34]

If Don Mills presented the neighbourhood unit of single-family housing as an instrument of suburban growth, the adjacent Flemingdon Park offered mixed housing and designed open spaces: an innovative, integrated cluster of apartment blocks, garden apartments, and row houses, along with community facilities. Based on British and Scandinavian precedents, but said to derive from Le Corbusier,[35] the unique combination of available land and suitable road connections made this project possible; it also showed that rental accommodation and access to central-city employment were compatible with planned suburban development.[36] These stood in contrast to Don Mills's pattern of homeownership and unrealistic intentions to provide substantial local employment. Given the sweep and energy of greater Toronto's growth, Flemingdon Park's success, like that of Don Mills, was never in question. And it, too, "would be imitated later in many cities."[37] It also presaged a time when, with families becoming smaller, alternative forms of housing would be needed. Flemingdon Park thus represented a major experiment in suburban design, even as Habitat 67 was, a decade later, in urban living.

Manufacturing relocation, retail reorganization and expansion, and more comprehensive forms of local government were pointing to a new future for Canadian suburbs. Major new initiatives in shopping centres were being undertaken, such as Park Royal and Oakridge in Vancouver, or on the Jockey Club grounds in Hamilton, to name but a few. Reorientation of consumer provision in both inner and outer suburbs pointed to planned "town centres." Theories of suburban centrality envisaged town centres based on com-

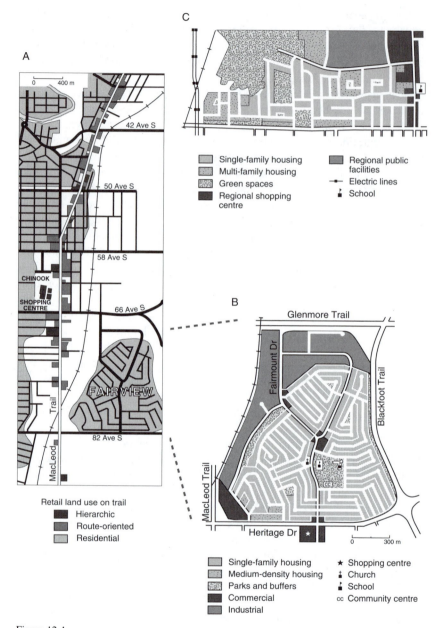

A

0   400 m

42 Ave S

50 Ave S

58 Ave S

CHINOOK

SHOPPING
CENTRE

66 Ave S

FAIRVIEW

82 Ave S

Trail

MacLeod

Retail land use on trail

- Hierarchic
- Route-oriented
- Residential

C

- Single-family housing
- Multi-family housing
- Green spaces
- Regional shopping centre
- Regional public facilities
- Electric lines
- School

B

Glenmore Trail

Fairmount Dr

Blackfoot Trail

MacLeod Trail

cc

Heritage Dr ★

0   300 m

- Single-family housing
- Medium-density housing
- Parks and buffers
- Commercial
- Industrial
- ★ Shopping centre
- ‡ Church
- School
- cc Community centre

Figure 12.1

(A) Suburban expansion in Calgary showing sectoral growth; (B) placement of a new, planned neighbourhood unit, Fairview; (C) a neighbourhood unit with a modified grid in Laval, suburban Montreal. *Sources*: A and B: Harasym and Smith (1978); C: Divay and Gaudreau (1984).

mercial cores in metropolitan areas.[38] Carver's *Cities in the Suburbs* (1962) complemented these ideas with concern for the "social completeness" of such centres.[39] Public planning agencies were to provide a broad framework for large-scale urban or metropolitan development, based on a spatial hierarchy of centres within the metropolitan area. Details would reflect local conditions and involve both private and public participation. Transportation planning would be crucial. Suburbs thus became both integral to the functioning of the whole unit and more important in their own right.

Perhaps one of the clearest applications of this approach occurred in the Livable Region plans of the Greater Vancouver Regional District, first formulated in the mid-1970s, with much public participation, and revised in the late 1980s. Elements of these plans are gradually being implemented, most notably with construction of Metrotown as a "new downtown" in Burnaby. Such developments were parallelled across the country by new hierarchies of corporate outlets, such as those created by the principal department store chains and chartered banks.[40] Indeed, as shown by construction of the suburban West Edmonton Mall, the corporate presence can now affect dramatically the nodal structure of the whole metropolitan area.

TOWARD AN UNDERSTANDING OF
THE MODERN SUBURB

*Households, Individuals, and
Community Settings*

If much of our understanding of suburbs has been a by-product of their inclusion in metropolis-wide studies, analyses of the larger urban complex will not fully reveal their individual characters. Case studies from the points of view of households and individuals provide alternative perspectives.

Rapid household formation has increased the demand for housing and encouraged both redevelopment of established residential areas to higher building densities and creation of new suburbs. The forms of housing have also changed in response to metamorphoses in household structure and shrinking household size. Two-person households, for example, increased at about 4 per cent per annum between 1951 and 1976, whereas all households increased at 3 per cent and the general population at only 2 per cent.[41] This rapidly increasing group deserves attention for the complexity of its social arrangements and effect on housing needs, although husband-wife combinations remain dominant and tend to have the highest incomes. At the other end of the income profile, parent-child, two-person households usually comprise a single mother and child. They affect housing in inner cities perhaps more than that in the suburbs. But tensions concerning "illegal suites" in single-family houses in the suburbs of Vancouver and Toronto highlight problems of income and housing affordability; such "informal" transformation of the built environment is a response to rapid formation of small, limited-income households which need small dwelling spaces.[42] For all this, the size of dwelling occupied by younger two-person households, and by young families, especially those with both partners working, tended to increase during the 1970s and early 1980s, a period of real income growth.[43]

The dual-earner household is defined by labour-force participation rather than by household size, although it coincides to a degree with the two-person household. Dual-earner families in Montreal, for example, have been differentiated in a general way as "home-oriented" and inclined to live in the suburbs or "consumption-oriented" and gravitating to central areas.[44] The occupational status of the wife is the key to this distinction. Home-oriented suburban wives, who also work outside the home for wages, tend to hold jobs with less prestige, responsibility, authority, and pay than do their husbands. Wives in consumption-oriented marriages tend to have equal or nearly equal occupational status, and the couples tend to live closer to the city centre where their jobs are located. For suburban households, the need for a second income helps to account for the fact that wives are working for wages,[45] and a large proportion of such women work in local suburban areas near their homes, as demonstrated in Scarborough.[46] Recent evidence suggests, however, that suburban women tend to advance in status when they hold centrally located jobs[47] and that women's incomes tend to increase with commuting distance.[48] Such findings may reflect higher levels of education and training, life-cycle stages when women feel that they can leave the house all day, a tendency for professional couples to move to certain suburbs in order to invest in a principal residence, and ownership of or access to a car.

If the status of women in marriages is indeed changing, and if this is related in turn to further locational differentiation of daily household activities, many other factors, in addition to earning power and income, also affect residential choice. Car ownership, the convenience of public transit, clustering patterns of suburban neighbourhood units, school busing, and central-city livability will all have an effect, as will the roles assumed by the marriage partners. In general, however, residential location seems related most strongly to the husband's pattern of commuting, while the wife's job search is constrained spatially by the residence as point of origin.[49] The working wife who is also the mother of a young family has a complex and mutable set of issues to consider, especially those involving children. The apparent sacrifice of pay, prestige, responsibility, and authority that might accompany a job may represent a conscious choice in order to devote time and attention to the raising of children.

Indeed, caring for children was the critical consideration in Coquitlam, a middle-class Vancouver suburb.[50] While the occupations represented by women in this study, apart from wife, mother, or caregiver, were quite variable, they tended to influence the way in which the commitment to children was expressed rather than modify the fact of such commitment. Thus the daily round, articulated in time and space, proceeded from the core concern of securing "safe spaces" where children not only would be protected but also could be nurtured. Provision of a variety of childcare arrangements, accessible within an appropriate distance, thus becomes important in suburbs as well as in central cities.

The mere increase in childcare opportunities, however, does not fully indicate the level of household planning required in suburban dual-earner households. For both husband and wife, the journey to work takes on a new and qualitatively different significance; the times, distances, directions, and destinations of daily travel must be co-ordinated in relation to the duration and location of childcare. Access to workplaces also takes on greater significance and cannot be expressed in an isochronal map focused on the central city, important as that is for general routeway and transit planning. Companies understand this, and many locate in the suburbs in order to take advantage of "labour pools" there. While this creates accessible employment, it also points to suburban diversification and "social completeness."

More and more dual-earner households are relocating toward the suburbs, a move related specifically to the difference in occupational class between husbands and wives.[51] Some of these households do not have children, although some may move in anticipation of raising families. As discussed later, the apparent adequacy of both publicly and privately supplied services are factors in the decision to relocate, as would be the decision to invest in a principal residence. Childcare outside the home will become a more important suburban issue and, to the extent that it becomes available, will allow larger numbers of qualified women to join those whose income opportunities increase with distance travelled. Inadequate childcare may well confine women to more locally available jobs in their own suburbs, and, even though some would still prefer to choose these, others would find their options curtailed. Thus the degree to which residential selection tends to reflect husbands' commuting may alter in future, the present evidence being that "women's work-trip lengths are becoming less a function of their marital status."[52]

Emerging Community Structures:
The Example of the Public Household

Childcare outside the home, however, points to broader and profound changes in service provision. At any moment one of many such social needs might come to the fore, but in general "forms of support and protection which adults and children traditionally secured through family households, kinship networks and belief structure are increasingly available within the general community."[53] Coquitlam mothers searched for these sources of assistance, which must have been factors in the moves of dual-earner families toward the suburbs of Montreal. To the extent that "support and protection in the advanced welfare and market society exist in theory for all ... who possess *public access rights*,"[54] provision of such amenities allowed emergence of the "public household."[55] This has many possible elements and forms, perhaps including both publicly and privately supplied services; but its impetus would appear to stem from the co-operative ideal.

But if the post-war suburb was based on child-oriented, largely self-sufficient private households situated within the neighbourhood unit, changing household forms, age profiles, labour-force participation, and patterns of property ownership and tenancy are redefining these characteristics. As a suburb matures, its children require and demand more opportunity and stimulation, and the simple structures of neighbourhoods and local centres become increasingly inadequate. Its growing elderly population faces decisions concerning property maintenance or disposal[56] and the need for more services.[57] Indeed, the location and size of property and the activity involved in its maintenance are directly related to the health of the suburban elderly population.[58] Further, property owners at all stages of life seek to protect their principal residence as an investment. To do this they must remain alert to local development and may involve themselves in municipal politics, often as a conservative element attempting to deflect or constrain construction. Such processes, which balance individual and collective interests, would influence the continuing shape of a public household.

Ethnicity, Class, and Housing

It is unclear, however, how such a publicly based conceptualization would account for other social dimensions such as ethnicity, class, and individualistic or even "deviant" behaviour, dimensions generally considered characteristic of inner cities rather than suburbs. But there are signs that these forms of social stratification are gradually making suburbs more diversified. Social transformation tends to occur along sectors, and, with minorities such as the Italians, the direction of movement is predominantly outward from the inner city (Figure 12.2).

But rural ethnic migration has strongly patterned suburban development. As in the Toronto of a generation ago observed by Clark, in contemporary Winnipeg the distribution of suburban ethnicity corresponds to historical rural patterns in the province.[59] And the only sizeable concentration of francophones in British Columbia, first settled as mill workers soon after the turn of the century, continues to shape the identity of suburban Coquitlam outside Vancouver.[60] More recently, non-European communities have de-

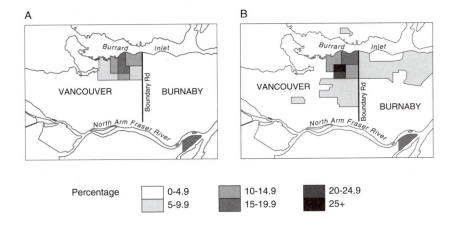

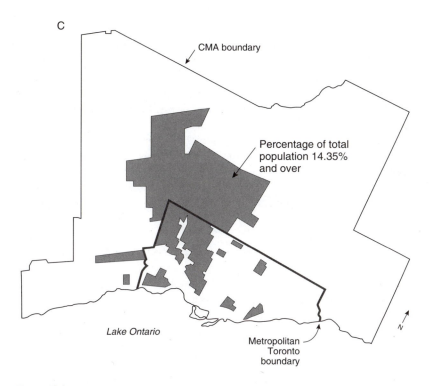

Figure 12.2
Population of Italian ethnic origin: (A) by residence, Vancouver, 1961; (B) sectoral growth, Vancouver and Burnaby, 1971. *Source*: Collett (1977). (C) Population of Italian ethnic origin: sectoral growth, Toronto CMA, 1981. *Source*: Bourne et al. (1986).

veloped as outliers in suburban areas, initially concentrating, in the case of the Chinese, around commercial establishments.[61] Sikh or Muslim communities have sought residences close to centres of employment and religious worship,[62] the latter sometimes located at highly accessible points in the road network. For Canadians of Japanese ancestry, a high degree of social and locational assimilation since the Second World War is now modified by self-conscious reassertion of identity, as exemplified in the building of a major community centre in suburban Toronto and in the resettling of Steveston, formerly a largely Japanese fishing village at the mouth of the Fraser River, but now a growing suburban district south of Vancouver.

Social class and housing conditions display further patterns of segregation.[63] The processes that underlie these patterns have been explored mainly at the scale of the urban area as a whole, with emphasis on housing more than on class.[64] Homeownership rates vary generally with class across the urban area but are usually higher in the suburbs. Landscape preferences and ethnicity have also been shown to influence housing location.[65] Affordability, however, is a crucial determinant of ownership, reflecting in the suburbs household income and its origin with one or two (or more) wage or salary earners and the trade-off between housing costs and journey to work. Suburban properties have been comparatively accessible in the market and, at least in Montreal, have produced "a veritable ownership boom."[66] Elements that condition the market accessibility of housing include availability of credit, access to privately held capital, and government assistance.[67] But access to new housing involves not only class and financial capability but also family background, intergenerational social mobility, career path, and quality of housing and location.[68] While suburban areas may thus become mutually segregated through differential affordability of housing, the benefits of ownership apparently cut across lines of class; ownership itself differentiates households along lines of political expression, if not formal political affiliation.[69]

Communities of Interest and Places of Identity

While housing affordability, residential mobility, changing gender relations, and journey to work are dynamic conditions that shape suburban communities, none refers to the community as a whole. How does a completely new suburb develop a distinctive community of interest, a focus that the inhabitants themselves will recognize? Further, what happens when a pre-existing, possibly agricultural community is complemented or even overwhelmed by suburbanization? How would the "old" community survive in the face of change, and how would the new developments take root in established patterns of behaviour, institutions, and landscapes which had grown earlier, in an entirely different socioeconomic context?

In a traditional but rapidly suburbanizing community in Surrey, east of Vancouver, the long-established population perceived the territory of their community as compactly shaped and based on the principal institutional and shopping foci, along with the major access routes. In contrast, newcomers tended to hold spatial images emphasizing the principal road alignments and had much less comprehension of the areal basis of the community.[70] These findings are consistent with a later study north of Toronto, where social network analysis showed that newcomers tended to have fragmented patterns of contact, many extending well beyond the immediate community.[71] Further, any local networks generally were initiated and maintained by women, pointing again to the importance of gender roles in suburban community formation. In contrast, the networks of long-established rural residents emphasized the immediate locality, kinship patterns, and complex linkages throughout all levels of the local community.

The pressures of development found in such suburban areas contributed to the need to reorganize local governments, as noted earlier. But the lack of fit among social linkages, behaviour, and associations of a changing population may lead to rootlessness and lack of commitment in the formation of a new community. In the raising of a new generation, deliberate creation and fostering of places of identity and belonging can be helpful. Local government, the legitimate corporate form of the community, can foster such efforts by investing in community symbols. Mississauga's dramatic new civic centre, for example, with its city hall tower rising above a prosaic suburban landscape, establishes a community focus, in line with goals set out in the official plan.[72] In time, such a focus will subsume but not necessarily destroy the numerous smaller identities already giving social texture to the municipal region. This attempt to maintain a link with the past is, however, tinged with irony and pathos; the municipality has preserved a sealed-up farm-house located in the centre of an encircling expressway interchange.[73]

The evolving identities of suburbs are reinforced (or not) by their inhabitants' perceptions of them. Perception studies based on the analysis of "mental maps" in Hamilton and Ottawa provide a glimpse of such self-images.[74] Distinctive places emerge in relation to perceived locational status, social behaviour, and job orientation. Certain suburbs of Hamilton, located in close mutual proximity, are "worlds apart" socially, with the separation being etched into the landscape by boundaries – heavily trafficked routeways that have come to serve as social divides. Conversely, more distant suburbs may be socially quite close, with much mutual interaction.

In the prairies, where large-scale metropolitan growth is well developed but locationally dispersed, perceptions of satisfaction with quality of life, as compared between Winnipeg and Edmonton, have been found to correspond in part with overall urban growth rates and associated conditions.[75] A comparatively narrow social focus is common in faster-growing Edmonton, reflecting recent suburban experience, with its emphasis on family life. Satisfaction with life in slower-growing Winnipeg is related more to "neighbourhood and friendship," although family life in general is perceived to be equally satisfactory in both places. The finding that the older, slower-growth city exhibits wider and more complex social networks is consistent with the results of the studies set north of Toronto and east of Vancouver, already noted, where social networks of long-established residents are locally more complex than are those of newcomers.

The study of the social geography of Canadian cities has generally ignored the complex and dynamic issues of community development concerning aboriginal populations. While Native Indians and Métis tend to locate in inner cities, especially in the west, their control of reserve lands in growing suburban areas gives them a major role in the development of selected areas of major cities. The Kahnawake on the South Shore of the St Lawrence River opposite Montreal, the Sarcee on Calgary's southwest side, the Musqueam at the mouth of the Fraser, and the Squamish Band's reserves on Vancouver's north shore are high-profile examples.[76] As these communities gain in confidence as well as numbers of people, pressures of socioeconomic development, compatibility of land uses with neighbouring areas, land claims, and crowding on the reserves will affect surrounding areas. But, with their territorial bases of identity, these communities are profoundly segregated, even though they may exist in close proximity to the wider suburban society.

PROSPECTS

In this interpretation of the suburbs we have emphasized the perspective of the household, the individual, and the community, but complementary studies are needed to provide a larger picture. If more manufacturing, office, and shopping locations decentralize, for example, how do these trends blend with other patterns of suburban development and community formation? Suburbs are apparently diversifying across a broad range of social, economic, and political characteristics, but to what extent and by what processes are they becoming mutually differentiated in space? Further, how do interrelated cultural, ethnic, gender, and class identities cause these patterns to appear and evolve? Given the present emphasis on the household or individual, perhaps the conceptualization of the public household has merit. It is holistic, points to formation of a self-conscious community of interest, and requires a geographical resource base that must in part be fashioned by the community itself. But there are complications. If different groups of people share space but express cultures that are incongruent in varying degree, or, conversely, if cultural traditions are held in common by groups inhabiting geographically fragmented territories, then application of the concept in these cases would be moot. Alternatively, the concept, if extended, might apply at a higher level, incorporating more than one public household in a plurality, and perhaps representing interactions among more than one such collectivity.[77]

It is generally assumed that suburbs are absorbed over time into the city. While this may be true in the evolving forms of the townscape or in the occasional imposition of a hierarchical rationalization of local government, it is also clear that local identity is an elemental force in creating new places on the edges of the city. Whether fostered by planned public and private development, as illustrated by Mississauga and Don Mills, or simply emergent, as studies of Toronto and Vancouver demonstrate, if local identities were to multiply at rates corresponding to newly developing suburban areas, we might ask whether the suburbs would constitute the geographical centre of society. In contemporary cities the "central city" would thus be "reduced" to a specialized neighbourhood, as already recognized in such characterizations of it as the commercial core, the executive city, the tourist enclave, or the place of public spectacle. Inner-city redevelopment notwithstanding, if "where the people are" is increasingly in the suburbs, then the social geography of Canadian cities will increasingly be that of the suburbs. And that will be a geography of many centres and much variety. "Des banlieues ... pas si plates!"[78]

CHAPTER THIRTEEN

The Social Geography of Small Towns

J . C . E V E R I T T A N D A . M . G I L L

Although life in Canadian small towns is a recurring theme in literature,[1] sometimes presented favourably and sometimes less auspiciously,[2] the social geography of small towns has been neglected,[3] even though small towns and villages are "the 'basic building' blocks of our settlement system models."[4] Moreover, many smaller settlements throughout Canada are evolving as they become more integrated into the urban system of their surrounding regions, which in turn reduces their independence.[5] Small towns, however, still play an important and discrete role in Canada, for although they vary widely in function, structure, and location, the characteristics that they share make them distinct from larger cities.[6] Their general neglect cannot be easily understood,[7] although the diversity of subject-matter (particularly when looked at nationally) and a bias in research toward larger urban areas may be partly responsible. Further, there is a shortage of readily available data sources,[8] and the analytical tools are often inapppropriate, designed as they were for studying larger urban units.

In this chapter we explore the changing role and social character of small towns with particular reference to recent research by geographers. In addition we shall show why, despite the real and perceived problems associated with living outside large urban centres, many smaller settlements continue to grow. We shall concentrate mainly on small towns in the prairies (Figure 13.1), as well as small resource communities (usually located in the north).[9]

WHAT IS A SMALL TOWN?

As Williams indicates,[10] "small town residents and advocates make much of the difference between urban and rural life." Although size is important, and economic functions must be allowed for, this distinctiveness lies in a poorly defined element of the social environment, which several authors have acknowledged as central to the persistence of Canadian small towns.[11] Yet

Figure 13.1
Aerial view of Killarney, Manitoba, 1988. *Photo*: courtesy Manitoba Pool.

these places still share many attributes with big cities,[12] and in some ways
"the traditional intellectual division between rural and urban is rapidly losing its meaning."[13] One definitional solution has been simply to look at population size in the settlement hierarchy. Thus many urban studies have, as a
matter of convenience if not philosophical conviction, begun with cities of
100,000 people or more, where data sets are more complete, whereas research on small towns has generally been limited to settlements of 10,000 or
fewer. For places in between these two thresholds an identity along the
rural-urban continuum remains ambiguous, although people living in settlements of up to 40,000 may often regard themselves as small-town dwellers,[14] and settlements of 15,000, though sometimes officially designated
cities, function in essence as small towns.[15] Towns in the resource sector are
defined variously, not by size but by function.[16]

Another definitional difficulty occurs because these settlements are not
closed systems but integral parts of their rural regions.[17] Delineation of these
hinterlands is also not clear.[18] Small towns have become more complexly integrated with their surrounding regions; counter-urbanization tendencies
noted in the 1970s encouraged some population dispersion in rural areas.[19]
This demographic resurgence does, however, vary regionally, reflecting at
least in part the differential development of urban systems. Although the
rural resurgence may have begun in many Canadian small towns in the early
1970s, it has had greater effect in southern Ontario than elsewhere. As

Dahms points out, to "fully comprehend the development of cities and towns we must also appreciate their places and functions in systems of settlement that have evolved over time and space."[20] Coffey and Polèse have suggested that "central" areas within 100 km of a "large centre" have been affected more than "peripheral" areas beyond. Indeed, Ontario-based research suggests that some small settlements may, depending on their distance from large urban centres, have become part of a "dispersed city." In this new spatial format, small groups of settlements function in the local region as economic entities largely independent of any core urban centre, aside from access to the highest-order functions. Residents in many rural communities, such as Wroxeter in Ontario, are today less dependent on the agricultural economy of the area and may, in fact, shop, work, and live in spatially separate settlements.[21] Thus the transportation revolution that undid numerous small towns has now led to revival of many smaller urban places which are interconnected by their inhabitants' choices and activities.[22] Although this development is most pronounced in Ontario, Meredith, working in Saskatchewan, has posited a "Prairie community system" in which communities are systems incorporating several trade centres (small towns, villages, and hamlets) and the open country around them.[23] Such systems may, however, lead to the decline of some communities, as residents search for the best possible price and variety of goods and services within the entire region. This is particularly likely if critical functional units, such as schools and post offices, are closed down and consolidated into larger urban places.[24]

We do not attempt to settle on one quantifiable delineation of "small town" but rather define smaller settlements in a behavioural context, as places where most areas of the town are known, at least to some degree, by most of the residents. We exclude places where patterns of social interaction have more in common with larger communities, where there is more opportunity for formation of distinct neighbourhoods.

FUNCTION AND VITALITY OF SMALL TOWNS

The functional classification of cities provides a convenient framework for revealing the variety in urban forms and processes, and the same argument holds true for smaller settlements. Small communities, however, may be more unstable in their position within a functional classification than are large cities, with for instance former farming towns changing to become, perhaps, minor centres of industry (such as Taber, Alberta), dormitory towns (such as Abbotsford, British Columbia), or alternatively even ghost towns (such as Bannerman, Manitoba) over a relatively short period.[25]

Although smaller settlements may be multifunctional, a single function often dominates and may be used in classification. A simple, but commonly

Table 13.1
Functional classification of small towns

Type	Examples
Location- and amenity-dependent	
Regional service centres	Ville Marie, Que.
	Dauphin, Man.
	Rosetown, Sask.
Dormitory towns	Verchères, Que.
	Carleton Place, Ont.
	Airdrie, Alta
	Coaldale, Alta
	Abbotsford, BC
Specialized manufacturing	Bridgetown, NS
	Taber, Alta
	Brooks, Alta
Near-urban recreation	Bracebridge, Ont.
	Collingwood, Ont.
	Grand Beach, Man.
Retirement centres	Hamiota, Man.
	Kelowna, BC
Specialty retailing, life-style, and culture	Acton, Ont.
	Elora, Ont.
	Ganges, BC
Resource-dependent	
Non-renewable	Thompson, Man.
	Tumbler Ridge, BC
Renewable – fishing	St George's, Nfld
	Lunenburg, NS
Renewable – forestry	Cornerbrook, Nfld
	Hinton, Alta
Renewable – tourism	Jasper, Alta
	Penticton, BC
"Ghost" towns	Bannerman, Man.
	Moore Park, Man.
	Sheridon, Man.
	Bevan, BC

recognized classification divides small towns into two major groupings (Table 13.1). First, some settlements exist because of their amenity and relative location, functioning perhaps as central places, small manufacturing centres within range of larger markets, commuter dormitories near urban recreation destinations, and retirement centres. Second, resource-based

Table 13.2A
Population growth in small incorporated prairie communities, 1976–86

Time period	Small-community* population change (%)			Total provincial population change (%)			National population change (%)
	Alta	Sask.	Man.	Alta	Sask.	Man.	
1976–81	+35.3	+6.6	+3.5	+21.7	+5.1	+0.5	+5.9
1981–86	+3.8	+2.8	+1.9	+5.7	+4.3	+3.6	+4.0

* Up to 10,000 people.

Table 13.2B
Population growth in small incorporated prairie communities, by settlement size, 1976–86

Province and years	0–499	500–999	1,000–1,499	1,500–2,999	3,000–10,000
Alberta					
1976–81	−9.1	−8.0	+3.5	+21.0	+39.8
1981–86	+4.3	+2.0	−16.7	−16.3	+10.8
Saskatchewan					
1976–81	−6.7	+5.0	+28.2	+2.2	+41.9
1981–86	−0.5	+1.8	−2.3	+3.0	+18.7
Manitoba					
1976–81	+8.3	−0.9	−12.5	+8.4	+3.4
1981–86	−18.6	−0.3	+13.6	−8.8	+1.9

Sources: Census of Canada, 1976, 1981, and 1986.

communities serve both renewable (forestry, fishing, tourism) and non-renewable (mining) industries and are especially vulnerable to factors that affect this single function.[26] Function and location are also linked. Whereas most service centres are located in agriculturally productive areas, resource towns are frequently isolated and situated beyond the main ecumene.

However, there are regional and sub-regional variations in terms of growth.[27] Settlements in the prairies, for instance, may be divided according to population size and/or the number of central-place functions offered, and some authors have suggested population thresholds above or below which growth or decline can be predicted.[28] However, as Table 13.2 indicates, the situation is not this clear.[29] From 1976 to 1981 (Table 13.2A), small-town growth led provincial population growth, whereas from 1981 to 1986 it lagged behind provincial figures. In addition, growth patterns of different community groupings were inconsistent (Table 13.2B), with categories experiencing increases and decreases in no obvious fashion. For instance, from 1976 to 1981 settlements of 1,000–1,500 people grew in Alberta and Saskatchewan but decreased in Manitoba. Conversely, from 1981 to 1986 this same category increased in size in Manitoba but declined in the other two prairie provinces.

Recent years have seen steady growth of the retirement population and its preference for small-town life. This factor is likely to increase in importance – though with fluctuations – as the population aged 65 and over grows from 10.7 per cent in 1986 to a figure close to 20 per cent in 2031.[30] Consider, for example, the village of Hamiota, Manitoba, with a population in 1956 of 690, of which 19.4 per cent were 65 years old and over; by 1986 the population had risen to 816, with 39.2 per cent of the population over 64 years old.[31] Several factors explain the migration of retirees to this village from the surrounding countryside: development of facilities (including health services) to serve them, proximity to family and friends, and the relative affordability of housing. Other settlements in Manitoba show similar results, with communities of under 5,000 people having over twice as many residents 65 years old and over as urban areas.[32] Nevertheless the population is also "greying" in regional centres. Brandon, Manitoba, with a 1986 population of 38,708, has 14.1 per cent of its residents over 64 years old, which is above both the national (10.7 per cent) and Manitoban (12.6 per cent) averages. The city is clearly functioning in part as a retirement centre, much the same as the surrounding small towns and villages, and the cohort of people over 64 years old is increasing both absolutely and proportionately.[33]

Extension of the "urban fields" of large cities has encompassed previously more "rural" locations,[34] and as a result the specific location of an individual's residence is no longer as critical as it used to be. Consequently the vitality of small towns can no longer be explained simply by the number of central-place functions; such communities may have declined as central places but not as places of residence. Economic revitalization is not occurring, however, in all smaller settlements. In the prairie provinces removal of rail lines and closing of grain elevators, post offices, and rural stores have been called "the death knell of rural towns."[35] In addition, mechanization of agriculture has led to a loss of rural population, which has further undermined many services, and this "vicious circle can only be broken if the small-town urban economy can be weaned away from its dependence on rural servicing,"[36] perhaps into potential growth areas such as manufacturing and/or tourism. Such centres as Banff, Alberta, Whistler, British Columbia, and Peggy's Cove, Nova Scotia, have nationally or internationally renowned natural or cultural attractions which are enhanced by well-developed infrastructures to serve tourists. Other communities have developed tourism through their locations near large urban centres – for example, the many communities in the "cottage country" north of Toronto, where towns such as Orillia, Bracebridge, and Huntsville serve the region's dispersed tourism activities. Recreational opportunity and accessibility also account for the rapid growth of, for example, Labelle, Quebec, situated north of Montreal.[37] Special themes or attractions account for the success of other tourist towns. Thus the Shaw Festival which attracts theatre patrons to Niagara-on-the-Lake has stimulated growth. Drumheller, Alberta, has also developed a sizeable tourism industry with opening of the Tyrell Museum of Paleontology, which in 1988 attracted 600,000 visitors,[38] and Chemainus, British Columbia, a dying pulp-mill town which revived its economy by painting murals on public buildings, attracted 350,000 visitors in 1988.

Problems of decline also haunt resource towns which often depend on resources that are controlled by multinational companies and global markets. "Boom-bust" cycles are often a fact of life, and while some communities may go through several such phases, others never recover.[39] Relatively little attention has been paid to the winding down or closure of resource communities, and such eventualities are rarely planned for.[40] Bradbury and St Martin did, however, document the decline of Schefferville, Quebec, where the iron ore mine and concentrator closed in 1982. This followed a five-year period of winding down in which the population declined from around 3,500 to 250, accompanied by restructuring of the work-force, desinvestment and relocation of capital, and the company's withdrawal from housing, municipal affairs, and public services.[41] Although such communities often seek to diversify, their options are usually limited. Chemainus and Kimberly in British Columbia are examples of resource towns that did find new economic growth (in tourism), but these are unusual, if informative, exceptions to the norm.

SPATIAL FORM AND HOUSING

Despite regional differences in the spatial form of Canadian small towns, these settlements – be they the Acadian street villages of New Brunswick,[42] the small towns of Saskatchewan,[43] or the resource towns of the north[44] – all have some variation on a "Main Street." Although Main Street may not be "dead" in Canada, as Hart claims is the case for the US Midwest,[45] it has often been dramatically transformed.[46] In the Maritimes, for example, "rigid colonial grid-iron" plats reflect former, distinctly British outposts, which contrast dramatically with the "seemingly organic and strikingly vibrant street villages" built by the Acadians in the same area.[47] Similar street villages dominate rural Quebec, defining characteristic French long-lot land settlement. The dispersed settlement patterns of the Irish in eastern Canada reflected traditional forms in Ireland,[48] whereas most prairie settlers did not have the same range of options. "Neighbourly, village-like arrangements of farmhouses were not ... characteristic of Western settlement" because the section, township, and range survey system was not designed to allow for it.[49] Exceptions, such as the Mennonites and Doukhobors, and later the Hutterites, included relatively few people and only a small number of settlements.[50]

For the most part, rigid geometrical principles held sway in the west, and the settlements were built on a grid plan.[51] Although some did evolve into a more irregular and distinctive form, the fear articulated by Rupert Brooke – that prairie towns might never mature and acquire "that something different, something more worth having"[52] – commonly proved to be the case.[53] In Ontario and the Eastern Townships of Quebec as well, grid-like patterns were often imposed and small settlements grew to serve the surrounding agricultural areas.[54]

Although the spatial form of resource towns built early was more haphazard, those constructed since 1945 have usually been comprehensively planned, reflecting the thought of the day.[55] Commonly, they have a nucleus of commercial and institutional uses in the town centre, with high-density, multiple-family housing nearby. The majority of housing stock, often put up by the resource company, is single family, organized into neighbourhood units, and includes mobile home developments. Most newer resource towns are located some distance from the industrial activity on which they depend. The resulting lack of industrial land use distinguishes them from larger Canadian cities.[56] They have often been characterized as "wilderness suburbs" – transplanted southern communities in isolated northern settings.

During the 1970s environmental sensitivity led to innovative designs considered more fitting to the northern environment, with Fermont, Quebec, and Leaf Rapids, Manitoba, being the best-known examples.[57] Planning and design have centred around "quality of life," with the overall objective of reducing labour turnover. One of the most recent and widely publicized mining towns in Canada, Tumbler Ridge, on British Columbia's northeast coalfield, reflects these changes, including increased input into experimental design and planning from environmental psychologists and sociologists.[58] The physical design, which was guided by a set of social principles, reflects more established small towns in southern Canada; the town centre is based on a "Main Street" theme, with winding streets, canopied sidewalks, and, at the focal point in downtown, a post-modern town hall (Figure 13.2). Such features are intended to give the residents a sense of history, identity, and permanence, although other problems have still not been resolved.[59]

In rural settlements "almost all households, rich and poor, are homeowners," and this percentage has increased in recent years; indeed, the presence of affordable housing is an important attraction.[60] Despite lower-quality housing, development has been "vigorous and impressive" over the past two decades.[61] Recently implemented government programs have improved infrastructure features.[62] The "Main Street" initiative of the Heritage Canada

Figure 13.2
Main Street, Tumbler Ridge, British Columbia. *Photo*: A. Gill.

Foundation, which recognizes that a smaller settlement has a geographical "heart and soul," has revived a number of Main Streets across the country as "meeting grounds" and "social spaces." It has thereby helped to "preserve the living past" of many settlements.[63] Such endeavours have been particularly effective in the smallest places (under 500 people) with weak and unstable economies, especially those in isolated, often northern, locations.[64]

Housing can be used as "a potent lever capable of uplifting alienated individuals and stirring stagnant communities."[65] Housing in small towns is considered as shelter, but usually not as an investment, because there is no conventional pattern of supply and demand.[66] Consequently, the only chance for community revitalization comes from government programs or smaller, more personal solutions, such as the building of "self-help" houses. In resource communities, where housing markets do not function "normally" because of uncertain economic conditions, there is considerable debate over whether the company or the homeowner should assume the risk of lost equity. Historically, company involvement has evolved from rental of company-constructed housing to use of financial subsidies (such as interest-free second mortgages) to assist employees to purchase homes on the open market,[67] and many companies have had to introduce "buy-back" policies to encourage house purchase.[68] Lack of choice is often a problem in newer communities where there is no existing older and cheaper housing stock, and new residents often have to assume the risk of purchasing a new house. A persistent problem has been access to housing for residents in the service sector, who are usually ineligible to purchase company-constructed accommodation. Housing in these communities also differs from that in most other small towns in its social connotation. As Bradbury suggested, "where ownership of the means of production is vested in an external class, little opportunity exists for social mobility through the ownership of local housing property."[69] Indeed, the normal class distinction that exists in many communities between owners and tenants is overridden by work-related status, and in some situations management personnel, especially those in more mobile positions, may rent rather than own their houses.

DEMOGRAPHIC STRUCTURE OF SMALL TOWNS

The demography of small towns reflects their functional diversity. Thus the "population pyramid" of a prairie agricultural centre differs from that of an exurban dormitory village, a northern mining town, or a resort community (Figure 13.3). Resource towns usually have generally younger households, with a high proportion of children. In some older towns, such as Flin Flon, Manitoba (established 1938), there are three generations of some families, and places such as Thompson, Manitoba, and Kitimat, British Columbia, are now exhibiting more "mature" population profiles, with increasing numbers of more elderly residents.[70]

Resort communities such as Banff, Alberta, Collingwood, Ontario, and Whistler, British Columbia, also exhibit distinctive demographic features. However, the majority of small towns across Canada that depend on tourism are in effect service centres for a tourist region, rather than major tourist destinations themselves. Part-time residents affect the social fabric but are not enumerated in the census.[71] Second-home residences may become permanent retirement homes, if community services, especially hospital facilities, are accessible.

In contrast, many small prairie settlements (such as Hamiota, Manitoba) commonly have ageing populations, although small towns within the commutersheds of large urban centre may deviate from this general rule. In Ontario many smaller centres like Wroxeter and Ripley have been dramatically changed by an influx of exurban families,[72] and Maritime age structures are often more mixed because regional variations in economic opportunities and support systems leave some areas in virtual poverty and others in relative affluence.

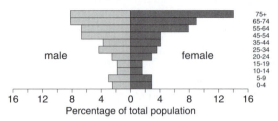

Hamiota, Man., 1986
total population: 815

male   female

75+
65-74
55-64
45-54
35-44
25-34
20-24
15-19
10-14
5-9
0-4

16 12 8 4 0 4 8 12 16
Percentage of total population

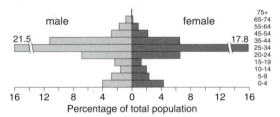

Whistler, BC, 1986
total population: 2,000

male   female

75+
65-74
55-64
45-54
35-44
25-34
20-24
15-19
10-14
5-9
0-4

21.5   17.8

16 12 8 4 0 4 8 12 16
Percentage of total population

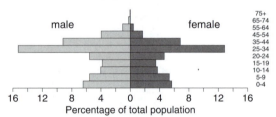

Tumbler Ridge, BC, 1986
total population: 4,390

male   female

75+
65-74
55-64
45-54
35-44
25-34
20-24
15-19
10-14
5-9
0-4

16 12 8 4 0 4 8 12 16
Percentage of total population

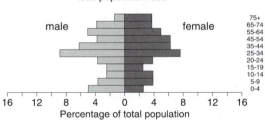

Wroxeter, Ont., 1986
total population: 395*

male   female

75+
65-74
55-64
45-54
35-44
25-34
20-24
15-19
10-14
5-9
0-4

16 12 8 4 0 4 8 12 16
Percentage of total population

*Data for the enumeration area containing the
unincorporated settlement of Wroxeter
(population 317)

Figure 13.3
Population pyramids of selected small towns, 1986

LIFE-STYLES IN
SMALL COMMUNITIES

Increasingly, studies of smaller settlements demonstrate that residents choose location primarily because of kinship, familiarity, and quality of life.[73] Unemployment and particularly underemployment are often higher than in larger urban areas,[74] and seasonal fluctuations are also likely to be wider. A study in the Central Plains region of Manitoba revealed, however, considerable within-regional variation in employment, reflecting environmental, historical, and contemporary forces at work – and demonstrating the differentiation among these entities. But even in the absence of available work in their home region, a large majority of respondents would not leave to seek jobs elsewhere – even though many small-town residents believe that their community infrastructure is deficient in a number of ways.[75] For example, Manitobans in several types of small settlements considered a wide range of community services, such as sewage systems, daycare, recreational facilities, and shopping services, to be deficient. They valued, however, the friendly character of the local communities, the "quiet small town atmosphere," the regional recreation facilities, and the wide variety of stores available in the region as a whole.[76]

The transience of a resource community differentiates it from other small towns, slowing the development of friendships and exacerbating loneliness and depression, especially among women. However, unlike more established towns, where community leadership and involvement tend to be more entrenched and newcomers find it difficult to break into the "inner circle," there are always opportunities for those who wish to become involved. The workplace is the dominant influence on social organization,[77] and status in the resource company is translated into status in the community. In older resource communities, managers and hourly staff were segregated, but in recently constructed resource towns heterogeneous housing developments attempt to integrate workers of differing status. However, as Gill has demonstrated in Tumbler Ridge, such planning seems to have little effect, and friendships are related primarily to the workplace. Workers on shift find community activities inaccessible because of the scheduling of events; in Tumbler Ridge, involvement was much higher among non-shift workers.[78]

Social problems in resource communities include insufficient employment for women – although entrepreneurial activity has recently increased among women, particularly in the creation of small businesses. The needs of children and adolescents have been acknowledged but rarely addressed.[79] Unlike smaller southern communities, where secondary education, if not available locally, is available at a larger, accessible regional centre, resource towns are often perceived to offer inadequate educational choice. The high proportion of "professional rookies" which Redekop has found in health and social services is also an issue in the educational system.[80] And older people, especially those over 55 years of age, face particular problems during periods of economic downswing; they tend to be relatively immobile because of overly specialized skills, lower educational levels than younger workers, substantial equity in their homes, and an entrenched social position in the community.[81]

THE PERSISTENCE OF SMALL COMMUNITIES

A question frequently raised with respect to small towns is "Why do they persist?[82] A more revealing issue is what Corbett has called the "perceptual paradox," that is, why so many small communities not only persist but grow, given the commonly expounded disadvantages of living in them.[83] One line of argument has observed that since 1960 basic public provisions such as police protection and health services have been universalized; this has enhanced the local economy, brought in officials and professionals, and linked peripheral areas to the mass society. An increase in local entrepreneurship has also stimulated this process.[84]

An interpretation beyond that of economic viability is clearly needed, and it seems to lie in the subjective judgment of "quality of life," which is intricately bound up with "small-town stability."[85] The quality of life is judged on such factors as environmental issues, including cleaner air and outdoor opportunities, and social dimensions, including friendliness and the slower pace of life. The "good life" now seen to be offered by many small communities has attracted new residents, many of whom choose to live there even if it means commuting elsewhere for employment. It seems clear that attempts to "foster the growth and economic development of small towns probably will flounder unless they are based upon the realization that these places have changed" their role.[86]

At the same time, not all smaller settlements are alive and well and popular as dwelling places. Undoubtedly economic viability is an important consideration.[87] In the us Great Plains, and arguably in the Canadian prairies, decline in smaller settlements has been termed the "collapse of small towns."[88] Overall, however, we argue that sometimes, and perhaps often, social factors are at least as important as economic ones to the survival of settlements. "Planning efforts directed at small towns must be every bit as cognizant of the subjective factors which constitute development as they are of the more familiar objective criteria."[89]

PART FOUR

Needs: Social Well-being and Public Policy

CHAPTER FOURTEEN

Social Planning and the Welfare State

J.T. LEMON

Through this century, while most Canadians urbanized and experienced increased prosperity, many have worked for an inclusive social welfare system to protect against threats to well-being and to widen opportunities. Largely through government auspices, they created a thick welfare net of many strands. While most of the policy decisions and the funding of social services now rest with the provincial and federal levels, municipalities and independent non-profit agencies have continued to ask social questions, to develop organizations responding to needs, and to engage in social planning.

As historians of welfare point out, prior to the twentieth century, social thought and action were imbued with the "residual" mode of thought: dealing with the poor as a last resort when the market fails, using means tests, and distinguishing between deserving and undeserving poor – the latter capable of work but morally unwilling and so subject to "least eligibility." The social sphere was restricted; private philanthropies, religious charities, and local municipalities, following the model of the Elizabethan Poor Law, primarily and parsimoniously helped those incapable of working. By the mid-twentieth century the notion of social planning had emerged, promoting the goals of coherence, comprehensiveness, and co-ordination among a wide range of social goods equitably distributed among all citizens. The concepts of a "social minimum" and even of "universality" were added to the lexicons of policy makers. In the 1950s, grand land-use planning paralleled social planning and pointed to high expectations of fulfilment.[1]

Even though the idea of universality has become widely accepted in Canada, social planning activity (or, in some quarters now, "social development") still has been concerned mainly with protecting and, it is hoped, improving the lot of lower-paid and unpaid people, suggesting still the notion of charity, or at least that some people are seen as disadvantaged or less able. Since 1975 the "residual" mode has been reasserted, universality questioned, and government financial support for those in need cut back in tandem with a declining commitment to the Keynesian style of demand management that lay behind much policy thinking between 1940 and 1975.

Under discussion here are many of the major social developments of this century. Health, education, housing, social assistance, community services, and, indeed, full employment have been on the agendas of social planners. In the context of provincial and federal action and of shifting economic conditions, I shall emphasize the role that urban municipalities and other local organizations have played. Local social planning bodies continue to undertake research and advocate policies and programs, especially community services. Since they are independent of government, they can be critical of official policies. Much of what is said has to be speculative because few social policy historians have dealt with the urban dimension. By contrast, the literature on federal-provincial relations has grown enormously. Most of the urban aspects discussed below concentrate on Toronto, which has experienced, over the past decade, the most severe growth pressures. I assume that the types and qualities of service do not vary greatly among Canadian cities. Finally, I shall deal with the issue of universal social assistance through the "long wave" and the "social limits to growth" debates. The social and environmental "spillover" effects of urban life have become more obvious as the century has worn on: cities cannot solve problems on their own, but perhaps their citizens' groups and governments can lead the way to solutions, as they sometimes have in the past.[2]

MUNICIPAL ORIGINS OF THE WELFARE STATE TO THE 1930S

The long march to our present social welfare systems was initiated largely by urban dwellers through voluntary organizations and their local governments, though rural Canadians also had a strong hand. Public responsibility for the handicapped, indigent, and unemployed began about 1600 in England through the Elizabethan Poor Law. This goal was transported to Canada formally and informally, and so municipalities, counties, and their property taxpayers were held responsible, assisted by churches and other charities. After 1850, municipalities created public services, such as free and compulsory education, public health measures, hospitals, parks, and libraries, to serve a much wider swath of the population, as the middle class came to see their value.[3]

Between 1900 and 1912, the strongest period of relative urban expansion in the country's history, action became dramatic. Municipalities were riding high: development meant higher assessments and so expanded tax revenues. Local politicians came to believe that they could handle all problems from their own property tax base. There were grounds for this expectation: expenditures by Toronto, for example, exceeded those of Ontario between the 1880s and 1913, with no direct provincial support. In 1913, municipalities collectively spent more than the federal and provincial governments on health and social welfare. Municipal ownership of utilities, or at least tight regulation of some utilities by other levels of government at the cities' insistence, became widely entrenched across the country. Citizen boards, composed mainly of business people, built and ran municipal hospitals, now available for all citizens, not only the poor. School boards built technical and commercial high schools, thus dominating these fields of education, and by 1920 schooling to age sixteen was compulsory. In 1914, Toronto's mayor, Horatio Hocken, could boast to an Ottawa audience about his city's wondrous social undertakings. Boosterism went beyond the buildings and economic base of the city: social welfare was part of the equation. Toronto's leadership in social matters was paralleled, if perhaps less vigorously, in other Canadian cities.[4]

But the bloom faded, as economic and population growth turned down sharply in 1913. In 1915, Mayor Tommy Church complained: "Toronto should not be expected to look after the unemployed of the whole province, and be made a dumping ground." By 1918 he was arguing that education was the "natural work" of the province. Grants from Queen's Park to the city for schools and hospitals were "trivial." A major social shift was thus in the making; cities were now crying for provincial and federal financial help.[5]

Other levels of government responded modestly to the calls for help. On health and social welfare, the total spent in Canada rose from $15.2 million in 1913 to $100.2 million in 1920–21. The municipal share dropped from more than half to less than one-fifth, as the federal portion jumped by twenty times and the provincial seven. In Ontario, total provincial gross spending tripled between 1900 and 1910, at the same rate as Toronto. After 1912, the province surpassed Toronto, raising its expenditure by a whopping seven times while the city increased its by only 60 per cent and its per capita spending actually fell.[6]

Thus, a transfer of responsibility, however tentative, had moved some of the burden of social welfare financing from the municipalities to other levels of government. Ottawa and some municipalities even dabbled in public housing construction at the end of the war. In the 1910s provinces introduced workers' compensation and child welfare bureaux; in 1920 they launched mothers' allowance, especially for widows. Although the municipalities had to pay half of the mothers' allowance, at least in Ontario "it marked the first time that a category of the poor was paid a permanent living allowance." This was clearly a step in a new direction.[7]

In the 1920s, as the income of many increased, welfare improved slowly, or even retreated, though the federal government began to recognize the need for the funding of unemployment relief, if only "temporarily," and in 1927 old age pensions were introduced. In 1930 an Ontario royal commission recommended a stronger provincial financial commitment to reducing the still heavy reliance of the poor and disadvantaged on municipalities and on private charities. This suggestion was surprising, given the slow pace of reform, the dominance of business principles, and the power of the voluntary philanthropy sector.[8]

Private secular and religious charities had expanded slowly after 1880. By the end of the First World War, professional social workers dominated much of the activity. Settlement houses catered to poor immigrants. In Toronto the Neighbourhood Workers' Association and the Child Welfare Council were formed. Spurred on by groups such as the Bureau of Municipal Research (founded in 1914), the first comprehensive fund-raising scheme, a precursor of the United Fund, was established. In Montreal and Ottawa some coherence was achieved through Councils of Social Agencies. But a great deal of confusion prevailed through the 1920s as various groups, public and non-profit, religious and secular, vied for power in dealing with the poor. Ward and neighbourhood ratepayer associations and home and school organizations also emerged after 1900, the former dominated by men and protective of home and property values, the latter controlled by women but limited to a marginal role in the quality of schools and excluded largely from curriculum issues. These groups rose apparently as a counterweight to the increased power of the professionals, at city hall and in schools and on school boards. Citizens' participation had taken on an organized form, especially in more affluent districts.[9]

THE DEPRESSION AND STEPS
TOWARD UNIVERSALITY, 1930–56

If the initiatives of the early twentieth century were tentative, those of the 1930s were not, as the Depression established a more lasting base for collective social action and planning. But the legislative results were slow in coming. Even though more than one-quarter of the labour force was out of work in April 1933, governments and their business allies were either befuddled or reluctant to sacrifice the notion of individual and family – and hence limited public – responsibility. Public works were tried for a while in the early 1930s but were replaced by a low level of financial relief. Dismal work camps were set up for single unemployed males, who probably suffered the deepest indignity, in being told that their condition was their own fault. Fathers felt shame in not finding jobs, and mothers endured beatings at home as a result. "Go back to the farm," was another refrain, an outdated echoing of nineteenth-century agrarian and frontier ideology. That could hardly succeed with a population of which well over half was urban.[10]

Initially the financial burden of relief fell disproportionately on the municipalities. Many places were themselves hard pressed, and some even went into receivership, including some of Toronto's poorer suburbs. Only the most affluent could provide a modicum of social generosity, such as Toronto, which borrowed heavily for relief during the early 1930s. If conditions improved in Ontario and Quebec after 1933, they still remained bleak on the prairies. The Maritime provinces had not fallen as far as the west, largely because chronic depression had set in after 1918.[11]

Late in the decade, as the pressure of unemployment receded, the provinces and the federal government assumed a larger share of expenditures. While the total social cost had doubled (in real terms, because inflation was nil) from 1930 to 1940, the municipal share dropped from about 20 to 15 per cent, after temporarily rising to 25 per cent. Pressure from a wide variety of interests brought about this change. Architects and home builders lobbied for national housing acts. Although the acts of 1935 and 1938 did little for low-income people (except for a housing renovation scheme initiated by Toronto), some construction workers got jobs. Of many reports issued during the period, the League for Social Reconstruction's *Social Planning for Canada* and the Report of the Rowell-Sirois Commission were the most conspicuous. Both called for a much stronger federal presence. Farm-support measures of the mid-1930s and unemployment insurance, instituted in 1940, were major national achievements arising from the Depression experience.[12]

Locally, in the late 1930s welfare councils and social planning bodies (separate from business-dominated funding bodies) were established, and so were social housing and planning advocacy groups. In 1939 the Toronto Welfare Council published *The Cost of Living*, a numerical analysis modelled on a similar study done by the Montreal Council of Agencies over a decade earlier. The Toronto council argued unsuccessfully for a "social minimum," a guaranteed income. Had stronger leadership been shown in Ottawa, more would have been accomplished; W.L. Mackenzie King was as reluctant to act as his predecessor, R.B. Bennett.[13]

Midway through the Second World War, Canadians began to worry that depression conditions might return after victory. This concern was reflected in the rise of the Co-operative Commonwealth Federation (CCF) as a federal electoral threat in 1942, its 1943 near-win in Ontario, and its 1944 victory in Saskatchewan. Deepening awareness of Keynesian demand management also precipitated action. Federally and provincially mandated bargaining rights for industrial workers (following the US Wagner Act of 1935) helped increase real wages and consumption. The creation of the Central (later Canada) Housing and Mortgage Corp. (CMHC) fostered housing development through low interest rates and guaranteed mortgages. The universal family allowance (the baby bonus) put more money in household budgets. Between 1942 and 1945 the federal share of all health and social spending jumped from 45 to 72 per cent, the provincial fell from 39 to 21, and the municipal from 17 to only 7 – a far cry from 1913, when municipalities provided over half. The total amount spent had increased enormously.[14]

Yet the "residual mould" was not yet broken; the social safety net remained thin and fragile. Following the Beveridge Report in Britain, Leonard Marsh's 1943 *Report on Social Security for Canada* provided a blueprint for comprehensive social programs, though many of its proposals would not be enacted for two decades. In largely rural Saskatchewan, the CCF government set the stage for eventual change through its innovative social programs. But Ontario and Quebec prevented an expansion of the welfare net throughout Canada by blocking a federal initiative in 1945 for universal health, old age pensions, and higher unemployment assistance. In fact, cities once again shouldered assistance for the able-bodied unemployed.[15]

At the local level, voluntary and municipal activity was notable through the 1940s, as energy flowed out of the war effort. Local branches of the Community Planning Association, growing from earlier housing groups, sought to combine social and land-use objectives in planning; in response, provinces enacted strengthened planning acts, requiring official plans for municipalities and attention to social and economic matters. Community planners also pushed for metropolitan governments, an idea mooted earlier. Beginning with Toronto in 1953, such governments, of varying types, brought greater tax equity in the urbanizing regions of Winnipeg, Montreal, Ottawa, Hamilton, and other cities. Neighbourhood planning made its appearance, and, closely related, social workers and researchers fostered the creation of tracts for the 1951 census, hoping through scientific analysis to provide data to bolster arguments for the elimination of poverty and slum conditions. When Toronto voters approved the Regent Park project in 1946, Canada embarked on a modest public housing program, more than a decade after the United States and well behind Europe. Although Toronto politicians and activists had to struggle for provincial and federal support, and the design was less than hoped for, the first phase of Regent Park was a socially advanced notion; people were displaced with minimum disruption. Later urban renewal schemes were to be less sensitively pursued, as the goal became less housing for families of working people and more for welfare recipients. By the mid-1960s urban renewal would be indicted nearly as much as the "federal bulldozers" in the United States that destroyed more housing units than were built.[16]

TOWARD UNIVERSALITY AND COHERENT SOCIAL PLANNING, 1956–75

Over the two decades up to 1975, Keynesian demand management combined with a political urge for greater social security led to extensive development, to some a "flowering," of the service state. Spending by all levels of government rose as a proportion of gross national product from 26 per cent to 36 per cent, much of it for social purposes. Of this, the provinces' share of expenditures rose from one-quarter to one-third, through transfers from Ottawa and higher provincial tax revenues. Municipalities ended up where they started, at 19 per cent, though until the late 1960s (as since 1945), their share virtually equalled that of the provinces. But by 1972 the provincial portion in spending for health and welfare had jumped to two-fifths from one-quarter, whereas the municipalities had fallen from 5 per cent in 1956 to only 2 per cent.[17]

The Federal-Provincial Tax Agreement of 1956 created a negotiated system of transfer payments (and, as of 1962, taxes collected by Ottawa on behalf of the provinces), taking into account as never before disparities among provinces. Begun as temporary expedients in the 1920s, conditional grants became a permanent method of revenue sharing – at least until 1977. The provinces increasingly transferred funds to municipalities, hospitals, and other boards, thus reducing the burden of property taxes. Beyond that were direct payments, mainly federal, to individuals. Besides, expenditures on public infrastructure contributed to a "highly equalizing" system, according to a British observer, when compared to those of other federal states. Provincial power, including Quebec's through its "quiet revolution," was crucial in altering the much more obvious centralized power established during the Second World War. Yet because the grants to the provinces remained conditional on federal measures of efficiency and standards of delivery, Ottawa remained dominant.[18]

Widespread policy initiatives were also pursued. The unemployment insurance system was revamped in 1956 and 1971. Regional economic programs, with social implications, were begun and then expanded with high expectations. Spending on education jumped enormously for a decade after 1959, including 50-per-cent federal contributions to provincial funding of higher education. The relief system was improved with the Canada Assistance Plan of 1966, Ottawa providing one-half, the provinces 30 per cent and the municipalities, at least in Ontario and Nova Scotia, 20 per cent. In the same year, federal pensions for those over sixty-five were dramatically increased and, like the family allowance, made universal. A Guaranteed Income Supplement helped the low-income elderly, and Canada and Quebec pension plans guaranteed more security for most retired people.[19]

Medical insurance on a nearly universal basis was introduced in 1968. Pioneered by Saskatchewan and initially limited to hospital care, medicare was rapidly put in place across the country. In contrast to Australia where universality was not achieved, provincial power was a major factor in its introduction in Canada: once it was established in one province, pressure from the public in others was too strong for their governments to resist. While the federal government was directly responsible for many of the cheques sent to individuals, provincial spending on education and medicare is what mainly drove expenditures to new heights, something like a factor of ten in nominal terms over two decades. Certainly, even though critics pointed out that equity had not been achieved, either regionally or individually, it seemed that Benthamite "less eligibility" was receding and the "residual mould" was breaking apart.[20]

If the municipalities' monetary share for social services fell, their activity hardly did, nor did that of voluntary groups. In Ontario, municipalities (or later counties and some regional governments) continued to administer relief, by then called "welfare." Some municipalities, such as London, took to the task only reluctantly. Major demographic shifts occurred with the end of the baby boom, including rising numbers of elderly to be cared for, and a rapidly increasing number of single mothers with dependent children. The latter were becoming the most needy group within the social welfare net, amounting by 1975 to one-third of those assisted.[21]

Municipalities and provinces also had to respond to a rising tide of community activism in urban matters. These events had considerable relevance for social policy. Led largely by central-area professionals in Vancouver, Winnipeg, and Toronto especially, resistance to American-style urban development reached a peak in the early 1970s. Canadian central cities differed sharply from many older American cities where poor blacks came to dominate numerically, if not yet politically. While Americans had torn up many of their inner residential areas with expressways, Canadian activists stopped them in several cases. Similarly, high-rise private and public apartments came under fire. Unlike the situation in most US cities, many private but few large public housing projects had been built in Canada's central cities. From the mid-1950s on in Metro Toronto, many of these public projects had been located on new land in the suburbs but became near-ghettoes of the poor nonetheless. The reformers promoted social housing which mixed designs of medium density and was occupied by people with varied incomes within diverse rental, freehold, and co-operative tenures.

These events thus had a strong social planning incentive, that of improving the quality of residential life in the central city. Residents' groups, which expressed concern more over the public environment than over simply protecting property rights, replaced or modified ratepayers' associations, previously defined by ownership. Tenants, increasing in numbers in the central city, could participate fully in neighbourhood life. Besides, and more important to them, many tenants in large buildings organized themselves into associations.[22]

Social planning, as a legitimate activity for volunteers and non-profit and non-governmental professionals, effectively emerged in the mid-1950s. As in the United States, the Community Chest was re-created, as the United Community Fund, to be more comprehenive, drawing on contributions from businesses, workers, and households for voluntary agencies. Unlike in the United States, the Social Planning Council of Metropolitan Toronto (SPC) was separated from the funding body, following the path of the Toronto Welfare Council. Social planning, promoted by the left in the 1930s, was now accepted by the business community – as long at *its* members kept control. The strong presence of business people in the SPC slackened in the late 1960s, possibly because by 1968 Canada had achieved the major stated social policy objectives and "Metro" had taken over welfare administration from local municipalities. After some controversy, the council involved itself more in community participation and dropped its role of deciding who should get United Way funds.[23]

By 1957, researchers and advocates had refined their tools and with access to much sharper data – such as for census tracts – published five times as many reports as earlier. The SPC commented on a wide range of social issues in Metro Toronto. Community services for young people and the elderly and family support for the disadvantaged continued to loom large. Daycare appeared in 1964. Through the 1960s, help for immigrants, still settling largely in the central city, received a lot of attention, as did urban renewal and public housing. The cost of living was addressed from time to time, though not consistently through budget guidelines until the late 1960s, when inflation was increasing dramatically. Employment issues were not central, though in the early years of the council school dropouts were a perennial concern. For a number of years, area bodies within Metro Toronto were maintained, as were a few direct central services such as an information centre and a volunteer bureau. By the mid-1970s, the SPC had become much more clearly a research, public-education, and advocacy body, rather than a direct service agency, but it still sought to build coalitions with emerging groups.[24]

CUTBACKS WEAKEN SOCIAL POLICIES AND PLANNING, FROM 1975

About 1975, cutbacks, privatization, and, if less blatantly than in, say, 1875, blaming the victim emerged from the shadows. Various reports on government productivity concluded that the voluntary sector should do more, government less, in social action. In 1975 Maxwell Henderson's report recommended that income support in Ontario be cut back and that incentives to work be built in. Soon afterward, the notion of "workfare," based on the hoary concept of "least eligibility," was promoted by the Ontario Economic Council. By then social supports had begun to decline. In the early 1980s government ministers were openly espousing and advocating such ideas, and through the decade calls for restraint became more strident, though hardly to the same degree as in Reagan's United States or Thatcher's

Britain, largely because Canada is less centralized than either and less divided by race or class. The ideology of free enterprise was, nonetheless, translated into increasing disparities of income.[25]

Besides, intergovernmental finance has become, it seems, a matter of passing the buck. Under Established Programme Financing of 1977, the federal government no longer gives 50 per cent for health and higher education, and so responsibility has fallen more on provincial shoulders. More recently, the Canada Assistance Plan has been weakened. In turn, the provinces have passed on the costs to municipalities and so to property taxes. For several years before 1991 Ontario cut its grants for Metro Toronto. This process had gone on for over a decade, despite cries from Metro Toronto that the financial partnership was "crumbling." The taxation system has become more regressive, with higher sales taxes, user fees, lower corporation taxes, and reduced marginal rates on the affluent.[26]

The long-term income trend of Canadians in poverty and those in affluence hardly differs from that in 1951, after rising and then falling. In 1989, 15 per cent of Canadians lived below Statistics Canada's poverty line, while the Canadian Council of Social Development says that a quarter lived in poverty. Depending on their household status and region, poor Canadians had to get by in 1989 on incomes that were 25 to 80 per cent below the official poverty line. Without government transfers, plus modest tax abatements, the lowest quintile of the population would be getting much less, indeed hardly more than a third of the poverty-line threshold. Between 1973 and 1986 the poverty gap between employment income and total income widened by 50 per cent! Thus dependence on transfers has become deeper and deeper for those with low incomes. Food banks and increased homelessness for some, as Dear and Wolch argue in chapter 16, suggest that the safety net is strained, and not working as well as it once did. Canada is well behind northern European countries in providing market employment, social support, and a tax system that takes a much larger share of the incomes of the affluent.[27]

Who the poor are has changed markedly. First, the working poor, especially those who live alone or are the single breadwinner for a family, and those never unionized, have seen their real incomes decline. The minimum wage for individuals effectively fell by 30 per cent in the 1980s. Many of the working poor, even if few are literally homeless, are paying such a large share of their income on rents that they are as dependent on food banks as are employable welfare recipients. Many industrial workers laid off through restructuring have also lost ground.[28]

Two other striking changes have been the decreasing numbers of the elderly in poverty and the rising numbers of single mothers and their children. Largely because of indexed pensions and the Guaranteed Income Supplement, the share of households with heads over sixty-five in poverty fell during the 1980s from over 40 to 14 per cent. Among the single elderly, disproportionately female, the share dropped from over 60 to nearly 40 per cent.[29] Single mothers and their children have not done nearly as well, nor have families with many children. Even though women generally have increased their incomes, a disproportionate share of this gain has gone to second earners in already financially capable families. (This has also cut into the pool of potential volunteers.) But four in ten poor families are headed by single women, and only one in ten by single men. Obviously it does not pay to be without a spouse, especially for women. Also, childless couples are far better off than those with three or more children. In fact, even by Statistics Canada's standards, about one million children grow up in poverty, which, as in the 1930s, is reflected in poor nutrition. The more children born to a family, the harder it is for mothers to join or re-enter the work-force.[30]

During the 1980s the country experienced an even sharper regional differentiation of economic growth and accumulation of wealth, despite transfer payments. Although the levels of poverty are almost equal for various categories of urban places and rural Canada, according to the National Council of Welfare, employment opportunities are very limited in smaller places. Regional development schemes, the great promise of the late 1960s and the 1970s, have done little to create employment in eastern Canada. The Canadian Council of Social Development (CCSD) has pointed out that in 1987 nearly one-quarter of Quebec's families, second only to Newfoundland, lived below its poverty line of 50 per cent of the average income for a family of three. Even in Ontario, 13 per cent of families struggled to survive below the CCSD's poverty threshold. Government spending constituted from nearly 40 per cent of Ontario's economy to over 70 per cent of Newfoundland's and Prince Edward Island's. Yet when all accounts, implicit (such as tax abatements) and explicit, are totalled, those in the top income levels derive more benefit from this spending than those on transfer payments at the bottom. Hence, richer areas have done better.[31]

One could argue that the situation could have become worse for the poor had it not been for the advocacy of social planning bodies and similar groups. The medical insurance system was actually improved, though danger increasingly looms over financing. Many physically and mentally handicapped are probably better off than they ever have been, through integration into community life, though injured workers and the deinstitutionalized mentally ill would hardly agree. Another ray of hope for low-income single people is the right to earn some income without losing all benefits, as advocated by the 1988 report of the Ontario Social Assistance Review, *Transitions* – the first comprehensive social study in twenty years. It has been a useful antidote to the free-wheeling market ideology of the last decade and a half. As in Vancouver, through the initiative of the Downtown Eastside Residents' Association, permanent living accommodations for homeless singles are replacing hostels. In Ontario, Quebec, and Nova Scotia, rental buildings have some protection from conversion to condominiums. Not all central-city neighbourhoods have been turned over to the upper middle class, even though property costs have favoured the affluent.[32]

Since 1975, social planning bodies have become more defensive because of cutbacks but have expanded nevertheless, at least until 1990. All the large cities have independent organizations, most supported substantially by the United Way. Some independent groups still provide direct services, as in Vancouver, where the Social Planning and Research Council of British Columbia (SPARC) runs the parking program for the disabled. The city of Vancouver itself does social planning, Toronto has incorporated social planning goals into CityPlan 91, and in Halifax the Community Planning Association largely fills the role. Montreal formerly had councils in both language groups; now a committee of Centrade (United Way) has a limited role, largely because of strong provincial activity. Groups in some cities rely less than formerly on the United Way, drawing in research funds from other levels of government. Besides, tension between funders and councils is less obvious than earlier in the century. That Metro Toronto now contributes 20 per cent of the SPC's budget suggests an emerging sense of partnership, reflecting a convergence of interests stimulated by the province's decision to place more social costs on local governments.[33]

The Metro Toronto SPC has been very active, with a budget in 1989 of over one million dollars and a core staff of about fifteen. Continuing many earlier initiatives, it has analysed the community services of the post-war suburbs, which are gradually becoming more similar socially to the central city. Strengthened and new local advocacy bodies linked together in a Metro Toronto–wide group have emerged. The SPC has been extending that study into the regions beyond Metro Toronto. Housing problems have been highlighted, as have minority rights. Equally important, the council's researchers pointed to the danger and short-sightedness of privatizing social services. Attempts to integrate social and economic policy, formerly kept at arm's length, through studies on unemployment and labour-market structures, imply a shift to western European ideas. In British Columbia, SPARC has been moving in similar directions. Perhaps these and other activities persuaded the Ontario Social Assistance Review Committee to promote waged work without the complete loss of welfare payments.[34]

EQUITY AND THE POWER OF THE PURSUIT OF STATUS

We have traced the course of social planning in this century at the local level within the context of expanding policies and financing. We have seen that action by municipalities and voluntary groups has remained significant, that municipalities and citizens' groups have argued for greater support, that a welfare state of considerable strength involves all levels of government, and that it has been eroding over the past decade and a half. Here I offer some tentative reasons for these patterns. First, urban Canadians have maintained local power in several regions, most strongly in Ontario. Municipalities in Nova Scotia and Ontario have responsibility for many social services. Of course, they can no longer exclude the poor "belonging" to other municipalities under the Poor Law and cannot sustain the same degree of civic initiative found around 1910. Periodic restructurings of local governments are positive attempts to redefine the locus of social action. In Australia, by contrast, all state governments run far more municipal services, partly through special boards, so that municipalities remain as weak as they did in the late nineteenth century. In Britain under Margaret Thatcher, Westminster attempted to take more control, though the class balance may be shifting once again. Then, too, unlike the United States, with its fragmented urban governments, Canadians have not suffered from as sharp central city–suburban splits, the denial of minority rights, and anti-urban sentiments. To put the case positively is less easy: perhaps it is the American experience against which Canadians measure themselves, so that maintaining the quality of the urban environment still remains a major goal. Although municipalities in the United States account for more of all government spending than do those in Canada, there is a world of difference in how urban areas are run. Provinces, unlike most states, hold a vital stake in urban life.[35]

Canada also participates in the course of Western history. For decades the magic of economic growth has been drummed into our heads: only it can guarantee a supply of social goods. For a while it was believed that private-sector growth would be helped along by Keynesian demand management. In Canada, since the mid-1970s, economic growth has slowed and became increasingly concentrated in fewer regions. With that came the strident reassertion that only the private sector is productive. New office towers providing more and more space for executives are somehow productive; people who sell necessities to those on welfare are apparently not so.

A partial, if speculative, argument possibly lies in the wave-like quality of Western economic life. Since at least 1790 (and probably earlier), over a more or less fifty-year cycle, absolute material productivity rises, stabilizes, falls relatively, and then collapses. Before the fall and even into its early years, rhetorical faith in the market is more frantic than usual, as in the gilded age of the early 1890s, in the speculative late 1920s leading into the Depression, and apparently in our times. A growing inequality of income and wealth, with the affluent taking more, seems both a symptom and a cause. Given that most people refuse to hear predictions of bad news and that precise forecasts are impossible, each generation repeats (in a way) the mistakes of the past.[36]

Another partial answer possibly lies in the concept of "social limits to growth." Fred Hirsch has argued that as societies become increasingly affluent, economic growth gradually slows. As more and more people are blessed with more and more discretionary income, the hunt for status becomes more widespread and acutely felt. The first Jaguar on the street inspires envy; soon everyone has to have one. But what is next? As more people increase their real incomes, the social quality of goods and so of status declines.[37]

But rather than falling, striving for status intensifies, particularly, it seems, near the downward end of a long wave. More than in the upswing, zero-sum games are played: people can succeed only by outdoing others. When the pressure is greatest, that is, in the last days of long cycles, as in the late 1920s and in the 1980s, the pursuit becomes desperate. Of late, promises of the best of all possible worlds from business have led to the piling up of both private and public debt. Governments have thrown money at businesses in a random and unplanned fashion through tax concessions, while the rich and the corporations have, since the late 1960s, drastically cut their contributions to the United Way and other social welfare bodies, so that even more of a burden is forced onto local taxes. The rising expectations of those paid from the public purse (including educators) have yet come under what might be termed a "social discipline" parallel to the market discipline imposed on industrial and service workers during the current economic turndown.

Most fundamental, the universality of some social programs – certainly a significant achievement – is threatened as we are subliminally persuaded to beggar our neighbours. Canadians may like to pat themselves on the back as having social policies superior to those of the United States and, nowadays, to those of Britain. Canada's stronger sense of collectivity is the legacy of an intricate federal system: it has not been easy to institute social welfare programs, but once established they are difficult to dismantle. Some would like to believe also that it is a legacy of public school systems that are much closer to being universal than those in other English-speaking countries. Yet this time around through the long wave, Canadians are burdened not only with massive global economic restructuring, but also with increasingly scarce and expensive material resources on which they have come to depend. As a result, pursuing social equity and economic growth in the twenty-first century may not be as straight forward as in what Wilfrid Laurier proclaimed as "Canada's century."[38]

CHAPTER FIFTEEN

The Meaning of Home, Homeownership, and Public Policy

R. HARRIS AND G.J. PRATT

The home has multiple and complex meanings that provide clues to a whole range of social relationships. In this chapter we review how the meaning and spatial organization of homes in Canadian cities have evolved and how variations persist among different social groups. Because ownership is intimately bound up with the meaning of home in Canada, we discuss the historical trend toward increased urban homeownership, assessing its causes and implications. While some authors assert a universal desire for property ownership,[1] we stress the social foundations of this desire, including its roots in public policy. In the final section, we emphasize the role of public policy in enabling or constraining housing options available to urban Canadians.

THE MEANING OF THE HOME

In Canadian cities, the use and meaning of the house have changed over the last century in at least three respects. First, there has been a long-term shift of paid employment away from the house.[2] Second, with increased separation between home and work, the home has taken on new meanings for both children and adults. It has become a haven for a family life protected from the stimulation and threat of the city. For the adult, it has become a refuge from an alienating and exhausting world of work, a place of security, privacy, and personal control. Third, it has become an important status symbol, a measure and symbol of personal success.

The most significant change influencing the meaning of the home has been the removal of many types of work and growing geographical separation from paid employment. In mid-nineteenth century cities, for many people the house coincided with, or was at least fairly close to, the workplace. The growth in scale of capitalist industry, extensive construction of street railway lines, shortened hours of work which allowed time for commuting, and the noxious externalities associated with larger industrial establishments encouraged people to live further and further from places of employment.[3]

The past century has seen less domestic production of goods and services in the home and increasing reliance on the market. In Montreal in the mid-nineteenth century, for example, working-class families supplemented their wages by keeping pigs, tending vegetable gardens, and renting to boarders.[4] Although the incidence of boarding varied from place to place with fluctuations in the local economy,[5] it remained quite common into the period between the two world wars. In Canada, the 1930s seem to have been a critical transitional period. Although boarding probably increased in the early years of the Depression, over the decade 1931–41 the proportion of Canadian urban households with lodgers fell from 20 per cent to 11 per cent.[6] In the post-war era, single people have been far more likely to establish their own residence than to board with families, and more than one family are unlikely to live under the same roof.[7] In 1961, 79 per cent of all Canadian families lived alone (that is, they shared a residence with neither a boarder nor another family), but by 1981 this figure was 88 per cent.[8] The extent to which Canadian households obtain informal revenue through letting out rooms has dramatically declined (though this may be changing – a theme we return to in the final section), producing a very significant recomposition of the household. Both factors have affected the meaning of the home.

In contrast, the home has retained and perhaps even acquired a new moral force for the development and stability of the private family.[9] Social reformers of the late nineteenth and early twentieth centuries saw the increasing incidence and worsening conditions of urban slums in some eastern Canadian cities as a threat to personal morality and family stability.[10] By the 1920s, if not before, the appropriate solution took the form of an alternative suburban environment, centred on the family and shielded from the stimulation and prosmiscuity of the city.[11] (This is not to say that suburbanization was a simple result of this intense association between family life and the suburban home; this association was merely one factor stimulating a desire for suburban housing.) The changing design of the suburban house also reflected this concern to support and control the social development of children. For example, a 1955 issue of *House Beautiful* recommended inclusion of an extensive range of recreational opportunities within the private home in order to keep children away from the influence of broader society.[12] Some called the house a "third parent": "The house in which children grow up is almost as much a parent to them as a father or mother. With its all-pervading influences – both good and bad – a house helps shape values and set standards for the younger generation. In this respect a house is really a third parent."[13]

The home can be seen as a protective environment for the adult as well. Many have argued that the home has become a refuge, a "haven in a heartless world."[14] This meaning is usually attributed to the single family–owned suburban home, although Rainwater has also developed this idea in relation to poorer households living in public housing projects in the United States in the 1950s and 1960s.[15] In environments where safety is a real concern, individual housing units are especially valued for their security. For the suburban homeowner, this "retreat" may be better thought of in terms of both the house and the local neighbourhood. Richard Sennett interprets the retreat into the family and a privatized suburban residential environment as a response to, and escape from, the social disorder and diversity of early industrial cities.[16] Other writers emphasize changes in the work world, arguing that as people have lost control over the work process, they have found compensating satisfactions in their control of their home sphere.[17]

A variation of this second hypothesis has been explored in the Canadian context. Pratt has examined whether homeowners living in a suburb of Vancouver (along with respondents to a national survey) showed evidence of this retreat into the community at the expense of class-based or workplace politics, by examining attitudes toward social welfare and patterns of political participation.[18] The influence of homeownership on social and political attitudes seems to depend on occupational class; even when age, incidence of children, and household income are controlled, homeowners in white-collar professions are more conservative than white-collar renters. The politics of skilled blue-collar workers is, however, unrelated to whether or not they own their home and seems to be shaped by their union experiences; interviewees demonstrated little evidence of an ideological or practical retreat into the home or community. This study suggests that any withdrawal into the home is partial, competing with other meanings and other experiences.

Third, the home conveys messages about status and identity. Ownership is clearly one aspect of the status message, but so too is neighbourhood, house style, and interior and exterior decoration. It may be that the home has become more significant as a status object in recent times. In assessing increased residential differentiation in Britain's Victorian cities, Ward contends that it became more important to express distinctions of status through the home and residential address, as differences established through work, in terms of income and skill, disappeared.[19] The Duncans attribute the growing role of the home as a status marker to changing patterns of social networks.[20] They argue that the fixed features of the house tend to communicate identity when social networks are open and fluid, which is more common in modern, as opposed to traditional, societies.

Of course social and geographical variations may obscure the historical trend. Within North America, it has been argued, homeownership measures status more for upper-middle-class men than for upper-middle-class women (the latter tending to conceive of the home more as an emotional refuge)[21] and for middle- as compared to working-class households.[22] Among those who use the house as a status object, different reference groups may value different landscape aesthetics, and this can be a source of conflict. Holdsworth has documented the competing landscape tastes of Portuguese and middle-class Anglo-Canadian families living in downtown Toronto.[23] Portuguese families have actively transformed existing houses to suit their aesthetic tastes and personal needs – for example, by installing summer kitchens and wine cellars and removing Victorian gables. Few such changes conform to the taste of middle-class Anglo-Canadians; they effectively remove the transformed homes from the latter's housing market.

Variations occur within social classes as well as between them. High-income households in Vancouver convey canons of taste through their house interiors. Members of a new elite group studied in West Vancouver created interior landscapes filled with eclectic, contemporary furnishings (one family brought a ten-foot totem pole into its living room!) and sought to impose their own personalities and creativity. A more established Shaughnessy elite chose traditional interior decorations and used domestic landscapes to signal membership in a closed social group.[24]

Links between family life and the home are not consistent. Some groups do not view the suburban single-family home as the ideal family environment. Many families prefer inner-city living, in either single-family housing or high-rise apartments. From a large, longitudinal survey of families living in Metropolitan Toronto during the early 1970s, Michelson identifies a subset that chose downtown high-rise apartments; access to employment and public transportation influenced this choice.[25] Female-led families in the inner city seem to value the rich network of readily available resources.[26] In some cases, women in two-headed families also seem less enthusiastic than men about a suburban location, especially if they lack access to a car.[27]

Social theorists, including feminists, have questioned the desirability of the family home, sequestered from the full range of urban activities.[28] Feminists have long argued that residential environments separated from work opportunities rigidify the existing gender division of labour. American feminists have for decades tried to redesign the home and neighbourhood so as to lessen women's domestic responsibilities and reintegrate home and work – for instance, by providing possibilities for co-operative housekeeping or space for home-based businesses.[29] In Canada, feminist alternatives for urban living are well represented in co-operative housing.[30] Women's co-ops have been built in at least eight Canadian cities, and some design aspects reflect feminist principles.[31] For example, all co-ops offer a range of units, appropriate to a diversity of family and household forms. The Constance Hamilton project in Toronto has five unit designs, including arrangements for two or three women sharing, multi-generational families, and two-single-parent-family households. The living areas (living room and dining/kitchen) have been split between floors to create more semi-private social spaces. In Toronto's Beguinage co-op, bedrooms are the same size, to denote equal status among all family/household members. Funding constraints have prevented provision of enterprise spaces, which would allow integration of paid employment and home, and more extensive communal areas.

Mills argues that inner-city market condominium housing also facilitates "post-patriarchal" family arrangements. Intensive interviews in Fairview Slopes, a gentrified inner-city neighbourhood in Vancouver, revealed that residents value the relative ease with which they can combine work at home and in paid employment; the house form requires little maintenance and the location is close to a range of job opportunities.[32] Mackenzie[33] finds that in many Canadian cities and towns, women are creating paid jobs within the home and local community. Dyck's ethnographic study in the Vancouver suburb of Coquitlam documents women's extensive use of neighbourhood networks to combine work at home and paid employment even within conventional suburban environments.[34] The concept of the home as refuge and separate sphere is, then, being challenged and reworked.

Urban Canadians invest dynamic and multifaceted meanings in their homes. One should not assume that residents living in one particular type of housing, such as single-family homes, have a richer set of meanings than those living in other types, such as multi-family units, or that homeowners always value their home more than do renters. In some instances, owners of single-family housing may have little attachment to home. A recent British study of affluent homeowners demonstrates that executives may self-consciously choose to minimize attachment to the house in order to lessen the pain of residential mobility.[35]

Numerous studies, however, have documented the pain that residents of areas designated as slums have undergone when forced to relocate.[36] The black community of Africville in Halifax was uprooted and relocated between 1964 and 1969. The majority (75 per cent) of residents had lived there all their lives, and many had family ties to the area dating to the mid-nineteenth century. Only a handful of families could establish legal title to land, many claimed squatter rights, and some rented. Most residents were extremely poor, and the area was designated as a slum. Yet when interviewed in 1969, 54 per cent of relocatees reported returning often to Africville.[37] In 1988, a woman whose family was moved from Africville into public housing in the 1960s still claimed: "A lot of people, if it was possible to get their land back somehow, would live back there again. I don't blame them. Home is home is home."[38]

It is nevertheless often assumed that owners are more attached to their homes than are tenants. There are a number of reasons why this may be the case: owners move less often than tenants, and their homes are therefore likely to harbour more memories; the owners' home embodies not only personal meanings and sweat equity, but also a large capital investment. So important is the issue of ownership that it is sometimes used to distinguish "homes" from mere "dwellings." As a Vancouver realty firm put it in 1911: "The house you live in is not a 'home' if you don't own it. If you are paying rent you are living in someone else's home, not your own home."[39] This type of statement denigrates the very real meaning of the home to the tenant. But it does point to the fact that homeownership has long had a special meaning to Canadians.

OWNING THE HOME

Most Canadian households attach considerable importance to owning their own home. A survey carried out in the 1970s found that in Toronto more than four-fifths of all households would like to own their home if they could afford to do so.[40] There is no reason to believe that the proportion was any lower in other parts of the country, or that the situation has changed in recent years.

This commitment to homeownership, like the meaning of the home itself, is the result of historical circumstances. The desire to own property has been especially strong among first-generation immigrants, eager to put down roots in a new world.[41] People from many backgrounds have worked unusually hard, lowered their present standard of living, and even neglected the education of their children, in order to acquire property.[42] Many have sacrificed privacy by taking in boarders and lodgers; within particular social classes, the incidence of boarding has usually been higher among immigrants than among the native-born.[43] In some cases two families have shared the use and ownership of a home until one or both could afford its own.

But Canada's history of immigration is only part of the story. The desire for ownership among the native-born is almost as strong as among newcomers. In one sense, this wish represents a search for social prestige.[44] Until quite recently, the higher status of owners as opposed to tenants was reflected in municipal franchise restrictions that discriminated against the latter. In Kingston, Ontario, in the early 1970s, for example, only tenants who had been resident at their current address for more than twelve months were allowed to vote. Owners could vote in any ward where they owned property (thus, up to seven times).[45] Because owners are more likely to get involved in local politics, they continue to exert a disproportionate influence within their community.[46]

Owner-occupiers typically are raising children. Families with children often have difficulty finding suitable rental accommodation: landlords in multi-unit buildings discriminate against them, and these days barely 10 per cent of single, detached home are available for rent.[47] Largely for that reason, families with children are disproportionately concentrated in owner-occupied housing, while homeowners are especially likely to have children living at home.

Symbols of social status often reflect an underlying economic reality, as is the case with homeownership. In economic terms the owner-occupied home is – in most situations – a valuable investment unavailable to the tenant.[48] Most consumer items, including "durables," lose value after their initial purchase. Homes typically do not, and for two reasons. First, technology and labour productivity in the residential construction industry have changed quite slowly in recent decades.[49] The cost of construction has therefore risen in comparison with that of producing most other necessities, including food and clothing. Second, the cost of land has risen in absolute and relative terms throughout most of the twentieth century. It makes up a large and increasing proportion of the cost of housing and has helped to guarantee that house prices have risen at or above the overall rate of inflation.[50]

In Canada, especially in the larger and more diversified centres, owning a home has been a profitable and reliable investment. Even when households have lost equity in their present home, their faith in ownership has remained largely unshaken. A large majority in a sample of Calgarians interviewed in 1983 intended to buy when they next moved, even though all were recent home buyers whose purchases were worth less than when they had bought.[51]

The Growth and Incidence of
Urban Homeownership

Household incomes have increased faster than the cost of living for most of the twentieth century. As a result, rates of urban homeownership have risen from about 30 per cent at the turn of the century to 57 per cent in 1986 (see Table 15.1).[52] Decline occurred during the Depression, when unemployment was high and the incomes of many households fell, and also in the 1920s and 1960s, when many young people with few savings were looking to set up their own households. As a result, after the buying boom that ended in the early 1920s, the increase in ownership has been quite modest.

Differences in affordability account for the broad geographical variations in the ownership of homes. Ownership rates have always been higher in the country than in the city, usually by at least 25 percentage points. In 1981, for example, when 56 per cent of urban households lived in their own home, 82 per cent of rural, non-farm households did the same.[53] Around most cities there is a tenure gradient: ownership rates increase steadily from the inner city to the exurban fringe. In Toronto in 1986, for example, ownership rates were only 40 per cent within the city, rising to 56 per cent in the rest of Metropolitan Toronto, and reaching an average of 74 per cent in the outer

Table 15.1
Homeownership rates in Canada since 1881

Year	Homeowners as a percentage of all households	
	Urban and rural	Urban Canada*
1881	–	c. 27
1891	–	c. 29
1901	–	c. 27
1911	–	c. 45
1921[†]	58	49
1931	61	46
1941	57	41
1951	66	56
1961	66	59
1971	60	54
1981	62	56
1986	62	57

Sources: Census of Canada from 1921 on.
* For 1881, 1891, 1901, and 1911, Canadian estimates are based on data reported for Toronto, Hamilton, and Kingston, Ont. See Harris and Hamnett (1987).
† Household rates estimated from reported data for families.

parts of the census metropolitan area.[54] This tenure gradient seems to have emerged in many cities in the late nineteenth and early twentieth centuries along with the growth of a non-residential central business district (CBD).[55] In Toronto, for example, it was not fully developed until after 1900 (Fig. 15.1). Today the slope of this tenure gradient reflects the variable cost of land[56], as well as differences in incomes and family circumstances of households, with a higher proportion of lower-income, female-headed, single and elderly households living within the central city. Development of condominium tenure, however, is confusing the picture, since it permits owner-occupation at high densities. Construction of condominiums very close to the CBD can raise ownership rates at the centre above those in the immediately surrounding areas. This had happened in Toronto even by 1981 (Fig. 15.2). It is likely that as many cities continue to gentrify, ownership rates in inner districts will rise and the tenure surface will become more complex.

In contrast to the national trend, Montreal has had an unusually low ownership rate which cannot, for the most part, be attributed to the cost of housing in relation to incomes. It has often been said that the distinctive tenure situation in Montreal reflects French-Canadian indifference toward property ownership. However, Montreal is the only city in Quebec that has an ownership rate that is significantly lower than that of English-Canadian cities of

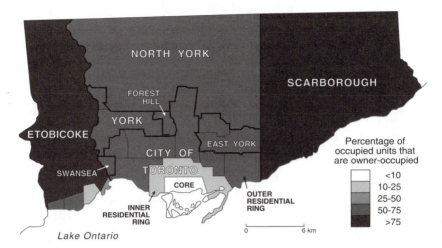

Figure 15.1
Homeownership rates in Metropolitan Toronto, 1961

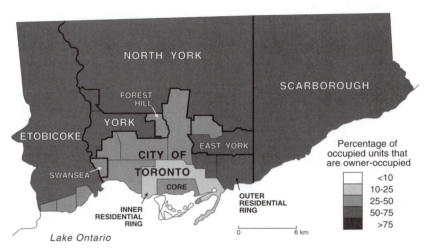

Figure 15.2
Homeownership rates in Metropolitan Toronto, 1981

comparable size. Paradoxically, it is the only major Quebec city with a high proportion of anglophones. Moreover, recent research has shown that francophone areas within Montreal have some of the highest ownership rates in the city. Altogether, the "ethnic" explanation seems weak. Instead, it seems that Montrealers developed their own culture of property, which crossed the boundaries of both class and ethnicity. The nature and origins of this culture, however, remain obscure.[57]

Table 15.2
Class and income differences in urban homeownership, 1931–82

Social class*	All households			Household income quartiles, 1982			
	1931	1974	1982	I	II	III	IV
Owners and managers	58	67	74	39	52	73	86
Self-employed	56	79	81	72	84	83	89
Middle class	41	50	59	25	33	64	80
Working class	38	55	61				
White collar	–	48	54	27	46	66	81
Blue collar	–	61	67	43	58	72	83
Others†	–	53	57	51	70	74	77
All households	46	57	63	45	55	71	83

Sources: Harris (1986). Calculated from 1931 census and Statistics Canada, 1974 and 1982 Household Income (1973 and 1981) Facilities and Equipment.

* Households are classified on the basis of the occupation of the (usually male) "head."

† Includes households with no members currently in the labour force.

Apart from such local differences in housing tenure, there are also quite substantial differences between the major social classes (Table 15.2).[58] Owners, managers, and the self-employed have usually been the most likely people to own their own homes, mainly because of their higher incomes. However, ownership rates among blue-collar workers have been consistently higher than among the middle classes. It seems that the desire to own property, coupled with an ability and willingness to build their own homes, has been especially strong among manual workers.[59]

A Crisis of Affordability?

In recent years the upward trend in homeownership has slowed in response to increases in housing costs. Between 1981 and 1990 in cities across Canada the average price of an existing home doubled, from $50,700 to $102,000 – new houses were more expensive yet. Except for Montreal, prices were higher still in the larger cities, reaching $154,000 in Vancouver and $204,000 in Toronto (Table 15.3).[60] For current owners, such increases underscore the profitability of the home as an investment. For first-time buyers, however, they seem to prevent attainment of the Canadian dream.

In fact, it is not clear whether homeownership has become more or less affordable in recent years. Since 1945, rising incomes have allowed people to occupy larger and better-equipped homes. They have also allowed many people, including an increasing number of single people, to form their

Table 15.3
Housing affordability in Canadian metropolitan areas

City	Average monthly rent for one-bedroom apartment Oct. 1990	Rank	Rental vacancy rate in Oct. 1990	Rank	Average sale price of existing dwelling* ($000) 1990	Rank	Index of average sale price of existing dwellings (1981 = 100) 1990	Rank	Proportion of one-family households spending more than 30 per cent of income on housing† (1986)
St John's	479	7	1.6	8	71	15	145	15	19
Halifax	485	5	3.6	14	84	13	160	13	19
Quebec	440	13	6.1	17	71	15	161	12	17
Montreal	440	13	5.9	16	91	11	202	8	20
Ottawa-Hull	511‡	3	1.2	5	105	8	210	6	16
Toronto	557	2	0.9	2	204	1	309	3	17
St Catharines	450	10	1.9	10	100	9	270	4	14
Hamilton	433	15	1:2	5	132	4	314	1	15
Kitchener	443	12	1.3	7	141	3	313	2	14
London	447	11	2.8	13	117	6	254	5	15
Windsor	485	5	2.2	12	85	12	207	7	14
Thunder Bay	476	8	0.9	2	83	14	169	10	12
Winnipeg	423	16	6.4	18	64	18	149	14	16
Regina	391	18	5.0	15	61	19	133	17	17
Saskatoon	363	19	7.5	19	66	17	132	18	20
Calgary	458	9	2.0	11	113	16	138	16	21
Edmonton	414	17	1.8	9	93	10	119	19	21
Vancouver	566	1	0.9	2	154	2	179	9	22
Victoria	488	4	0.3	1	128	5	164	11	22
Canada	–		3.3§		102§		200		–

Sources: CMHC, Canada Housing Statistics, 1990, Tables 28, 29, 76. Statistics Canada, 1986 Census of Canada.
* Single detached homes financed under the NHA.
† In the case of owner-occupiers, includes cost of mortgage, taxes, and utilities.
‡ Ottawa only.
§ Average for 24 CMAs.

own household instead of having to stay at home or to board with another family.[61] Even in the 1970s, when affordability again became a public issue, housing costs rose less rapidly than household incomes. Between 1971 and 1981, the increases were, respectively, 126 per cent and 190 per cent.[62] Since then it seems likely that prices have more than kept pace with incomes. But housing is still more affordable than it was in 1971. Such a statement is, however, misleading, since it pertains only to a national average.

Recent trends in house prices have, if anything, heightened local and regional differences in housing costs (Table 15.3). In 1981 an average home was about 30 per cent more expensive in Toronto than in metropolitan Canada as a whole. By 1990 it was twice as costly. In some places, notably Vancouver and Toronto where rents and housing prices are the highest in the country and where rental vacancy rates are among the lowest, housing is clearly not affordable for many. The national average figure also disguises the growing difficulties faced by lower-income households. Recent demographic changes have seen more households able to afford only the cheapest of accommodation. Of these, the most numerous are single-parent, usually female-led, families.[63] These groups have difficulty in finding affordable housing, especially when the cheaper inner-city housing that they might have occupied is being upgraded and gentrified. A growing number are ending up homeless.[64]

HOUSING AND PUBLIC POLICY

The problems of affordability and homelessness are matters of widespread concern, but in Canada, compared with most western European countries, governments play only a modest role in the housing market. Even so, Canadian governments have both reacted to and shaped the meaning and quality of housing. The range of housing policy issues is vast; we restrict our discussion to a few areas. First, we discuss the ways in which federal housing policy has helped to construct the meaning of homeownership, in part by increasing its attractions relative to renting. We then consider more fully how government has structured the other housing meanings discussed above. Government policy has nevertheless threatened one of the meanings attached to home – security and personal control – for some Canadians at some points in time; we examine citizens' resistance to public programs that have had this effect. We then examine some recent efforts by some Canadians to reconstruct the meaning of home, as well as some of the economic pressures that seem to force this process.

Federal Housing Policy and Homeownership

Perhaps the federal government's most significant (and expensive) housing-related activity has been to promote homeownership. The financial attractions of owning a home reflect government policy and the legal interpretation of the rights attached to different tenures.[65] Through the Dominion Housing Act (DHA) of 1935, Ottawa reduced the required downpayment from 40 to 20 per cent of the value of the house and extended the amortization period from roughly five to 20 years.[66] These provisions were extended in the 1938 National Housing Act (which superseded the DHA) and essentially remain in place today.[67] By 1950 almost 50 per cent of all new housing starts in Canada were financed with NHA assistance.[68] Federal policies have also tended to favour ownership, not taxing imputed rent and capital gains from owning a home. Dowler estimates that these tax breaks amounted to 76.8 per cent of all federal housing subsidies in 1979 – a substantial skew toward homeowners.[69] Since owners are, on average, more affluent, this bias was especially inequitable. Moreover, the greatest tax subsidies went to those with the most expensive homes and, presumably, the highest incomes. Though the situation has changed slightly since 1986, when limited capital gains exemptions were extended to other investments, the net effect of current federal housing policies, implicit and explicit, is certainly regressive.

There are various reasons why successive federal governments have promoted homeownership. Some analysts suggest that ownership of a single-family dwelling has been compatible with heightened consumption of mass-produced goods, and thus the economic benefits of the tenure and house form extend throughout the economy.[70] Others have stressed the socially integrative effects of ownership, arguing that it allows satisfactions within the home that dampen workers' unrest and currents of political instability. In the words of a spokesman for the Vancouver Real Estate Exchange in 1920, homeownership "teaches thrift and sobriety; it makes better and more contented citizens and eliminates that prey of unrest and radicalism – discontented rent-payers."[71] The federal Progressive Conservatives claimed in the election campaign of 1979: "A society in which a major proportion of the population owns its own homes is likely to be a more stable, settled and productive society."[72] But these hypothesized effects do not hold across all occupational classes. They seem more accurately to describe white-collar than skilled blue-collar workers.[73] In any cases, this view of social control through extended homeownerhip must be complemented with recognition that many working- and middle-class families desired and actively sought a home.[74]

The bias in favour of ownership has not gone uncontested. In response to pressure from tenants, most provinces instituted rent controls in the 1970s, as well as provisions for increased security of tenure for renters.[75] Implementation of such measures changed the relative attractiveness of renting versus owning and redefined the meaning of renting in Canada (i.e. its security vis-à-vis ownership). This can be an uphill struggle, however. In British Columbia, for instance, most of the programs that the provincial NDP government (1972–75) put in place to benefit renters have now been abandoned: rent controls have been slowly phase out, the Rentalsman office has been dismantled, and the renter's tax rebate has been cancelled.[76]

Public Policy and the Meaning of Home

Government has helped create homogeneous residential environments which make the home a retreat, a status symbol, and a locus for a family-centred life-style. The federal government has done this by targeting mortgage funding and insurance to suburban areas and to large developers who were prepared to construct "appropriate" suburban environments.[77]

Even more significant have been the actions of local governments. One of their most powerful regulatory tools is that of zoning. While less marked than in the United States, exclusionary zoning does occur in Canada and creates perceptible problems. A number of jurisdictions just outside Metropolitan Toronto are legislating large-lot zoning which effectively excludes low- and moderate-income households; Vaughan Township is a case in point. As the city of Toronto continues to gentrify, an increasing proportion of low-income households will probably concentrate in the inner suburban areas within Metro, with unfortunate fiscal consequences for these municipalities.

Zoning that increases the social homogeneity of suburban residential areas may enhance the meaning of the home as status symbol, retreat, and support for the social development of children within traditional family settings. However, Relph has argued that, in other ways, it potentially strips the home of meaning. Relph sees modern suburban landscapes as outcomes of excessive planning and as environments in which the resident is prevented, by various regulations, from making anything but the most trivial alterations. "In modern suburban landscapes a small but important freedom is therefore denied. This is the freedom to demonstrate a commitment and responsibility to a place by engaging in the making and maintenance of its landscape in something other than a cosmetic fashion. The process by which this is denied can be called 'benevolent environmental authoritarianism'."[78]

Nevertheless, such control is not monolithic. In contemporary Canadian cities there are pressures to defend old, and create new, housing meanings. During the 1960s, pressures for inner-city redevelopment became very strong in many of the larger centres and provoked a defence of existing neighbourhoods.[79] The most militant forms of community opposition were reserved for publicly sponsored redevelopment. In both Vancouver and Toronto, proposals for urban renewal and downtown highway construction proved catalysts in the emergence of a local urban reform movement.[80] In Vancouver, the Chinatown expressway was stopped. In Toronto, the Spadina expressway was halted in mid-course. More important, the opposition of residents to urban redevelopment in cities across the country altered the whole thrust of federal urban policy, away from wholesale renewal toward more environmentally sensitive programs for neighbourhood upgrading.[81]

People's attachment to their homes had shown itself a major factor in urban politics.

Today, community opposition to neighbourhood change may be less evident than it was in the 1960s. It would be wrong, however, to conclude that people are any less attached to their homes. Changes in the political climate have reduced the overall level of activism, on both the urban and the national scenes. At least as significant, developers and local governments have learned a lesson. In recent years the major urban redevelopment projects have involved vacant or industrial sites, such as Vancouver's False Creek or the St Lawrence, Railway Lands, and Harbourfront projects in Toronto. They have not been free from controversy, but they have avoided the opposition of displaced residents. The most common form of displacement today occurs through gentrification.[82] Since it is perceived of as a private market process and typically occurs one dwelling at a time, it generates much less co-ordinated political opposition than the "bulldozer" approach of the 1960s.

Not all residents are equally likely to get involved in community politics. Middle-class professionals are more active than blue- or white-collar workers. For this reason, local activism is more common in middle-class areas of the city than in working-class districts. In Vancouver in the 1970s, for example, conflicts over land-use change were more common on the (middle-class) west side than on the (working-class) east side, even though urban redevelopment was proceeding at a similar pace in both areas. Professionals often take over neighbourhood organizations even in quite mixed areas.[83] Such was the case, for example, in the late 1960s in Kingston's Sydenham Ward,[84] which contained a mixture of students, professionals, and many lower-income workers. Owner-occupiers, in comparison with tenants, tend to get involved in local politics; owners are somewhat more likely to be middle class, and housing tenure itself is important. This is shown by the experience of Kingston's overwhelmingly working-class North End in the late 1960s.[85] When a residents' association was formed in the late 1960s, in part to resist a proposed cross-town expressway, it was immediately dominated by working-class homeowners. Tenants, who made up a large majority of the district, played little role.

Creating New Housing Meanings

Along with attempts to maintain existing neighbourhoods and housing, there are pressures to create new housing meanings, which have provoked creative policy responses from government. For example, in Vancouver, artists challenged the contemporary conventional separation between home and workplace and have been granted zoning allowances from local government to authorize live-work artists' studios in certain areas of the city.[86] A similar negotiation is ongoing in other Canadian cities (e.g. Winnipeg and Toronto). Some women's housing co-operatives have, as yet unsuccessfully, challenged the Canada Mortgage and Housing Corp. to rethink its regulations to allow for "enterprise" or work space within residential housing,[87] thus allowing for reintegration of home and paid employment.

Some of these challenges to existing housing meanings are forced; they follow from problems of housing affordability and from ongoing restructuring of the work process. Considering first the problem of affordability: the proportion of gross household income that, on average, is spent on housing has risen in stages in the twentieth century and currently stands at about one-quarter. Housing affordability is not a major problem in all Canadian cities (Table 15.3), depending on the balance of housing costs and wages. Writers such as Goldberg have argued that there is no general housing crisis, only a series of specific problems.[88] This implies the need for specific programs, possibly income supplements, targeted to particular groups, not a major intervention in the housing market. This view, however, ignores or understates the adjustments that many households have had to make in order to buy or rent their home. They may have had to send into the labour force a second adult, who may have gone reluctantly, and the situation may have created stress.[89] Increased house costs and sustained desire for homeownership have therefore forced fundamental reorganizations of family life. There have been adjustments in housing arrangements as well. Some new home buyers take in lodgers, or rent out basement apartments, in order to pay for their mortgages. In Vancouver it is estimated that one out of every four houses contains an illegal or secondary suite.[90] Clearly, even today the home is not a place used simply for consumption. We may, in fact, be returning to the housing arrangements typical of the early twentieth century. Further, single parents, who must rely on one income, have been forced into smaller housing units in cheaper, but less convenient, locations, while scrimping on food and clothing.[91] Others have been forced to "double up" through homesharing arrangements or reliance on relatives, again signalling reversals of long-standing trends.

Changes in the organization of work are also forcing renegotiation of the meaning of home. Old forms of homework – for example, sewing on the "outwork" system – seem to be flourishing.[92] A US ban placed on homework in 1941 in the women's apparel and six other industries is currently being renegotiated and at least partly removed. (In Canada, homework was never banned altogether but, rather, is closely regulated by provincial governments.) Changes in microtechnology make it likely that many clerical and professional jobs will also be moved, in part at least, into the home. Some observers are wary of the consequences of these developments: while potentially offering flexible accommodation of domestic responsibilities and paid employment, lower wage rates and stress – engendered by trying to carry out two tasks simultaneously (i.e. childcare and paid work) – are equally likely outcomes for some households. Existing evidence suggests that women and men have different reactions to homework: men often choose it for its advantages (freedom from commuting, flexibility of schedule), while, for women, it is forced by the constraints of coping with domestic work and paid employment. Qualitative studies of home-based workers find that most women would prefer to go out to work if they could.[93] It is important to keep in mind the very different experiences of men and women engaged in homework, especially at the present time, when major policy decisions need to be made around this issue.

We see then that the meanings attached to the home and homeownership are dynamic. They vary across time and space and are continually renegotiated, by choice and of necessity. At the present time the meaning of the home seems to be under intense renegotiation. The analytic separation between home and work within social scientific discourse has always been artificial;[94] it becomes even more so as various forms of paid employment are brought back into the home. The contemporary fluidity of family and household structure, with increasing divorce rates and numbers of single-parent families, suggests a plurality of housing needs and preferences. Perhaps most troubling, for many Canadians the home as place of security and personal control is intermingled with stress, related to lack of affordability and insecurity of tenure. Growing homelessness in many large urban centres represents the extreme experience of such insecurity; ethnographic reports suggest that it touches and transforms the core of personal identity.[95]

We end where we began: the meaning of housing and homeownership reflects the societal context and, as in the case of high housing costs necessitating two wage earners, can change social relations. We have also noted the increasing spatial variation in problems such as housing affordability; programs needed in one city may be irrelevant in another. In general, federal policies have been shaped by the situation in the larger centres, while the particular needs and resources of smaller towns and rural districts have often been overlooked. If housing markets vary so much in character from place to place, geographers, trained to be sensitive to such variations, must inform and help shape public policy.

CHAPTER SIXTEEN

Homelessness

M.J. DEAR AND J. WOLCH

Canadians have traditionally viewed theirs as a caring society. Far distant from the rigidities of its British and French heritages, and resistant to the example of its southern neighbour, Canada developed a relatively comprehensive "welfare state" in a pragmatic and piecemeal manner.[1] The social safety net reached its zenith during the 1970s. Since then, there has been increasing retrenchment,[2] but Canada's enduring commitment to social welfare remains fundamentally unshaken. The present crisis of homelessness therefore comes as something of a surprise to many Canadians. Since the early 1980s, newspaper accounts have documented a tale of growing human misery and deprivation.[3] National attention was focused on homelessness in 1987, International Year of Shelter for the Homeless. Only then did it become clear that the current crisis reflected deep-seated changes in Canadian society that had been under way for as long as two decades.

The presence of the homeless in Canada's cities and towns is exaggerated by their high visibility, as well as the apparent intractability of the problem. Some downtown sidewalks are cluttered with the homeless and their belongings, including shopping carts, bed rolls, carrier bags, and cardboard boxes. Downtown, too, is where the services to help the homeless are congregated, acting as magnets to attract still more homeless people. Heightened awareness of the homeless has had contradictory effects. On the one hand, a cry to help them has been raised; on the other hand, a small but distinct community backlash has been noted, from people who want the homeless removed (forcibly if necessary) from their neighbourhoods. The politics of homelessness has been aggravated by the persistent refusal of some federal, provincial, and local politicians to concede the need to devise policies to combat the problem. In addition, there has been disagreement on which levels of government should act.

In this chapter, we shall argue that the problems of homelessness are complex in origin and will undoubtedly be with us for many years to come. Urgent direct attention, over an extended period, is therefore vital if homelessness is to be eradicated. And we shall show how an understanding of the geography of homelessness can help in devising realistic and humane solutions.

WHO ARE THE HOMELESS?

No one has a clear idea of the numbers of homeless people in Canada. We lack even a comprehensive but practical definition of the homeless condition. At its simplest, "homelessness" is defined as the lack of a stable residence where one can sleep and receive mail. But this would allow a rooming-house or a hotel to count as a residence. Other, more inclusive definitions have encompassed social dislocation – for example: "Homelessness is not simply the lack of a stable shelter; it is life in disarray. The homeless person's existence is a public existence – there is no privacy. It is a day-to-day question of basic survival."[4] Even more comprehensive is the following, almost totally impracticable definition; "The absence of a continuing or permanent *home* over which individuals and families have personal control and which provides the essential needs of shelter, privacy and security at an affordable cost, together with ready access to social, economic and cultural public services."[5] So the problem of estimating numbers is exacerbated because the homeless are notoriously fugitive. They are highly mobile and have no fixed address at which they can be contacted. As a consequence, they often fall between the gaps in the municipal social welfare network, which is concerned principally to look after its "own" people.

Estimates of the numbers of homeless thus vary widely. One early survey, based on head counts at emergency shelters and soup kitchens, put homeless Canadians at between 20,000 and 40,000.[6] But these figures were quickly questioned: Ontario alone was judged to have over 10,000 homeless; and Toronto's figure was "guesstimated" to be as high as 20,000.[7] The most reliable data to date appear to have been produced by a survey of shelters undertaken by the Canadian Council on Social Development (CCSD) in 1987. The survey focused on people who did not have a secure home as well as those whose housing was grossly inadequate.[8] It revealed that 102,819 different people had been sheltered across Canada sometime during 1986. Shelter had been provided out of 472 shelters with a bed-capacity of 13,797. Taking account of those who were unsheltered, the CCSD estimated the homeless in Canada at between 130,000 and 250,000.[9]

One CCSD finding (confirmed by many other studies) was that the homeless population is very heterogeneous.[10] One report identified members from the following groups:[11] refugees, mentally disabled, Native Indians, seniors, youth, mentally ill, women, single-room occupants, farmworkers, transients, and the physically disabled. Since the demographic structure is so diverse, we might anticipate that the genesis of homelessness, and any search for solutions, must be complex.

The homeless population is no longer dominated by the middle-aged, white, male alcoholic (those whom many regard as homeless "by choice"). It can be characterized as follows. There are *more young people*. Twenty-five years ago, the average homeless man in Canada was between 36 and 44 years old; today he is 18–34.[12] Almost one-third of Toronto's homeless are between 18 and 24.[13] And fully one-half of Montreal's 10,000 homeless are young adults.[14] There are *more women and children*. Children are the fastest-growing group among the homeless, and one-quarter to one-third of Toronto's homeless are women (often the victims of domestic violence).[15] In Montreal in 1986, there were approximately 3,000–4,000 homeless women in a city that had only 77 shelter bedspaces for women.[16] There are *more mentally disabled*. As many as 60 per cent of Winnipeg's homeless have some form of mental disability.[17] The average for Canada as a whole is 30–40 per cent.[18] These include former patients at psychiatric hospitals, as well as those never institutionalized. There are *more substance abusers*. One-third of the shelter residents surveyed by the CCSD reported problems with alcohol; and 15 per cent had a drug problem.[19]

More people are also experiencing multiple episodes of homelessness, and these episodes are becoming longer and longer.[20] These statistics merely paint a broad picture and do not take account of local variations. Hence, in some parts of Canada, Native Indians comprise a disproportionate share of the local homeless population.[21]

The overall numbers of homeless people in Canada remain relatively small. However, many more Canadians today find themselves on the margins of homelessness. How did such a situation arise? And how long is it likely to last? To answer these questions, we now turn to consider the factors on the "path" to homelessness. The answers we devise, as a nation, will tell us a great deal about ourselves and our future.

THE PATH TO HOMELESSNESS[22]

Homelessness is the end state of a complex social process. For convenience, we may begin by identifying a range of changing conditions that have altered the demand for and the supply of housing (Figure 16.1). These conditions have made homelessness possible in Canada. Then we explore those conditions that allow homelessness, once initiated, to become a permanent condition for many people. Our objective is to identify the general path to homelessness, rather than to detail local variations.

Demand for Housing

First, one of the most potent forces underlying social change within the last two decades has been a fundamental global economic restructuring. This has

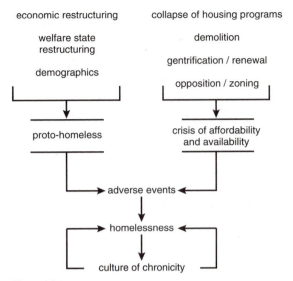

Figure 16.1
The path to homelessness

generally taken the form, in capitalist nations, of large-scale "dein-dustrialization" – the decline in long-established manufacturing industries, such as steel and automobiles. It has been paralleled by a rise in service-sector jobs, which are extremely diverse but, on average, involve less skill and hence are much lower paid. Many of these new jobs have been taken up by part-time employees (e.g. in fast-food stores). The Canadian economy has also been strongly harmed by world market conditions for its raw mate-rial exports, especially oil. Moreover, the incidence of these changes has been extremely uneven across the country.

The net result of these economic changes has been a recession which dur-ing the early 1980s increased poverty and unemployment among Canadians. In 1984, an estimated 4.2 million people lived below the official poverty line. In some cities, unemployment reached unprecedented heights (in Ed-monton's inner-city, as high as 44 per cent).[23] After a brief respite, the renewed economic slump of the early 1990s has sharpened the economic marginality of many Canadians.

Second, economic change has been accompanied by significant restruc-turing of the welfare state. Deinstitutionalization had its roots in the 1950s, took hold in the 1960s, and by the end of the 1970s had changed the char-acter of mental health care in Canada.[24] In a move away from institution-based care, Canada's psychiatric hospital bed capacity was reduced from 47,633 in 1960 to 7,935 by 1983.[25] The intention was to replace hospital beds with small-scale, community-based services. Unfortunately, the dollars did not follow the ex-patients into the community. In Ontario's 1986 budget, for example, $235 million was allocated to psychiatric hospitals, but only $48 million to community mental health programs.[26] In Manitoba, 90 cents of each mental health dollar still goes to hospital-based services.[27]

Community mental health programs were overtaken by the desire of governments to cut back on welfare expenditures. The deteriorating economy of the 1970s and 1980s caused politicians to retreat from the high levels of expenditure and service provision of the welfare "boom" of the 1960s and early 1970s. Programs were cut back or eliminated; eligibility criteria were tightened. For example, in Alberta, the shelter component of social assistance for single adults was cut from $290 per month to $150; the cheapest accommodations in the city cost $240 per month.[28] In many provinces, restructuring found its primary expression in the move toward "privatization."

Third, simple demographic change has been exacerbated by changing social attitudes and life-styles. The well-documented "greying" of the Canadian population has increased demand for affordable housing to accommodate single-person households – the weakest sector in most metropolitan housing markets.[29] Immigrants and refugees face special problems of adaptation, including access to housing.[30] Even more serious problems beset Canada's Native populations.[31]

Taken together, the changes in economic and social welfare conditions, plus the pressures of demography, have created a class of people who live in marginalized housing conditions. The proto-homeless exist on the fringes of the housing market, only one step ahead of homelessness.

Housing Supply

The proto-homeless face a rapidly diminishing supply of affordable housing in most Canadian cities. Gentrification and urban revitalization, conversion, and renewal have reduced the quantities of affordable housing. Many cheaper rental units have been lost to both public and private programs during the past two decades. Within the last ten years, for instance, Toronto has lost an estimated 10,000 units to rehabilitation and condominium conversion[32] and over half of the city's core-area rooming-houses.[33] Growth in central-city employment, especially in commercial offices, has intensified competition for the diminishing supply of cheaper downtown accommodation. This has increased demand for lower-paid service workers (janitors, etc.), who have begun to take over the rooming-house accommodation previously available to the service-dependent populations. Other factors have further diminished supply, as in the notorious Vancouver example of evictions of the poor to provide more hotel space for visitors to the World's Fair in 1986.[34]

In addition, government subsidies in the housing market have fallen.[35] In 1986, the federal government, for instance, allocated only $1.2 million to housing assistance for the very poor.[36] Major initiatives to increase affordable housing have been frustrated: only one-third of non-profit and co-op housing units have been directed toward low-income households, and most single people (except seniors) are not eligible for subsidized housing.

Efforts to provide housing for special populations have been slowed by community opposition and zoning practices. When news of an impending homeless shelter, group home, rooming-house, or low-income housing project leaks out, an intense community outcry is frequently raised – "not in my backyard" or "not on my street."[37] Most people readily concede the need for such facilities, but they do not want them in *their* neighbourhoods. As a consequence, many residential service facilities have typically been "zoned out" of neighbourhoods by planners seeking non-controversial locations.[38] Most of Metropolitan Toronto, for instance, requires special rezoning to allow rooming-houses.[39] Even within the relatively tolerant city of Toronto, there is gross maldistribution of residential facilities, which tend to be confined within and around the downtown area.[40]

In summary, the proto-homeless face a housing market in crisis. Supply is drastically diminished, and is more and more concentrated geographically. For the poor and service-dependent, housing is unavailable, unaffordable, and unsuitable. As early as 1982, Canada Mortgage and Housing Corp. estimated that the "core housing needs" of over half a million renter households in Canada had not been met (i.e. they lived in adequate or crowded dwellings, unable to improve their housing conditions without paying more than 30 per cent of their income).[41] In the city of Toronto, vacancy rates in the late 1980s were close to zero; real estate prices rose 30 per cent in one year (1987), and the average house price is now over $200,000.[42] In 1984, welfare payments in Toronto were $286 per month (for an unemployable single person), plus a shelter allowance of $115 maximum, for a total of $4,812 per annum, just less than half of the $9,800 identified by Statistics Canada as the official "poverty line" for a single person in Toronto.[43]

Adverse Events

The proto-homeless have only a precarious hold on an increasingly inaccessible housing market. For many of them, some adverse event in their everyday lives will be sufficient to propel them into homelessness. Such events typically include eviction, domestic conflict, or loss of a job or welfare support.[44] Individuals commonly experience these events sometime during their lives; but for the proto-homeless, they become the "last straw." The adverse event is often enough to tip them over the edge and out onto the street.

THE CULTURE OF
CHRONIC HOMELESSNESS

The descent into homelessness is not the end of our story. For the homeless themselves, it is often the beginning of a life of repeated episodes of homelessness, during which the times spent as a "homed" person become progressively shorter. For those who do not escape from homelessness, a "culture of chronicity" can quickly develop. The experience of being homeless itself perpetuates homelessness.

It is easy to overlook how destabilizing and disorienting the loss of a home can be. Appearance and cleanliness become a problem; privacy is almost totally sacrificed; rest and undisturbed sleep become a rarity; cold nights become an enemy. Exactly how well one stands up to such depredations depends on one's inner resources, both mental and physical. But individual coping is also linked to experiences on the street.[45] For example, our research has shown that those homeless who retain even rudimentary social support networks tend to fare better on the street than those who lose such support. Others face unaided the problems of depression, poor physical health, undernourishment, harassment, mugging, and the ravages of drug addiction. The problems facing homeless women are especially acute.[46] They have significantly fewer shelter opportunities (in Toronto, female shelter tenants were reported as being "troublesome"); they are much more vulnerable to assault; and they have fewer opportunities to earn money, frequently being reduced to "subsistence prostitution" (i.e. the exchange of sex for food, accommodation, and/or protection). When a woman has children to care for, the problems rapidly multiply.

THE SEARCH FOR SOLUTIONS

The complexities in our "explanation" underscore the profound difficulties facing those who seek to devise solutions for homelessness.[47] Policies will differ according to local conditions, including variations in local needs and stage in the homelessness cycle. Two general obstacles are particularly noteworthy. First, the problem is not going to disappear in a hurry. Economic recovery, as it occurs, is not reabsorbing many workers who now find themselves outside the newly reconstructed service economy. For example, the fifty-year-old, recently redundant auto plant worker is not going to find, or want, a position filling orders at a fast-food outlet. Even those who do find work are imperilled by the prospect of a further recession. Moreover, a large proportion of the service-dependent homeless will always remain outside the labour force.

Second, since the process of homelessness is extensive and multidimensional, corrective interventions must recognize that different policies will be needed at different stages of the process. Hence, they must combat the long-term effects of deindustrialization and the deskilling of the labour force, as well as providing emergency shelter for those who find themselves on the sidewalks in sub-freezing temperatures.

The absence of adequate policy responses, in Canada and elsewhere, is causing significant deterioration in the lives of the homeless and proto-homeless: the lack of action is pushing more and more people into homelessness. Social unrest around the issue may spill over into our streets, our neighbourhoods, and our political institutions.[48] The principal response appears to have been re- or trans-institutionalization (i.e. shifting people between institutions, as between hospital and jail).[49] More and more homeless people are being placed in, or in some cases returned to, institution-based care, including asylums and jails. In 1986, for instance, as many as 16 per cent of the inmates of Ontario penitentiaries were mentally disabled or mentally retarded.[50]

We believe that other resolutions are available to combat homelessness. In the remainder of this chapter, we focus on one solution which constructively uses the geography of homelessness in its recommendations.[51]

The Service Hub Approach[52]

Policy responses typically concentrate on the immediate shelter needs of the displaced. Most analysts identify three phases in a shelter program: emergency response, in which the immediate need for temporary shelter is addressed; transitional arrangements, in which now-stabilized individuals are given assistance to enable them to return to everyday life; and a stabilization phase, in which the homelessness cycle is broken and the individual regains self-autonomy and long-term stability, including permanent housing. This model program assumes that the homeless need some kind of assistance (beyond mere shelter) if they are to escape from their condition. So community-based shelter networks need to be augmented by more formal services, including medical care, counselling services, job training, nutrition services, and childcare.

One important – but often neglected – point about shelter and support services is that they have to be geographically accessible in order to function effectively. Moreover, close proximity to other services (used by the population at large) is also essential in the daily lives of the homeless. These include retail outlets, public transit, parks, and recreation centres. Hence, a homeless schizophrenic may best be served by affordable accommodation near psychiatric services, transportation, shopping, and community activities. Such an individual may be poorly served by a remote suburban or rural location, or even an urban location if appropriate services are not nearby. "Ghettos" of service-dependent homeless have developed spontaneously in the core areas of Canadian cities. These districts usually have cheap housing and appropriate structures for services, which are often excluded from the suburbs. Although physically deteriorated and plagued by social problems, the service-dependent ghetto acts as a "coping mechanism" or support network. Within it, the homeless, including the mentally disabled, dependent elderly, and physically disabled, assist each other in finding jobs and accommodation, as well as friendship and support.[53]

The service hub concept draws on our knowledge of service-dependent ghettos and the needs of the homeless. It is a diverse collection: small-scale, community-based facilities so close to each other that interaction between them is facilitated. It typically consists of a heterogeneous group of services, including some generic functions, such as shopping and recreation. Hence, it can address the needs of a wide variety of client groups.

The service hub approach creates decentralized services and housing throughout an urban area, in all districts where need exists. It provides more choice in housing location, as well as encouraging a "fair share" of caring for the homeless. In essence, it calls for replicating the positive support features of the service-dependent ghetto – now generally confined to the downtown core – in other zones of the city, by creating a totally new service infrastructure or (more commonly) by "adding on" to an existing infrastructure the basic elements of a support network. Add-ons would typically occur in established communities, with existing services such as a shopping mall with good transportation, a community centre, and a library. If specialized services for the homeless could be added on (e.g. a sheltered workshop), then supportive community-based networks could meet local needs.

The purpose of the service hub is everywhere the same. It aims to provide the necessary level of support and choice in community-based care so that the homeless are able to regain, and to maintain themselves in, independent living. Needless to say, it is difficult to define a priori what each service hub will consist of. Resources and needs will vary by client needs, city size, availability of specialized care, structure of the voluntary sector, and local policy.

Community Opposition and
Community Planning[54]

It is intuitively obvious that groups in need often have multiple problems which are most efficiently served through a set of closely linked facilities, i.e. a service hub. However, while many pay lip service to the concept, there are few examples where these principles are being used to help the homeless. Community opposition is strong, particularly to residential services. When service providers attempt to open residential facilities, they typically face organized and angry neighbours who block issuance of planning permits. A common result is exclusion of services from the rejecting community. Some localities have even institutionalized their exclusionary policies by banning certain classes of facilities from their jurisdictions.

Community opposition is based on fear: of the homeless, of increased traffic and noise, and of decline in property value. In most instances, these fears are groundless. Many studies have shown that "spillover" effects are negligible – for instance, there have been no demonstrable property-value declines near community mental health facilities, and local effects of human services vary by type of facility.[55] The problem for service providers and city officials is how to overcome misinformation and opposition and get on with helping the homeless.

We believe that action by both municipal and provincial/regional governments is essential in resolving conflicts between the rights of communities to self-determination and their obligations to assist the homeless.[56] At the local level, several strategies are possible. For example, cities could create a "clearing-house" to track the location of various types of services for homeless and other dependent populations; prepare siting assistance to service providers; educate neighbourhoods about the needs of the homeless and the value of community-based service; and mediate conflicts over siting or management of facilities. Another strategy would be to incorporate human service facilities into the formal planning process, designating areas of the city in which development of service hubs would be encouraged and where some types of facilities would have "by-right" zoning privileges.

As things stand, each municipality in a region makes its own land-use policy. If some cities wish to exclude the homeless and their support facilities, they are usually at liberty to do so. Moreover, there is no incentive for one community to develop service hubs unless all cities in the region are required so to do. Without this requirement, communities that play host to homeless services are apt to have their neighbours "free-ride" on their goodwill.

Regional and provincial governments therefore have to help create an equitable shelter and service system for the homeless. By establishing "fair-share" policies for their constituent municipalities and counties, higher tiers of government can assist the establishment of service hubs across uran areas. Such fair share policies would determine the proportion of services that each municipality should host, based on region-wide projections of service need and in collaboration with local jurisdictions. Negotiating strategies could include incentives to develop service hubs – additional public services (such as parkland) to offset the "costs" of sheltering the homeless; the ability to trade off a large-scale facility for more, smaller facilities; or some provision whereby localities can substitute funds in lieu of service sites. Such approaches would encourage development of service hubs in areas best suited to the needs of diverse client groups.

CONCLUSION

The distribution of homeless people in Canada is geographically uneven; so is the availability of services to help them. In our explanation of homelessness, we have emphasized the facts of geography in the genesis of the current crisis and in the generation of solutions to aid the homeless. However, we have also underscored non-geographical factors which helped to create homelessness – most notably, lack of political will to attack the problem. Perhaps recent moves to encourage the homeless to organize themselves into a political constituency will give some impetus toward political action.[57]

CHAPTER SEVENTEEN

Geography of Urban Health

S.M. TAYLOR

The geography of health is high on the agenda of public issues. It is a matter of urgent public concern, media attention, political debate, and policy statement. This chapter examines three topics that illustrate the salience of the geography of health in Canadian cities today. The first is spatial inequalities in health, particularly within cities. The reduction of inequalities has recently been identified as the first challenge on the national agenda for health promotion.[1] This topic establishes that there is indeed a geography of health. Environmental, socioeconomic, behavioural, and biological factors create disparities that are manifest spatially and socially. Evidence presented from Montreal, Toronto, and Hamilton shows that health inequalities persist and may be widening.

The second aspect is the relationship between health and exposure to environmental contaminants. A long and growing list of substances poses a possible threat to human health (e.g. PCBs, dioxins, lead, radon, and sulphur dioxide). Many of these are concentrated in cities, given the presence of heavy industry, toxic waste sites, high traffic volumes, and other pollution sources. From a geographic perspective, environmental contamination is a negative spatial externality – an undesirable "spillover" effect of a facility or activity on a surrounding region and population. Negative environmental, social, economic, and health effects can occur at a range of scales from the individual to the regional. Determining the effects of contaminants on human health is complex because of difficulties in defining and measuring exposures and outcomes. Studies of the effects of air pollutants and exposure to hazardous waste in Toronto and Hamilton are described as examples of the type of research that has been attempted.

The third topic is a special case of the relationship between environment and health – the effects of the community environment on the chronically mentally disabled. This is one of several issues related to the geography of community mental health to have emerged in the wake of deinstitutionalization – the shift of patients from hospital to community-based care. This topic is also directly related to priorities on the national health agenda, which identifies "enhancing people's capacity to cope" as one of three national health challenges.[2] Its relevance for the mentally ill is recognized in a federal document, *Mental Health for Canadians: Striking a Balance*.[3] Research on factors affecting coping and satisfaction among the chronically mentally disabled in Hamilton is described to illustrate the issues involved and the contribution of geographers.

The three topics selected illustrate a more general convergence of geographic and public health research whereby geographers can participate with other social and health scientists in multidisciplinary studies of health problems in Canada. The Lalonde Report, *A New Perspective on the Health of Canadians* (1974),[4] shifted the emphasis from biomedical to socioecological models of health and from biological to environmental, social, and behavioural factors as determining the health status of Canadians. The more recent federal document, *Achieving Health for All: A Framework for Health Promotion*,[5] speaks of health as "a resource which gives people the ability to manage and even change their surroundings ... a basic and dynamic force in our daily lives, influenced by our circumstances, our beliefs, our culture and our social, economic and physical environments."

Some brief comments on the measurement of health status may be useful. Research on the geography of health depends on availability of relevant health information at appropriate geographic scales. Vital statistics provide mortality data which can be used to analyse spatial variations in life expectancy and in causes of death. For particular diseases, registry data compiled by health agencies may provide detailed information on mortality and morbidity (i.e. the incidence and prevalence of disease). For example, the Ontario Cancer Registry contains information on over 500,000 cases of cancer newly diagnosed in residents of Ontario between 1964 and 1985. Statistics on the use of health care services include hospital records and statistics compiled from data from provincial health insurance plans. Information from community health surveys can uncover morbidity not otherwise reported through health care use or other records.

Taken together, these sources would seem to provide comprehensive data from which one could characterize spatial and temporal trends in mortality and morbidity. In fact, geographically disaggregated data are much more limited than would at first appear. Confidentiality of health records often bars access to researchers. Geographical information is frequently incomplete or missing in health records. Compilation of health data by geographic units has received little priority. Thus, despite vast quantities of health information, much of it is neither available nor accessible in a form suitable for geographical analysis, particularly at a local scale. The most complete and comprehensive data currently available are the mortality information contained in vital statistics published by Statistics Canada.

In spite of these limitations, research on the geography of health and health care is expanding quite rapidly among geographers and related social and health scientists both inside and outside Canada. Evidence can be seen in the recent publication of four texts,[6] the proceedings of three international symposia on medical geography held under the joint auspices of the Association of American Geographers and the Institute of British Geographers,[7] and the growing number of articles published in such journals as *Social Science and Medicine*.

INEQUALITIES IN HEALTH

The Black Report (1980) in the United Kingdom[8] drew international attention to the persistence of profound social inequalities in health status. It brought into question the efficacy of the National Health System in reducing inequalities and challenged the system's role in improving the health status of the population. In the wake of the Black Report, similar, though less fully documented, findings on social inequalities have been reported for other advanced nations, including Australia[9] and Canada. As a consequence, reduction of inequalities is now a high priority in national health programs and is a primary concern in devising strategies for health promotion.

In Canada, studies have shown a strong socioeconomic gradient in health status.[10] Using 1971 mortality data for the 2,228 census tracts in 21 census metropolitan areas, Wigle and Mao[11] confirmed that mortality rates for all causes of death combined and for several individual causes varied substantially by income level. For males, life expectancy at birth in income levels 1 (high) and 5 (low) was 72.5 and 66.3 years respectively. For females, the corresponding figures were 77.5 and 74.6 years. Sex differentials in mortality by income level were the subject of a second analysis of the same data.[12] The results indicated a much greater differential between lower-income males and upper-income females than between upper-income males and lower-income females. The interaction between gender and income implied the effects of life-style and other factors not contained in the data. A geography of health in cities is implicit in these differences. Evidence from Montreal, Hamilton, and Toronto clearly demonstrates the spatial patterns associated with social inequalities.

A series of studies in Montreal showed the temporal persistence of social and spatial inequalities in health status.[13] More recent work by Wilkins[14] reveals that major disparities are increasing over time. Using census tract data on mortality for 1975–77, he showed a nine-year disparity in life expectancy between the "Golden Ring" of wealthy inner suburban municipalities and the disadvantaged "Lower City." When the census tracts were aggregated to correspond with the eleven planning districts used by the city's Service de l'habitation et de l'urbanisme, the difference between "best" and "worst" districts was approximately seven years, from 68 in

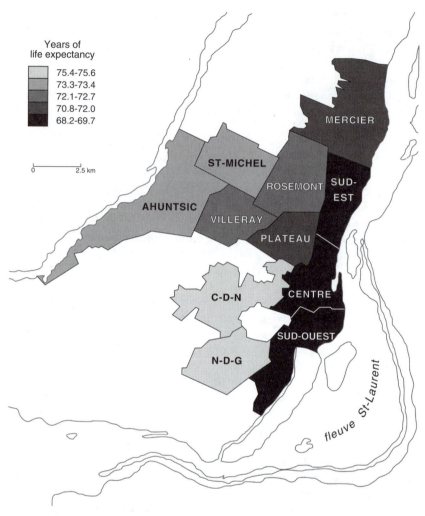

Figure 17.1
Social disparities in life expectancy in Montreal, 1976. *Source*: Wilkins (1980).

Sud-Est to 75 in Notre-Dame-de-Grâce (Figure 17.1). Between 1961 and
1976 life expectancy increased by 2.6 years in the "most advantaged" areas
but by only 0.3 year in the Lower City, a disquieting finding, given introduction of publicly funded hospital and medical insurance during the period.
Wilkins points to the need to think of public policy initiatives to change the
environment and life-styles and redress social inequalities in health status.
His comments are consistent with current thinking about determinants of
health and strategies for health promotion.

A   B

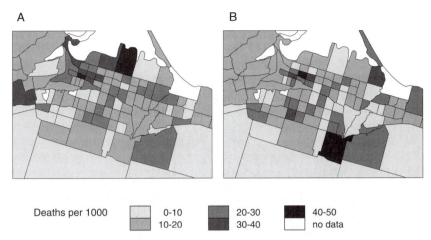

Deaths per 1000   [ ] 0-10   [ ] 20-30   [ ] 40-50
                   [ ] 10-20   [ ] 30-40   [ ] no data

Figure 17.2
The spatial pattern of annual death rates in the Hamilton-Wentworth region, 1980–82:
(A) males, 55–64; (B) females, 55–64. *Source*: Liaw et al. (1989).

A recent detailed analysis of intra-urban mortality in Hamilton[15], shows a situation comparable to Montreal's. Attention focused on mortality for men and women in the age group 55–64 for the period 1980–82. The census tract was again the unit of analysis. Three main conclusions emerged from spatial patterns of annual death rates (Figure 17.2). For both males and females, high mortality rates were found in the five or six low-income tracts close to the centre of the city. For males, a second concentration of high-mortality tracts coincided with the heavily industrialized "north end." For females, a more fragmented pattern may relate to the smaller number of female deaths and greater resistance to geographically differentiated mortality factors. An ecological analysis using the logit model estimated the relationships between mortality rates and seven socio-demographic variables (i.e. indicators of income, employment, and marital status) derived from census information. For men, income was the most important explanatory variable, accounting for about 40 per cent of the spatial variation in mortality. Income was also the strongest explanatory variable in female mortality, but its effect was weaker. Further analyses demonstrated that a negative mortality-income relationship persisted for males and females when death rates were disaggregated by major cause (i.e. cancer, circulatory diseases, and respiratory diseases), although the income effect was weaker on cancer mortality than for other causes of death.

Toronto provides another case study of spatial inequalities in health status. Mortality data compiled by the city's Department of Public Health and published as annual health status reports permit geographic analyses at the level of municipality, health area, and census tract.[16] Standardized mortality

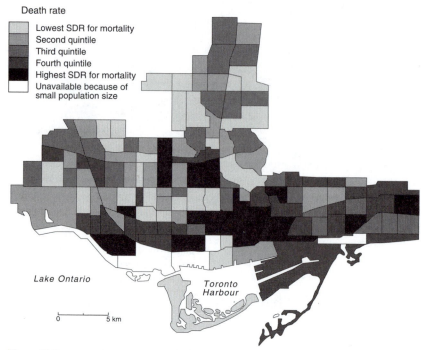

Figure 17.3
Directly standardized death rate (SDR) for all causes of death by census tract in the city of
Toronto, 1979–83. *Source*: City of Toronto (1985).

rates for males in 1982 show that Toronto ranked first (out of six metropolitan municipalities) for all causes combined, trauma (other than motor vehicle accidents), and suicides, and second for lung cancer and ischaemic heart
disease. The figures for females do not show such clear disparities. The results may reflect over-assignment of deaths in Toronto, but errors of this sort
would not account for the differentials in the rates for males. The mapping
of mortality rates (for 1978–83 combined) by census tract within Toronto
reveals more detailed spatial variation and suggests possible ecological correlates (Figure 17.3). For five of the six categories of cause of death examined, there was a large cluster of tracts with high rates in the south-central
area, adjacent to downtown. Deaths caused by cancer, however, showed
more even distribution. Tracts with lowest rates were quite dispersed but
were concentrated in the northern and western areas of the city. Many of the
tracts with the highest rates also had high unemployment and a large proportion of households below the poverty line. Although no statistical analysis
was attempted, this correspondence implies a relationship between income
and mortality similar to that reported in Hamilton.

The findings for the three cities show pronounced spatial inequalities in health status – or rather, in death – which correlate with variations in socioeconomic characteristics. Inequalities may be equally great for morbidity indicators. In this regard, Wilkins's innovative research combines life-expectancy and quality-of-life measures.

Wilkins and Adams[17] describe development of a health expectancy index which combines data on mortality from vital statistics, institutional data from long-term care facilities, and survey data on levels of disability in households. The index involves calculating life expectancy in each of six states of health: no disability; short-term disability only; minor activity restriction; restriction in a major activity; inability to perform a major activity; and long-term institutionalization. By assigning weights to each state, it is possible to calculate quality-adjusted life expectancy for groups defined by variables such as gender, income, region, and local area. The weights used by Wilkins and Adams range from 1.0 for each year of no disability to 0.4 for each year of long-term institutionalization. Wilkins[18] reports that inclusion of this quality-of-life factor leads to widening of social disparities.

Of particular geographical interest are the findings for different local areas of Montreal (Table 17.1). Standard life expectancy ranges from a low of 66.1 years in St-Henri/Petite Bourgogne to a high of 75.2 years in Westmount/Notre-Dame-de-Grâce Est, a difference of 9.1 years. Compare this with the disparity for quality-adjusted life years which ranges from 60 (St-Henri/Petite Bourgogne) to 71.4 (Montréal-Ouest/Notre-Dame-de-Grâce Ouest), a difference of 11.4 years. Both sets of life-expectancy figures generally correspond with variations in socioeconomic status, which fact constitutes further evidence of the covariance of social and spatial inequalities. In short, inequalities in quality of life appear to conform to the same pattern as those for quantity of life.

Why, despite generally improving health among urban Canadians, do such apparently profound spatial and social inequalities persist and may in fact be widening? The Black Report[19] found similar evidence on inequalities in the United Kingdom[20] and offered explanations that seem equally applicable to Canada. The first was natural and social selection: health status confers social status, rather than the reverse. Those who experience chronic illness are disadvantaged in terms of educational and occupational opportunity and thereby in economic and social advancement. This reversal of the relationships of social inequality to health inequality can be supported for certain types of chronic morbidity, especially mental illness.[21] Its wider application is less clear. It is also uncertain how this argument can account for differentials in mortality and life expectancy.

The second explanation is the behavioural or life-style hypothesis. Inequalities arise as a result of differential participation in health-threatening and health-promoting types of behaviour. Groups at high relative risk for

Table 17.1
Health expectancy and social-class indicators, by local area, Montreal, late 1970s

Area	Life expectancy* (years)	Disability-free life† (years)	Quality-adjusted life‡ (years)	Income 1970§ ($)	Professional occupations# (%)	Education < 9 years" (%)	Population 1976 (Total N)
Westmount/NDG-Est	75.2	64.9	70.8	14,500	18.9	13.8	61,000
Montréal-Ouest/NDG-Ouest	74.5	67.0	71.4	10,300	13.0	20.9	50,000
Metro	71.8	60.5	67.3	10,400	12.7	15.1	36,000
St-Louis/Mile End	70.8	60.1	66.3	6,600	2.6	53.0	55,000
St-Henri/Petite Bourgogne	66.1	51.6	60.0	6,500	2.6	53.9	25,000
Average	72.6	62.0	68.1	10,200	10.8	28.3	227,000

Source: Wilkins (1983).

* Health expectancy figures based on population data for 1976, death data for 1975–77, and non-institutionalized disability data for 1979–80; data on long-term institutionalization not included in this analysis. Social-class indicators based on data from 1971 and 1976 Census of Canada. Local areas refer to Local Community Service Centre districts of the Department of Community Health, Montreal General Hospital.

† Free of any health-related activity restriction (long-term limitations only).

‡ Weighting factors used were as follows: major activity impossible = 0.5; major activity restricted = 0.6; minor activity restricted = 0.7; short-term disability only = 0.5; no activity restriction = 1.0.

§ Average household income in 1970, including transfer payments.

Labour force engaged in managerial, administrative, teaching, and related occupations (occupation major groups 11 and 27) in 1971.

" Compared to population 15 years of age and over.

both morbidity and mortality commonly engage in health-threatening activities (e.g. smoking, excess alcohol and drug consumption) and infrequently participate in health-promoting behaviour (e.g. regular exercise, low-fat diet). By this argument, social and spatial inequalities arise through the congruence of social group and general patterns of health-related behaviour. This explanation has prompted strategies for reducing inequalities which identify the individual as responsible for his or her health, and it sees health promotion as supporting and encouraging people to exercise this personal responsibility.

The third explanation addresses inequalities from a structuralist perspective and challenges the assumptions that underlie the behavioural argument. It limits individual responsibility (which can too easily translate into "blaming the victim") and considers the structuring of life chances in society, which exposes certain groups to health hazards yet denies them access to high-quality health care and discourages participation in positive health behaviour. From this perspective, the higher mortality rates for men in the north end of Hamilton, for example, raise questions about exposure to hazards in the local workplace, a heavy industrial area. Equally, they may reflect limits in discretionary time and income as well as limited access to resources and facilities which would support engagement in positive health behaviour.

In their discussion of the Black Report, Townsend and Davidson[22] conclude that more of the evidence on inequalities can be explained by the third, or structural perspective. In short, health inequalities are rooted in class structure as it affects economic status, work conditions, and deprivation in the home, school, neighbourhood, and social environments. Strategies directed toward behavioural and life-style change may not necessarily fail, but effective prescriptions for improvement in health status ought to reflect the fundamental determinants of inequality.

ENVIRONMENTAL CONTAMINANTS AND HEALTH

The second aspect of the geography of health in Canadian cities to be considered is the relationship between exposure to environmental contaminants and health status. Although this is a high-profile issue, there have been relatively few studies of the effects of contaminants on the health of urban Canadians. There are some major methodological problems[23] in assessing both exposures and effects. There are problems, for example, in establishing the type, intensity, and duration of exposure. Consider a toxic waste facility where various types of solid and liquid waste have been dumped. There are unlikely to be detailed records from which to document the composition and amounts of the toxic agents present. Moreover, the chemical reactions between various agents could have created new and unknown compounds. In the absence of detailed monitoring over space and time, the intensity and duration of exposure in the residential community cannot be accurately determined. The problems of exposure assessment are compounded by mobility within the population – short-term movements outside the residential area to places of work and other locations and long-term mobility, as households enter and leave an affected area.

Complications equally impede the measurement of health effects. In most situations, the health outcomes that are plausibly related to specific exposures are not well defined. There may be uncertainties regarding type of exposure, but even if toxic agents are well established, toxicologic data may not define health outcomes, particularly to low-dose exposure. In addition, outcomes that typically generate most community concern, such as cancer and reproductive disorders, have a complex aetiology involving multiple causes, and even strong epidemiological designs cannot always separate the effects of environmental and other factors. Further complexities are the relatively few cases of a disorder likely to occur within an exposed population and the latency period between exposure and disease onset (15 or 20 years for certain cancers). One response to these last problems is to focus on short-term effects of exposure that are likely to show relatively high prevalence in an exposed group. Studies of the health effects of environmental contaminants from toxic waste sites, for example, have frequently used surveys to measure self-reported symptoms of short-term effects such as respiratory, digestive, and neurological disorders, although a sensitized population has a strong vested interest in the results.

This brief review of some of the difficulties surrounding research on the health effects of environmental contaminants underlines the complexity of the problems and the barriers to strong inference and definitive results. Study results, whether positive or negative, tend to contain a list of qualifiers such that the conclusion is equivocal about the degree of risk to human health and the results can compound rather than alleviate community concern. With these methodological problems in mind, we turn to four studies on the effects of environmental contaminants in Canadian cities. The examples are all drawn from southern Ontario, which is perhaps not surprising, given the juxtaposition of people and industries in this highly urbanized region. The first two studies deal with the effects of ambient air pollution in Hamilton; the last two relate to the effects of exposure to toxic chemicals from an industrial area in Toronto and a landfill site in Hamilton.

The first example is an ecological study of the relationship between air pollution and lung cancer in Hamilton.[24] Previous findings[25] reported a fivefold difference in lung cancer rates for 1967–71 among different areas of the city. The highest rates were in neighbourhoods adjacent to heavy industry with the heaviest air pollution, but age, sex, and other confounding factors were not controlled and so the effect attributable to pollutants could not be assessed. To improve on this earlier research, data on mortality from death certificates and hospital records were combined with questionnaire information on confounding variables.

The city was divided into four areas of air pollution level based on ward boundaries. Detailed pollution data for a representative set of monitoring stations were not available, and so the areal division has to be viewed as an approximate ordering of pollution level. The analysis involved comparing lung cancer mortality ratios for the four areas. Consistent with previous findings, the crude mortality ratio for males in the area of highest pollution was more than double that in the lowest-pollution area. Various standardized mortality ratios (SMRs) were calculated using the questionnaire data on age, smoking, and occupational histories. The difference between high- and low-pollution areas was greatly reduced once these confounders were controlled. For men, the SMR in the highest-pollution area was only 15 per cent above that for the low-pollution area. For women, the SMRs were not clearly associated with pollution level. The researchers conclude that there is evidence for males of some effect of air pollution on lung cancer, although its magnitude is considerably lower than previously suggested. They recognize the limits imposed by the crude definition of exposure – in particular, the possibility that specific point sources of pollution may result in considerable variation in exposure in each of the four areas. They make no claim to apply their findings to the risks imposed by current pollution levels, which differ from those in the past.

A second study of air pollution effects on the health of Hamilton residents[26] illustrates a different approach and focus. This study examined respiratory health among children rather than lung cancer mortality among adults. A prospective cohort design was used to monitor respiratory health in a random sample of over 3,000 elementary school children for a three-year period, 1978–81. The prevalence of respiratory symptoms was determined by questionnaire. Pulmonary function testing was conducted at the child's school. Outdoor air-pollution levels (suspended particulates and SO_2) at the school were estimated from measurements taken using a comprehensive network of monitoring stations distributed throughout the city. Data on indoor-air quality (parental smoking and gas cooking) were also obtained by means of the questionnaire. This allowed for investigation of the relative importance of outdoor and indoor air quality on respiratory outcomes. The study is noteworthy for its rigorous design, attention to confounding variables, and the quality control exercised in the measurement of exposures and outcomes. Several plausible influences on the children's respiratory health covaried with outdoor pollution levels. The prevalence of domestic smoking, parental respiratory symptoms, and gas cooking were highest in the industrial area of the city, which also had the hichest pollution. When these factors were controlled for, exposure to fine-particle air pollution was found to be significantly related to reduced lung function in the children.[27]

Current public concern about adverse effects of the environment on health are centred around exposure to toxic chemicals. Several health studies have been commissioned in response to public fears and demands for government action. Two examples are the Junction Triangle study in west Toronto[28] and the Upper Ottawa Street Landfill Site Health Study in Hamilton.[29] Geographers were included in the research teams in both cases. The main focus of concern in the Junction Triangle neighbourhood over several years was emissions from some twenty-four nearby industries. The district is demarcated by railway lines. Housing dates back to the 1920s, and many of the industrial plants have been operating for over twenty-five years. An informal residents' study in 1979 showed a high level of health concern related to frequent colds, coughing and sneezing, eye irritation, skin rashes, breathing difficulties, and fatigue. A 1981 study found higher absenteeism rates at the neighbourhood school compared with other randomly selected Toronto schools. A chemical spill in 1982 was the pretext for a major health survey which compared the health status of residents with that of people living in two other areas: a matched neighbourhood elsewhere in Toronto and the area surrounding the Junction Triangle.[30] The major finding was that the prevalence (i.e. all cases reported for the two-week period prior to the survey) in Junction Triangle children of "at least one cardinal symptom" (i.e. itching, burning or running nose not caused by a cold, itching or burning skin, throat irritation, tiredness or fatigue and itching, burning or watering eyes) was significantly higher than in the comparison areas (28 versus 15 per cent). A much higher proportion of Junction Triangle residents reported exposure to potentially hazardous risk factors (e.g. gas and chemical fumes), but physical measures were not available to assess these reports. Overall, the results were equivocal, but the Toronto Department of Public Health accepted the study's recommendation to assess further the health of children in the district to determine whether the increased reporting of symptoms was indicative of serious underlying health problems. Results thus far based on physical examinations have yielded negative results.[31]

The Upper Ottawa Street case in Hamilton closely resembles the Junction Triangle experience: mounting community health concerns culminated in demands for a health study. The culprit in this case was a municipal landfill site originally designated in the 1950s for solid waste disposal but used from the mid-1960s until its closure in 1980 for the illegal dumping of liquid industrial waste. During this period new residential development was constructed next to the site, thereby increasing the population potentially at risk. Two health studies were conducted. The first focused on the health status of municipal employees who had worked at the site and were therefore regarded as the group at greatest risk. The results of the workers' study were used to define the health outcomes to investigate in the subsequent residents' survey, which concentrated on respiratory, skin, sleep, and mood disorders. The self-reported symptoms of area residents and those of a matched control group were compared. The results confirmed a strong association between symptom reporting and exposure to the landfill site. The analysis applies methodological criteria (e.g. strength of association, consistency with findings from the workers' study, dose-response gradient, absence of recall bias) to assess the strength of the evidence for the association. The researchers recognized that the effects might be caused by perceptions of risk rather than actual exposure but conclude that on balance the evidence is stronger for the latter explanation.

In summary, the results described in this section show strong evidence of weak effects of various contaminants on different public health outcomes in the cases studied. The evidence is strong, given the methodological rigour in design and analysis. The demonstrated effects are relatively weak, however; not all outcomes showed significant differences between exposed and non-exposed groups, and not all differences reported were necessarily clinically significant. Typical of studies of this sort, the findings are therefore somewhat equivocal. They are not strong enough to prove or disprove whether exposure to contaminants represents a real threat to public health. Thus the conclusions can be used to support arguments on all sides of the debate over the need for stricter environmental regulation.[32] In the mean time, community fears about possible long-term health consequences of continued exposure may be compounded and may result in secondary health effects in terms of psychosocial morbidity.[33]

THE GEOGRAPHY OF
COMMUNITY MENTAL HEALTH

Deinstitutionalization, the move from hospital to community-based treatment of the chronically mentally disabled (CMD), has been the catalyst for two complementary avenues of research on the geography of community mental health in Canadian cities. The first has focused on community reaction to deinstitutionalized clients,[34] and the second on clients' reactions to the community. Attention in this section is restricted to the latter – specifically, the effects of the community environment on clients' well-being.

This third aspect of the geography of health is a special case of the relationship between environment and health but involves broader definitions of both than the examples in the previous section. By reason of disability and associated social and economic disadvantage, the CMD are vulnerable and typically have to deal with an amalgam of environmental challenges that would likely exceed the capacity to cope of many who enjoy full health.

Physical conditions include the quality of housing and the social situation – the nature and extent of social support networks and the attitudes and actions of neighbours and others in the community. Health status for this group obviously involves psychiatric diagnosis, but quality of life, which may well be linked to psychiatric outcomes, warrants special consideration in relation to environmental circumstances.

Advocates of deinstitutionalization assumed, without proof, that the community setting was more therapeutic and potentially more rehabilitative than continued confinement. In light of the deprivation experienced by many discharged patients,[35] the assumption now appears to have been naive and wishful. What factors within the community promote or retard integration, quality of life, and positive health outcomes? Geographers, in collaboration with other social and health scientists, can address these urgent issues. A recent study of the community experience of the CMD in Hamilton illustrates such a social geographic approach.

Kearns has proposed a general socio-ecological model of the community experience of the CMD based on his work with clients in three aftercare programs in Hamilton.[36] According to the model (Figure 17.4), positive (enabling) or negative (disabling) community experience depends on the interrelationships between the "outer world of shared experience" and the "inner world of personal experience." Contributing to the former are social support, employment, income, housing, community services, and attitudes and beliefs. Three dimensions of personal experience mediate the effects of these factors on clients: "identity, which incorporates biographical experience and characteristics brought to bear on personal experience; place, which includes both dispositions towards locale and experience of one's place in the world; and material existence which includes possession of and access to material opportunities."[37] Evidence to support the model was derived from quantitative and qualitative analyses of data obtained from sixty-six clients who were interviewed on two occasions six months apart.

The quantitative analysis[38] was based on an analytical framework describing hypothesized relationships between sets of client and community variables and objective and subjective indicators of clients' well-being.[39] The research developed subjective indicators of community experience. Clients rated their coping ability and satisfaction with respect to social situation, living arrangement, community experience, employment, and income. Composite indices of coping and satisfaction were calculated from these data. The relationships between client and community variables and these composite measures showed that clients coping well in the community tended to be more involved in gregarious activities, had enough to do, had more "significant others," and were more involved in mental health services. Those more satisfied tended to be older, had more "significant others," did not live in a boarding-house, were more residentially stable,

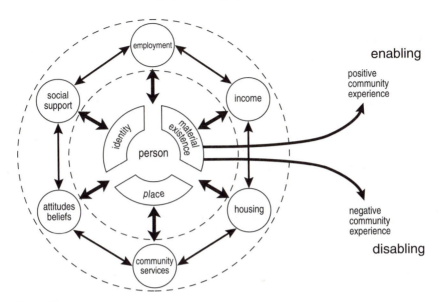

Figure 17.4
A socio-ecological model of the community experience of the chronically mentally disabled.
Source: Kearns (1987).

tended not to be on social assistance, and had enough to do in their spare time. The relationship between the client's housing satisfaction and ability to cope in the community was a focus of specific attention in a separate analysis.[40] Those more satisfied with their housing rated themselves as coping better overall.

The quantitative results were confirmed and elaborated by qualitative analysis of narrative accounts of experiences voiced by the clients during the interviews. Consistent with recent developments in social geography,[41] Lord, Schnarr, and Hutchinson have argued for use of qualitative methods in community mental health research on the grounds that they are naturalistic, holistic, and acknowledge the subjectivity of human behaviour.[42] Kearns organized the narrative data in two ways, by client and by theme. The two approaches poignantly convey the impact of the community environment on daily life experience, satisfaction, and ability to cope. Consider Larry's response to his housing situation in a single room in the central city: "I'm depressed because I'm not living where I'd like to be. My housing conditions are awful. Always have been in recent years. The rooms are always small. They're never well looked after. We're packed in like rats. There are mice and bugs where I'm living now. I never seem to be in a place I really want. I always get tired of places, so end up going to another one hoping it will be better. But it isn't, so I move on. It makes it worse because I'm used to better conditions. Right now I'm on Wellington. There's too much traffic and it's a slum. I just sit there and feel I'm just totally in the wrong place."[43]

When asked how he copes in the community, Larry responds with this impassioned indictment of the status quo: "There is no coping with the way it is. I have to work around knowing the wrongs of society and yet still live. You know it's pretty value laden to ask about coping. Best ask society how they're coping with people like me. We are what society's really like inside ... I'm a violation of the basic standard all Canadians should be living by. People think it's the standards of those at the top of society that make the quality of society. Well it's not. It's how the people at the bottom are doing that counts. And there are a lot of us."[44]

Taken together, results of the quantitative and qualitative analysis show the effects of the community environment on the well-being of the CMD in Hamilton. They support the components and relationships embedded within Kearns's socio-ecological model. Further work is clearly required to untangle the complex web of determinants that bear upon enabling or disabling outcomes for this group. The research mandate carries with it the challenge of translating the findings into strategies to improve the quality of everyday life for this vulnerable population and thereby narrow the gap between the intention and reality of community-based mental health care.

CONCLUSION

This discussion of the geography of health in Canadian cities has inevitably been highly selective in both topic and place. Attention has been restricted primarily to issues related to distribution and determinants of mortality and morbidity. A parallel chapter could be written emphasizing spatial aspects of the delivery of services. In the Canadian context, the work of Joseph and Phillips,[45] Rosenberg,[46] and Thouez[47] deserves particular note. The topics here were selected to show that geographical considerations are central to several major health issues in Canadian cities – an assertion supported by recent federal government statements on health priorities and health promotion strategies. Spatial analysis of intra-urban mortality and morbidity rates yields evidence of inequalities in health status, generates hypotheses regarding their ecological correlates, and provides a starting-point for further aetiological inquiry. Studies of the health effects of environmental contaminants are inherently spatial and coincide with several areas of geographical investigation, including spatial externalities, disease ecology, and hazard research. The complex relations between person and environment, fundamental to understanding the well-being of the CMD and other disadvantaged groups, can be profitably examined from a social geographical perspective, combining quantitative and qualitative approaches and methods of analysis.

There is good reason to suppose that the lines of inquiry illustrated by the studies described in this chapter will continue to be primary areas of investigation in future research on the geography of health in Canadian cities. With respect to the spatial analysis of intra-urban mortality and morbidity, access to new data sets will allow more detailed investigation of geographical variations in disease prevalence and incidence and more precise determination of the effects of ecological factors. For example, research is just beginning on geographical variations in cancer rates in Ontario, using the rich data set compiled by the Ontario Cancer Registry.[48] The effects of environmental contaminants on human health will continue to be a major public concern and a topic of intensive research. The health outcomes of interest will include both physical (e.g. cancer and reproductive disorders) and psychosocial effects (e.g. anxiety, depression, and interpersonal problems). Research in progress again in Ontario indicates the direction in which work on psychosocial effects might proceed.[49] Finally, the health of physically, mentally, and socially disadvantaged groups represents a priority for future research, consistent with the stated federal policy objectives of reducing inequities and enhancing people's capacity to cope.

CHAPTER EIGHTEEN

Changing Access to Public and Private Services: Non-Family Childcare

S. MACKENZIE AND M. TRUELOVE

In 1993, a young woman gives up her place in a university because she cannot find a place in childcare for her 18-month-old son. In Vancouver, a two-income family moves across the city in order to take advantage of better community childcare facilities for its two pre-schoolers. In Halifax, a professional woman changes from full-time to a less lucrative and less satisfying part-time job in order to "fit in" with the hours when childcare is available for her three-year-old, while a neighbour, the father of a family of four, quits his job to become a home-care giver, meeting his own and other families' needs for childcare while supplementing his family's income.

The bare figures on the need for childcare in Canadian cities are made up of thousands of such stories, of adjustments that Canadian parents and children are making on a daily basis. Adding up these tales of diverse forms of adjustment gives us statistics which indicate that most parents of pre-schoolers are now also wage earners; that 56 per cent of mothers of children under age 6 are in the paid labour force, 50 per cent of these working full time; and that 67 per cent of mothers of children aged between 6 and 15 are in the labour force. For families with a husband present, 90 per cent of fathers with children under 15 are in the labour force.[1]

These statistics, and the adjustments by parents and children that constitute them, are not just indicators of a specific social need which can be analysed in isolation. They are also components of fundamental changes in the nature of Canada's economy, society, and cities: alterations in family lives, parenting, relations to the paid labour force, and the roles of women and men. They signal and extend some significant adaptations in the places in which we live and work and in the ways in which we use our cities.

This chapter examines non-family childcare in the context of these social and environmental changes. We define non-family childcare as care undertaken for some form of remuneration on a regular basis outside the child's home by persons other than parents or guardians.[2] Building on a small but rapidly growing geographical literature, we outline a contextual framework for examining non-family childcare in contemporary Canadian cities. We then use this framework to examine the relationship between changing urban form and childcare, looking at how childcare needs have grown in Canadian cities over the past century and how people's responses to these needs are stimulating further social and economic changes. In conclusion, we suggest some of the implications of this framework for the provision of childcare and other human services in Canadian cities.

CHILDCARE AS A PUBLIC AND PRIVATE SERVICE

Much of the limited literature on non-family childcare in geography has tended to take the bald statistics on need as its starting-point. These studies have concentrated on formulating spatial models to maximize access to public childcare resources within a given urban structure.[3] While this work has made a valuable contribution to locational and urban planning, it has dealt with only a limited part of childcare. Not only has it concentrated almost exclusively on the formal, licensed sector, which in fact accounts for only about 25 per cent of non-family childcare for children under 6 in Canada,[4] but it has tended to treat patterns of childcare demand and provision as "independent variables," isolated from the historical context that led to these patterns and abstracted from the active adjustments of parents and children that produced them.

Some of the limitations of this literature stem from its underlying assumptions and limited objectives. It has analysed childcare primarily within the theory of urban public-facility location, itself an offspring of Pareto-optimal theory, which sees social outcomes generated through the decisions of optimizing individuals. These models have been criticized as inadequate for discussions of non–market-based urban facilities.[5] Attempts have been made to develop more realistic models through extending the variables considered to include political, non-market, and collective decision making[6] and by introducing social welfare criteria and equity as partners to efficiency.[7] Public-facility location theory has also been placed in a new context, following Castells's work on the collective means of consumption as the basis of urban life and organization,[8] or Dear's evocation of a "historical-hermeneutic" method.[9]

While these modifications have greatly extended the utility of public-facility location theory, they have done so by concentrating on the social and spatial outcomes, virtually excluding the content and activities of the home and neighbourhood. This orientation provided little or no basis for understanding the processes whereby needs and tastes are generated or altered. Needs and tastes appear, somehow, out of nowhere, or at best out of an un-examined and apparently inaccessible domestic and community sphere.[10]

Recently, influenced both by non-geographical literature on human ser-vices and by feminist methodology, which analytically privileges people's activity and experience,[11] some geographical discussions of childcare have taken a different approach. Rather than beginning with statistics on need and attempting to fit these into existing models, such work has started with the childcare experiences of parents and children and attempted to generalize and develop frameworks based in this experience.[12]

In this chapter, we wish to draw on both bodies of work, to go behind the statistics and spatial outcomes to ask how and why childcare needs and re-sponses are generated. We see non-family childcare needs, like needs for other human services, generated as people – women and men – live out their daily lives within the "private" places of the home and neighbourhood and the "public" places of wage work and political debate. The responses to these needs are located in the public sphere – for example, in the provision of state-regulated group childcare centres and in the growing politicization of childcare as a "national issue." Responses are also located in the private sphere of family and neighbourhood life – for example, in the provision of unregulated informal care in people's homes and in the juggling of time and resources within the family. Non-family childcare is thus neither "public" nor "private" but a set of needs and responses located at the intersection of public and private life.

We examine non-family childcare as generated by the interaction between family and neighbourhood activities and "public" economic and political life. Our framework thus incorporates the dynamic inherent in Pareto-optimal facility location models, the dynamic of relations between individ-ual actions and decisions made in the family and neighbourhood, and the aggregate economic, political, and spatial results. We place this dynamic in its historical and social context and take it as our starting-point.

CHILDCARE AT THE
INTERSECTION:
HOW DID IT GET THERE?

Caring for young children is a primary and universal human need. The way people have cared for children has varied tremendously over history and across the globe.[13] But childcare has consistently had certain characteristics: constant attention and instant decision-making in response to unpredictable demands. It is an intense, interactive, and very localized activity. Over his-tory, it has been primarily the responsibility of women, especially mothers, and because of its immediate and localized nature, it has generally been carried out within the household. However, the internal composition and social position of the household and thus women's position as part of that household have varied, both functionally and spatially.

In Canada, between the late nineteenth and mid-twentieth centuries, the household and its activities were defined as the core of "private" life. With increasing urbanization and the rise of wage-work as the predominant means of survival, the family lost many of its economic functions, and increasing numbers of households took the form of nuclear families, shedding extended family members and boarders.

This was a process reflected and reinforced in the changing spatial organization of the city and the home.[14] By the early twentieth century, extensive suburbs were developing in most Canadian cities. These suburbs were a new spatial form, part of a fundamental change in the structure of the early industrial city with its apparently chaotic jumble of workshops, offices, and transport facilities, interspersed with homes, shops, and schools. The suburbs, in contrast, were seen as healthy places, far from the congestion, pollution, and dangers of the downtown world of industry and commerce, a "private" place for recuperation and family life. Here, it was assumed, women would devote themselves to the care of children and husbands.[15]

The extent to which most Canadian suburbs and their resident households ever achieved the suburban ideal is questionable. But throughout this century, it remained the "taken-for-granted" basis of urban planning. For the most part, cities were built on the understanding that if sufficient sewers, gas lines, and housing were provided, the family – composed of a full-time domestic worker and a full-time wage earner together with their dependent children – would take care of the rest. Some services were provided – schools, the occasional shopping centre, frequent gas stations, and a few social agencies (often afterthoughts).

The provision of non-family childcare reflected these assumptions. Even though women in suburban households were less and less likely to have childcare assistance from extended family members or servants, there was nearly no formal provision made for childcare. Early-twentieth-century crèches and nurseries were run largely by charitable organizations for poor inner-city families. They freed low-income women, often sole-support mothers, for menial and domestic jobs. Following passage of the federal Mothers' Allowance Act in 1920, the availability of public childcare declined; in particular, agencies ceased to provide infant care. By 1933 there were only twenty day nurseries in all of Canada, serving about 2,500 children.[16] Aside from a brief flurry during the Second World War, childcare continued to be socially invisible.[17]

Yet by the 1960s, for more and more Canadian families, maintaining life in this private domain required two incomes, and a growing number of women with children entered the labour force.[18] These women did not stop caring for their children and husbands in the private sphere of the home; they added a new "public" job to their existing responsibilities. Every day, they moved back and forth between their private home workplace – designed on the assumption that they were there full time – to their public, paid workplace – organized on the assumption that workers had no other responsibilities.

Despite official recognition that women who worked for wages to assist the war effort required childcare, realization that this new movement of women into the wage sphere would throw up similar needs took a long time. This blissful political indifference, facilitated by the divorce of economic and household spheres in urban location models, was reinforced by a city divided into "public" economic-political and "private" household-community spaces. What women did at home was largely invisible to the world of industry and commerce and public policy makers. Employers and policy makers were able to assume that children were somehow being cared for somewhere else, while simultaneously taking for granted that their workforce had no home-based constraints.

Women, however, could not make such assumptions. The conflicts of the dual-role woman in a separated city found expression in a growing women's movement and, within geography, a developing literature on women and environments.[19] In these activities and analyses, the care of pre-school children was a central issue. Studies documented that access to childcare was one of the most vital issues in the daily life of urban women. The availability of childcare and its location, hours, and cost influenced the kind of wage work that women with children could do, the hours they could work, and the distance and direction they could travel to jobs. Childcare provision directly influences the time schedules and travel patterns of family members (see chapter 5). In addition, childcare provides money-earning work for women and an increasing number of men, in both formal group-care and informal home-care.[20] Childcare has also entered the "public" arena as a political issue. Throughout the 1970s and 1980s, politicians at every level have been confronted with arguments that childcare is a social responsibility and that adequate provision is necessary to economic efficiency.[21]

In short, childcare was not just a private family matter, nor was it just a "public service." It was a service that had implications for all aspects of working and family life, for employment, for travel patterns, for relations between men and women, and indeed for decisions about the number and spacing of children within a family. This is the context within which policy makers and individual parents and children have developed contemporary childcare services. In the process of so doing, these people were creating new kinds of spaces in cities, new patterns of movement, and new ways of working and living. The next section documents some of these responses.

CHILDCARE AT THE INTERSECTION: HOW ARE PEOPLE RESPONDING?

Childcare is pre-eminently a service where demands for government action are created directly from the adjustments that women (and men) have made to help maintain their private family life. Not surprising, as the number of mothers in the work-force grew to become the majority of mothers of young children, the response to these demands was both public – the establishment of formal childcare centres and programs – and private – the unregulated growth of a home-care sector to compensate for the inadequacies of public provision.

The Formal Sector

The origin of childcare as a charity for low-income families is still evident in contemporary policy with respect to formal childcare. The federal government does not have a comprehensive policy for funding of childcare services because fees for childcare are assumed to be a parental responsibility. However, a funding scheme begun in 1964, the Canada Assistance Plan (CAP), provided for equal federal-provincial cost sharing of a wide range of welfare services, including, since 1966, childcare funding for parents in certain circumstances. The terms of CAP allow the federal government to provide 50 per cent of provincial and municipal costs of childcare services for low-income families. Ottawa, however, has recently begun to impose ceilings on its contribution levels. An August 1991 decision by the Supreme Court of Canada confirmed its right to limit its funding levels.

Whether or not CAP is an appropriate way to finance childcare has been an issue ever since the plan was established. Its aim was to help those in need – a welfare service – and to have the federal government share more fully in the costs of social services; its aim was not to create a universal childcare system.[22] Many today are calling for a childcare program created on broader lines, with working parents included in the category of those "in need."

No direct funding is available to families not "in need," according to the regulations of CAP. However, a relatively large tax deduction under the Income Tax Act helps working parents meet childcare expenses. Since 1987 the maximum amounts claimable as a deduction for childcare costs have been $4,000 per child for children age 6 and under and $2,000 per child for older children. This deduction is available for formal or informal care, provided that receipts can be shown. In the 1988 tax year, 524,810 Canadians claimed $1.131 billion in allowable childcare expenses.[23] This deduction is a significant benefit for middle- and high-income Canadians. There is no

Table 18.1
Formal daycare spaces in Canada, 1973–90

Type of sponsorship	1973	1976	1979	1982	1985	1988	1990
Public	3,409	9,882	6,215	5,977	10,363*	10,911*	11,574*
	(12.7%)	(12.6%)	(7.3%)	(5.5%)	(6.7%)	(5.1%)	(4,6%)
Non-profit	10,850	34,380	36,319	59,075	81,433	112,066	145,061
	(40.4%)	(43.9%)	(43.1%)	(54.4%)	(52.4%)	(52.9%)	(57.3%)
Commercial	12,552	33,891	41,549	43,461	63,631	88,912	96,723
	(46.8%)	(43.3%)	(49.4%)	(40.1%)	(40.9%)	(42.0%)	(38.2%)
Total centre spaces	26,811	78,153	84,083	108,513	155,427	211,889	253,358
Family daycare spaces	1,560	5,367	9,769	14,427	22,623	30,839	38,159

Sources: Health and Welfare Canada (1973–90).
* Excluding Quebec's school-age daycare spaces.

quota on the number of people who can claim this deduction; but there are limits on the numbers of subsidized spaces for lower-income families. The working poor, and lone-parent families in particular, who cannot get a child-care subsidy may well have very serious problems in meeting their families' basic needs.

The location of childcare at the intersection of public and private spheres of life has resulted in the development of a variety of forms of public provision: publicly funded centres, commercial centres, non-profit centres – including community-run and co-operative programs – and licensed home childcare spaces. The growth of formal childcare in Canada is shown in Tables 18.1 and 18.2. Table 18.1 shows the growth in the number of spaces in the formal sector, in centres and licensed family homes, since 1973. (Most of these spaces are *not* subsidized via CAP.) Between 1973 and 1990 the total number of centre spaces increased by 845 per cent. Since 1978, availability of childcare has increased steadily, as a result of constant pressures and demands by community groups across Canada.

While the number of spaces has grown strongly, the three types of centres – public, non-profit, and commercial – have shown different trends. The number of spaces in public centres fell from a high in 1978 of 10,140, a total that was not reached again until 1985. The greatest growth was in non-profit centres, a trend that is expected to continue. Commercial centres have also shown extensive growth but at a rate somewhat behind that of the non-profit sector.

There are remarkable differences among provinces in the provision of daycare facilities (see Table 18.2). In 1990 Alberta and Ontario had 56.3 per

Table 18.2
Interprovincial comparisons of licensed daycare spaces, 1990

Province	Public centre	Non-profit centre	Commercial centre	Family daycare	Total
British Columbia	–	12,489	6,000	7,155	25,644
Alberta	1,000*	19,049	34,823	6,962	61,834
Saskatchewan	–	3,795	–	1,980	5,775
Manitoba	–	9,199	972	2,623	12,794
Ontario	10,574†	60,794	36,178	11,762	119,308
Quebec	29,107‡	30,574	10,638	7,273	77,592
New Brunswick	–	3,075	2,493	96	5,664
Nova Scotia	–	3,448	2,529	123	6,100
Prince Edward Island	–	876	1,037	35	1,948
Newfoundland	–	620	1,782	–	2,402
Northwest Territories	–	578	116	66	760
Yukon	–	564	155	84	803
Totals	40,681	145,061	96,723	38,159	320,624

Source: Health and Welfare Canada (1990); since 1988 this publication has combined public and non-profit centre spaces.

* Provided by Child Care Programs, Alberta Family and Social Services.

† Provided by Child Care Branch, Ministry of Community and Social Services, Ontario.

‡ School-age spaces only, operated by the Quebec Department of Education.

cent and 30.3 per cent respectively of their formal childcare spaces in commercial centers – but Alberta has few public centres while Ontario has 8.9 per cent of its formal childcare spaces in public centres. Only in Alberta, Prince Edward Island, and Newfoundland are the majority of spaces in commercial centres. Non-profit centres dominate in Saskatchewan, where new commercial centres have not been allowed to open since 1974, and in Manitoba, Ontario, and New Brunswick. Formal family home childcare is relatively undeveloped in eastern Canada. Clearly there are many different ways of providing childcare.

Why do these provincial variations exist? There are some strong political differences, with historical roots. Most provinces other than Ontario had no wartime history of childcare provision. There are also many approaches to the public-versus-private provision of social services; for example, Alberta used to have strong municipal childcare programs, but these have declined. British Columbia has not allowed infant care in centres in the past but has recently started some experimental programs.

The quality and quantity of care vary also within provinces, from city to city. In 1971 there were 104 centres offering full-time care in Metropolitan Toronto: 20 public, 29 non-profit, and 55 commercial. These centres had a total capacity of 5,774 children. Forty-nine of the non-profit and commercial centres also had "purchase-of-service" agreements with the metropolitan government: that is, subsidized as well as unsubsidized children could attend. In almost all cases, the municipal centre, owned and operated by Metropolitan Toronto, care for subsidized children only. By 1990 there were 498 childcare centres offering full-time care: 42 public, 289 non-profit, and 167 commercial. Two hundred and thirty-one of the non-profit centres and 131 of the commercial centres had purchase-of-service agreements.[24]

By mid-1990 there were approximately 21,500 full-time daycare centre spaces in Metro Toronto. If the most recent participation rates for the Canadian labour force are assumed to hold true for Metro Toronto, about 90,000 children under the age of 6 require childcare.[25] Even if half of their mothers do not work full time or prefer care by a relative or some other arrangement, parents of more than 45,000 children might potentially want to fill 21,500 full-time daycare spaces.

Moreover, these spaces are not equally distributed across the metropolitan area. Non-profit centres have been concentrated in the inner city, and commercial ones in the outer suburbs. This pattern seems to be the result of two principal factors. First, the outer suburbs were traditionally thought of as higher-income areas that would attract commercial centres, while the inner city was assumed to be lower income. This simple pattern is certainly not accurate today; governments have deliberately mixed housing types and income groups in many newer suburban areas. Commercial centres may also choose lower-rent locations on routes used for the journey to work. Second, the inner city is the traditional location of service and charitable agencies that would provide childcare and other services, while the outer suburbs have been seen as lacking a strong base of such voluntary agencies. These agencies usually require time to develop, yet families with young children abound in new subdivisions.

Policies and actions of some local governments and school boards have recently redefined these differences. The city of Toronto and the Board of Education, for example, have enacted policies to encourage the growth of non-profit centres only. York and North York have shown rapid growth in the number of non-profit centres, which are now in the majority in these two municipalities.

This location pattern has important effects on families' daily lives. The average distance travelled to a childcare centre in a sample of 1,619 children attending 43 centres full time was 2.9 km. Over 82 per cent of the children travel 5 km or less from home to childcare (see Table 18.3); indeed, over one-fifth (21.9 per cent) travel less than 0.5 km. In the sample, few children had to cover long distances.[26]

The average distance travelled is slightly lower for subsidized children (2.8 km) than for those paying full fees (3.1 km). For these distances, transit or car travel would be necessary; most low-income families would thus

Table 18.3
Distances travelled from home to daycare centre, Toronto study

Distances (km)	Total		Full-fee		Subsidized	
	Freq.	%	Freq.	%	Freq.	%
0–0.5	354	21.9	103	16.9	251	24.8
0.6–1.0	293	18.1	109	17.9	184	18.2
1.1–2.0	298	17.8	120	19.7	168	16.6
2.1–5.0	405	25.0	152	25.0	253	25.0
5.1–10.0	191	11.8	90	14.8	101	10.0
10.1–15.0	58	3.6	24	4.0	34	3.4
15.1–20.0	16	1.0	3	0.5	13	1.3
20.1–25.0	8	0.5	4	0.7	4	0.4
>25.0	6	0.4	3	0.5	3	0.3
Total	1,619		608		1,011	
Average distance	2.9 km		3.1 km		2.8 km	
Standard deviation	3.8 km		4.0 km		3.7 km	

Source: Truelove (1989b).

rely on transit, with the inconvenience and extra time usually involved. The most noticeable difference between these two groups is in the proportion of children travelling less than 0.5 km: 16.9 per cent of unsubsidized and 24.8 per cent of subsidized. This difference may be created by those childcare centres that have a large proportion of subsidized children in low-rent apartment buildings or very close by. Four of the five municipal centres in the sample are in public housing complexes. These heavily populated areas may have attracted childcare centers of all types: commercial, non-profit, and municipal. Also, despite the expansion of workplace childcare,[27] few new subsidized spaces have been made available because of provincial government fiscal restraint in the early 1980s. Thus the trips of subsidized children may not have lengthened, on average, as much as have those of full-fee children; again, this may reflect differences in mobility and car ownership.

Even the well-intentioned actions of government and non-government agencies may thus further limit the choice of childcare centre. At the same time, these public agencies influence families' private activities. Government agencies help decide the availability of formal care spaces and subsidies and thus affect the distances parents and children must travel and the price they pay for care. New policies can also alter the supply of and demand for childcare and thus influence the growth and use of informal home care.

The supply of formal childcare spaces in Metropolitan Toronto has grown rapidly from 1971 yet has not kept pace with demand. Must we conclude that the law of supply and demand does not apply to a service such as child-care? Why are people not coming forward to create centres for the families who state that they wish to use them? Or can quality childcare not be provided at the prices that parents are willing to pay? In Metro Toronto, the price of infant care in centres (for children less than 18 months old) is currently over $8,000 per year. In the past, few parents could afford formal childcare; now that the "market" has grown, perhaps the demand fluctuates in any neighbourhood or is considered too uncertain for operators to open centres.

Waiting lists should provide a measure of the demand for centre spaces. But a 1984 study of all centres serving pre-schoolers in Metro Toronto found that 4,638 children's names were on waiting lists at the same time that 1,696 vacancies were reported.[28] Such figures give only a rough estimate of actual demand; some centres do not keep waiting lists, while some parents submit their children's names to more than one centre. Recently some centres have asked for non-refundable deposits of up to $150 to list a name. The waiting lists of those who qualify for subsidized spaces, but who have not received a space, because of a limited supply, have grown very long in recent years when the Ontario government has provided very little additional funding for these subsidies.

For the whole country, the report *Status of Day Care in Canada* estimates than in 1990 the formal sector could accommodate less than 11 per cent of infants (age 0 to 17 months) who require full-time care, less than 20 per cent of children 18 to 35 months, less than 55 per cent of children 3 to 5 years, and less than 10 per cent of children 6 to 12 years. Childcare centres clearly best serve to 3-to-5-year age group.

In a variety of surveys a high proportion of Canadian parents have stated that they would prefer to use childcare centres, if available.[29] Governments at all levels are being pressured to increase their financial commitment to formal childcare; in the last two national elections, 1984 and 1988, childcare was a major issue. However, the attendant costs and the common belief that childcare is a family responsibility, have resulted in a great deal of political rhetoric but little action. Most non-family childcare is still provided by the unregulated, home-based, informal sector.

The Informal Sector

While non-family home care has been a consistent feature of Canadian urban life, its current growth and development are a direct response to the inadequacy of the formal sector. The lack of public provision has created a demand for alternatives and a supply of home-care givers, most of whom are mothers also caring for their own children.

Table 18.4
Personal characteristics of home-caregivers

	Kingston, Ont.	Trail-Nelson, BC	Total
Total sample	27	23	50
Number who			
were mothers	22	21	43
had children at home	17	19	36
had children under six at home	15	14	29
had husband at home	21	18	39

Sources: Interviews with home-caregivers, 1984–85 (see Mackenzie 1987).

Available research on this sector is limited, for obvious reasons. Researchers' concentration on public facilities, in a city divided into public and private spheres, has obscured the economic significance of work in the home and community. The following discussion is therefore based on a case study of fifty home-care givers in two areas of Canada: Kingston, a city of 60,000 people in eastern Ontario with a largely institutional economic base, supplemented increasingly by tourism and retirees; and the Trail-Nelson area in southeastern British Columbia, a region dependent primarily on declining primary extraction, smelting, and wood-processing.[30]

As indicated in Table 18.4, the basic sample included 27 care givers in Kingston and 23 in the Trail-Nelson area. The characteristics of the sample conform to the limited information available on home-care givers elsewhere: most were mothers, their education and income levels were slightly below the average for women in their areas, and they were geographically dispersed throughout each city.[31] These women, and others in the home-care sector, provided the largest portion of non-family childcare in their cities and in doing so were actively changing the nature of their homes and communities.

Home-caregivers were using the private environment of their homes and neighbourhoods in new ways in order to gain a livelihood. In converting these private spaces and skills into public ones, they were extending and often creating new resources in their households and communities. Their immediate environment – the home – was being actively redesignated as a workplace as well as a living space. As indicated in Table 18.5, it was often being redesigned as well. Caregivers, sometimes with the assistance of their families, were undertaking renovations: finishing basements as playrooms, fencing yards, taking safety measures. They were purchasing, or more often making, equipment: toys, outdoor playsets, cots, tables.

Table 18.5
Informal childcare networks, caregivers carrying out home renovations

	Kingston, Ont.	Trail-Nelson, BC	Total
Total sample	27	23	50
Number who carried out some renovation	20	17	37
Major structural renovation (adding room, major outdoor work)	6	7	13
Medium-scale renovation (setting aside room, building equipment)	15	12	27
Minor renovation (safety features, play corners)	25	15	40
Number who intended to do more renovation	15	13	28

Source: See Table 18.4.
Note: Numbers are not mutually exclusive.

These activities often looked like those carried out by other parents who did not earn money caring for children at home. Yet caregivers, unlike (other) parents, tended to see these alterations as productive investments as well as improvements to their living space. Many said that carrying out their job efficiently required specialized spaces: large, light, "childproof" rooms set aside for the hours when the children were in the home or large, well-equipped, and fenced outdoor play spaces.[32]

Most caregivers had also redesignated their neighbourhoods as working places. They tended to assess their environs in terms of its capacity to provide safe, accessible play spaces and public transportation for outings.[33] They also considered the number of local families requiring childcare, thus measuring actual or potential market demand.

As indicated in Table 18.6, the majority of caregivers used the wide variety of available community resources. Most of these had been developed and were sustained by caregivers or parents, sometimes assisted by local social service agencies or education departments. For example, there were toy libraries and resource centres in both Nelson and Kingston. In both cases, professional staff ran drop-in centres and educational seminars for caregivers, parents, and children and offered referrals and advice. There were also a number of play groups, run by parents and caregivers, funded by parents' fees with subsidies from local agencies in some cases. They were held in venues ranging from unused school rooms and church basements to people's living rooms.

Table 18.6
Home-caregivers, use of local facilities

	Kingston, Ont.	Trail-Nelson, BC	Total
Number using			
toy libraries	10	7	17
play groups	9	12	21
drop-in centres	16	6	22
informal meetings	21	14	35

Source: See Table 18.4.

Note: Numbers are not mutually exclusive.

All caregivers emphasized that these groups had to meet specific requirements. The most important was geographic accessibility, on bus routes or
within walking range for three-year-old legs. The majority of those who did
not use these groups cited inaccessibility as a major reason. Most felt that
small groups in familiar, existing structures such as schools and homes were
best able to meet their needs. All caregivers mentioned the value of such
groups as community foci and as extensions of their home workplace.

While redesignating private resources to meet a combination of public and
private uses provided an adaptation strategy for caregivers, it was one severely constrained by the city's being separated into "living" and "working"
spaces. Caregivers' ability to extend resources was constrained by a number
of barriers, including the quality of care possible in private homes, extremely low pay,[34] and seeming indifference or even hostility to the service
they provided on the part of statutory agencies and group-centre staff.

These problems stem largely from a common source: the use of a limited
set of resources designed for one purpose (for "private" family life and individual consumption), in order to carry out a related but different purpose
– the provision of a public service. The homes and neighbourhoods were designed and built so that almost all family needs would be taken care of
within the individual suburban home; efficient social communication and
community interaction were not primary concerns. There are few financial
or social support resources available to make homes and neighbourhoods efficient workplaces as well as living places. Similarly, there are few training
opportunities for those extending their parenting skills into the public business of making a living, and few subsidies to supplement the wages of those
who are doing for pay what other mothers are doing without pay.

The problems that caregivers are encountering in their attempts to provide
a public service in an environment designed for private life are symptomatic
of a more general social problem: failure to recognize the collective needs
that arise for family life in contemporary society.[35] This is a problem that af-

fects both public and private care sectors. In the concluding section, we suggest that the intersection between public and private activity may be the most important arena in understanding and responding to human service needs in urban areas.

CHILDCARE AT THE INTERSECTION: IMPLICATIONS

Over the past three decades, the question of who is caring for children and where and how they are being cared for has moved out from behind the walls of private residential units into the public arena. The discussion has moved largely because so many children have moved into non-family childcare settings. This shift reflects and extends a new relation between the family, the core of the private sphere, and the economy, the core of the public sphere. Social relations between family and economy are intensifying. Both parents in most two-parent families now have a lifetime commitment to wage work, the economy is increasingly staffed by dual-role people who have another job as domestic workers, and childcare outside the family becomes an inherent part of family life and of family social relations from the infancy of the child.

The initial policy on group centres in Canada in the 1960s and early 1970s was developed in a social climate of growing interdependence between the internal workings of the family and public policy. Progressive educators and social policy analysts, as well as growing numbers of parents, argued that raising children was a social responsibility and that adequate provision for their care was properly a social service as well as a family-based activity. The separation between public and private activities was more and more dissolving into the intersection – that area of social life and economic activity where individuals, using household and community resources, are attempting to make a living and to provide public services. This intersection has been actively developed, and given concrete form, by public-sector and home-caregivers and by parents and children who use childcare.

Many parents prefer formal childcare for their children, in licensed daycare centres or regulated and inspected private homes. For children aged 3 to 5 years, many parents prefer a school-like atmosphere, in daycare centres, some of which have developed in the empty classrooms of existing school buildings. In order for childcare to be fitted into a family's schedule, centres must be near home, near work, or on the way from home to work. Local governments, non-profit agencies, school boards, and employers have all responded, in varying degrees, to this growing need. This means that childcare is increasingly provided at workplaces and in neighbourhood facilities.

For parents whose childcare fees are subsidized, there is no choice of mode: subsidies are available only in the formal sector. The number of subsidies available is, in all provinces, far lower than the number of families who have applied for and qualified for a subsidy. Without subsidies, most lower-income parents cannot accept a job outside the home or finish their education. For parents with higher incomes, the cost of non-subsidized childcare is substantial. As a result, many centres provide care that meets only minimum provincial standards. The wages of childcare workers, in both the formal and informal sectors, are also a public policy issue. These workers have very low wages for very important work: caring for the next generation of citizens in their formative years.

Other groups have been advocating the development of universally accessible, comprehensive, non-profit, high-quality childcare.[36] Organizations such as the Canadian Day Care Advocacy Association point out that provision of childcare to those who want it, in non-profit centres, would assure good-quality care without means testing or income testing of families and would guarantee higher wages for childcare workers. However, many parents may not want childcare in the formal sector, and home-caregivers might not be included in such a system.

Childcare is now a very prominent issue in national politics. Its provision raises important and emotional questions that touch some of our deepest values: how to be parents, how to raise children. Differences concerning how childcare can help people resolve these questions satisfactorily will be settled only by long public debate and by careful study of childcare needs and responses. Meanwhile, governments at all levels are being pressured to increase their financial commitment to formal childcare, yet the federal government's contribution has recently been shrinking.

The intersection of public and private activities is not confined to childcare. Changes in wage work and family life may mean, for example, that parents working full time may have responsibilities for elderly or disabled people as well as for children. The pattern evident here – individuals providing human services through the conversion of family-based resources to public resources – is being applied throughout society, in fields such as care of disabled children and adults, public health, and care for the elderly. We suggest that it can form the basis of a general principle to develop social policies as well as geographical concepts that can contribute to such policies. While there is wide regional variation in the necessity and extent of these policies, the same forces that led to the adaptation strategies of home-caregivers are evident more broadly at the intersection of the wage economy and social welfare.

We also require better models in order to understand and respond strategically to these developments; some are emerging in childcare and related fields. At this point, however, general principles are difficult to enunciate. The needs, the strategies, and the required resources are small scale, are highly localized, and must vary with and adapt to local communities and individual household needs. There are two general requirements here. First, we must break down the conceptual and analytical barriers between public and private activities, between productive spaces and living spaces, in order to see the ways in which people are themselves actively breaking down these divisions in providing human services in cities. Second, we must look at the content of the "interface" between public and private spheres and extend our concept of social resources to include the emerging forms, networks, and skills being created in this "new frontier."

CHAPTER NINETEEN

Cities as a Social Responsibility: Planning and Urban Form

P.J. SMITH AND P.W. MOORE

Most of the previous chapters have treated the social geography of Canadian cities as the product of spontaneous forces working within Canadian society. Now we must reverse the point of view and consider how society has acted, deliberately and purposefully, to shape Canadian cities. The concept of planning refers, in this context, to the means by which society attempts to direct the processes of urban change and development for public or collective ends. Urban planning thus constitutes an integral part of the complex apparatus with which communities govern themselves. Its particular purpose, or social responsibility, is to ensure that the eventual forms of urban development, the "planned" forms, best satisfy the community's own goals and aspirations – or are the best than can be obtained in the circumstances.

Although the idea that society should exercise some control over urban form is an old one, establishment of official regulatory systems has come about largely in this century. Canada is no exception. Our first public planning organizations were created between 1909 and 1914, and the legal machinery of planning became quite highly developed over the next twenty-five years. It had little practical effect at the time, but useful foundations were laid for the years after the Second World War, when the great surge of urban growth began. As long-pent-up demands were at last released, and new expectations were given voice in the optimism of post-war reconstruction, there was a new willingness – desire, even – to plan for a better future. Building better cities was vital to that vision and so fed the growing conviction that urban planning was a socially responsible act.[1]

Over the past forty years or so, planning has become a regular institution of Canadian urban life – "a normal and necessary" public activity, as one author puts it.[2] But how effectively has it functioned under its post-war regime? Are Canadian cities indeed better than they might otherwise have been? Are they even different? These questions lie at the heart of this chapter, yet they are terribly difficult to answer. Canadian planning is not a monolithic structure. There have been broad similarities in planning instruments and procedures and in the major issues with which planning has been concerned, but these do not guarantee identical consequences. For that matter, it is not always clear what the consequences are, let alone how to judge them. No one has ever attempted a systematic assessment of what planning has achieved, or failed to achieve, across Canada, and we cannot possibly be that ambitious here.

We concentrate instead on two things. The first section discusses the general nature of urban planning as it has evolved in Canada, emphasizing two characteristics that account for the difficulty of assessing its physical outcomes – the politics of planning and the fragmentation of jurisdiction. The next two sections review two issues that have dominated Canadian planning practice: planning for urban expansion – including growth management, metropolitan-regional planning, and the planning of new suburbs; and renewal and modernization of the urban fabric – especially the changing context of, and public initiatives in, redevelopment. Throughout these latter two sections we draw on actual experiences in a selection of cities, from which we venture to offer some impressions of planning's overall impact.

CANADIAN URBAN PLANNING

Urban planning is a complex endeavour of extremely broad scope. It provides an institutional framework for addressing a great range of matters, from the structured pattern of settlements over large regions to the heights of fences between neighbouring houses. Foremost, however, it is a procedure for community decision-making, which gives rise to two characteristics that directly affect our ability to generalize about its effects: first, that planning is an intrinsically political form of social behaviour; and second, that authority for planning is highly fragmented, geographically and politically.

Planning as Political Process

As an instrument of democratic social action, planning is inescapably political. It is society's organized response to situations, affecting urban development, that cannot be dealt with by such everyday institutions as "the market" or "the law."[3] At the same time, there is an essential technical component to every planning decision. We can usefully think of planning as operating from two complementary perspectives – "process" and "activity."

In the process of urban planning, the central task is one of making decisions about future uses of land and the associated circulation of people and goods, a responsibility normally vested in elected bodies, such as city councils. Because everyone who lives in a city uses land and transport facilities, potentially everyone has an interest in planning decisions. Their interests are also likely to differ, which means that there will be conflict, in the social science sense of a difference over ends or objectives. Since "politics" is also defined in terms of competing objectives, conflict is ipso facto political. More particularly, politics is the interaction among parties in conflict as they attempt to influence those who have the legal authority to take decisions on behalf of the collectivity.[4] To the extent that planning decisions are arrived at in this manner, planning is a kind of politics – a political process, that is.

The activity of planning, by contrast, is the domain of professional planners. It focuses on the functional and aesthetic attributes of built environments and is concerned with laying out future development forms. Its central purpose is to determine what forms are best from a technical standpoint, based on such criteria as the efficient organization of land uses and service networks, and to advise the community's representatives accordingly. The final decisions, however, are not usually made on technical grounds alone. Planners' ideas of good form often conflict with the interests of others, and so planners cannot stand outside the political process, as disinterested experts, even when they are engaged in planning as an activity.[5] In the final analysis, planning is a method of choosing among competing ends and objectives, including technical ends and objectives. The process subsumes the activity; that is a democratic necessity, even if, as sometimes happens, the process then falls hostage to pressure-group politics.

The political realities of modern urban development cause profound difficulties for planning, philosophically as well as pragmatically. The city is regularly described these days as an "arena" in which competing interests play out their conflicts.[6] To varying degrees, depending on local circumstances, changes in urban form are the "negotiated outcomes" of interactions among these many interests, but that leaves planning in an ambiguous position. Under the adversarial conditions into which most development conflicts descend, planners frequently find themselves pitted against the communities that they are trying to serve, and even cast as villains. The theoretical response to this dilemma has been to portray planning as a process of "social co-operation," an ideal that is slowly coming to be realized in the best of Canadian practice as well. The core concepts of Vancouver's metropolitan plan, for example, were shaped through a mass exercise in community participation, about which the planning director of the day would later write, "The Vancouver experience has convinced me that effective planning of human settlements in western democratic society will come to depend more on human relations in the process of arriving at decisions than it will on the planner's science and art of preparing plans."[7] This does not imply that planners' technical skills are no longer needed, but it does mean that planning decisions should be a genuine expression of community purpose.

Fragmentation of Jurisdiction

Given the political character of the planning process, the effects of planning in any particular city will be strongly conditioned by the local decision-making context. An essential part of that context is the authority vested in decision-making bodies, which introduces the issue of planning jurisdiction – the legal right or power to carry out certain authorized tasks within some prescribed territory, such as a province, city, or regional municipality. In Canada, there are literally hundreds of planning jurisdictions, exercising various kinds of authority and often overlapping in confusing ways. Planning jurisdiction, then, is extremely fragmented, which not only contributes to the variability in physical outcomes but is a major source of conflict in the planning process, especially in metropolitan areas where several jurisdictions have to coexist in a close and often competitive political relationship.

Two factors chiefly account for the proliferation of planning jurisdictions. The first is the constitutional separation of powers between the national and provincial governments; the second is devolution of planning authority to municipal governments. The significance of the constitutional factor arises because planning is one of those matters that falls under provincial jurisdiction. This has two direct consequences. First, unlike many other countries, Canada has no national planning system, and there is no means of imposing uniform development policies on Canadian cities. Even when the government of Canada promotes a national policy measure, its implementation depends on the willing collaboration of provinces and municipalities. Second, planning is not a constitutional obligation. Rather, it is up to each provincial government to decide whether it wishes to have a planning system, and then to determine its scope and powers. The result, historically, has been a great deal of variation among the ten provinces, with the further implication that Canadian planning systems have differed considerably in their power to effect change in urban form.

The second factor, devolution, relates to the general belief that planning is treated best as a local responsibility, as something that urban communities should do for themselves. By and large, municipal governments have a great deal of freedom over the exercise of their planning duties, especially in terms of setting their own development goals and policies. However, no matter how great their autonomy under planning legislation, they always remain subject to the superior authority of their provincial governments, which may apply controls of their own over aspects of urban development. They may set limits on concentrations of potentially hazardous industries, for example, or prevent cities from expanding onto prime agricultural land.[8] Provinces also have financial power, which they exercise in two ways. The first is direct spending on their own public works projects, such as construction of highways and trunk sewers; the second is indirect spending in the form of grants to municipalities, which are frequently tied to particular projects or programs that the provincial government chooses to encourage. Thus the manner in which provincial and municipal governments share jurisdiction over urban development contributes to the variability of planning's effects.

PLANNING FOR URBAN EXPANSION

Planning in Canadian cities now comprises a massive, all-embracing system of controls over physical development. Potentially, its effects are everywhere, whether visible or invisible (in the case of development prevented). We can make sense of this potential by identifying two major issues that reflect the ongoing nature of urban investment: urban expansion and the redevelopment or re-use of urban land. Within these broad fields we can identify others that constitute the real substance of Canadian planning – regional planning, suburban neighbourhood planning, urban renewal, downtown revitalization, rapid transit, and expressways, all set against a backdrop of large-scale growth which has provided Canadian planning with its overriding issue of concern for most of the period since 1945.

The Concept of Growth Management

Since 1945, Canada's urban population has increased by about 13 million – an overall growth rate of some 200 per cent. Most of this increase has been accommodated on the fringes of our largest cities, causing them to expand, at a rough estimate, by as much as 500,000 ha. The planning implications of these trends have been immense, both because of their scale and because of the problems that they have generated. In the early post-war period, when most Canadian cities still had comparatively simple structures, the main object was to prevent "urban sprawl," or the premature subdivision of agricultural land and scattering of development throughout the rural-urban fringe. With more stringent regulations firmly enforced, the worst effects of sprawl began to come under control in the 1950s and 1960s, and planners turned their attention to the much larger issue of growth management – the general problem of controlling expansion of cities into their rural surrounds.

Growth management policies are commonly controversial, and just winning agreement on them can be a major achievement. Even then, there is no guarantee that plans will be carried out in their approved forms. Jurisdiction is important here, because the most effective control measures are vested in municipal governments, whereas the appropriate stage for growth management normally extends over several municipalities. In theory, the gap is bridged by regional planning, so it is useful to distinguish between two scales of planning for urban expansion: the metropolitan-regional scale, where comprehensive strategies for growth management are devised; and the suburban scale, where planning usually falls under the jurisdiction of single municipal governments.

Metropolitan-Regional Planning

The metropolitan form of regional planning was initiated by provincial governments in the 1940s and 1950s, primarily to provide an administrative structure for controlling urban sprawl beyond the incorporated boundaries of rapidly growing cities. The general approach was to create a higher tier of planning agencies empowered to practise large-scale land-use planning over several municipalities. In some cases, such as Toronto and Winnipeg, these agencies were closely associated with new forms of metropolitan government; in others, such as Edmonton, Calgary, and Vancouver, they were superimposed on an existing patchwork of municipalities. Whatever the institutional framework, however, regional planning agencies have generally found themselves in an ambiguous position in relation to the municipal governments below them and the provincial government above. The consequences are particularly well illustrated by Toronto's abortive experience with large-scale growth-management planning.[9]

The beginnings of regional planning in the Toronto area date from the late 1940s, but a more meaningful step came with creation of the Municipality of Metropolitan Toronto in 1953. As a federation of local municipalities, the metropolitan government was responsible for co-ordinating planning among the lower-tier units. It was also responsible for providing the major infrastructural services – trunk sewers, water-supply, arterial roads, and transit – that facilitated expansion into the suburban townships. With these powers, and the funds granted to it by the provincial government, the metropolitan municipality managed growth innovatively and effectively, as long as it was contained within the limits of its direct jurisdiction.[10] By the late 1960s, those limits were being overrun, an eventuality that had been anticipated in 1953 when the Metropolitan Toronto Planning Board was created. It was given jurisdiction over a territory three times as large as the new metropolitan municipality, but then the province, instead of expanding the municipal boundaries into the surrounding planning area as development advanced, reduced the planning board's territory and set up separate regional governments, with their own planning authority. Problems of co-ordinating development and services were clearly implicit in this division of jurisdiction; they came home to roost in the 1980s, as growth spread over an ever-widening domain.[11]

In the absence of a single local government with planning jurisdiction commensurate with Toronto's urban growth field, the government of Ontario stepped into the breach for a time. Government agencies produced a celebrated plan, the Toronto-centred region (TCR) plan, by which they hoped to structure the form of development over a radius of about 120 km from downtown Toronto. The plan recognized that growth would continue in Toronto's immediate environs and in the lakeshore zone toward Hamilton, but it also aimed to prevent development from coalescing into an undifferentiated mass. This was to be achieved by spacing communities along a "parkway belt" parallel to the lakeshore and separating them with secondary parkways or green belts. The land immediately north of the lakeshore zone was to be preserved for agriculture and recreation, while development further afield was to be concentrated in a small number of designated "growth points." A particularly important objective was to secure a more balanced distribution of growth within the region by stimulating development of key centres in the lagging areas east and northeast of Toronto.

The TCR plan was promulgated with great fanfare in 1970, yet it never was effectively implemented. Indeed, by its own actions the Ontario government did much to sabotage the plan. It built a huge trunk sewer into the "agricultural" zone north of Toronto, stimulating development there, and it dropped plans for three transportation projects that would have facilitated eastward growth, while continuing to build expressways to the north and west of Toronto.[12] All in all, it was a classic case of a plan being unhinged from the process required for its implementation. In making its decisions, the government was responding to public opinion, electoral messages, and other political stimuli from within the planning region, whereas the plan had been designed by technocrats who were not directly accountable to the many communities affected by it. In the upshot, the process effectively killed the plan.

The TCR experience was admittedly extreme, both in the scale of the plan's conception and in its utter collapse. Nonetheless, it serves to demonstrate the difficulties that bedevil metropolitan-regional planning and the ambiguity that attends the whole concept. In planning legislation and its underlying theory, metropolitan-regional planning has been accorded a prominent place in Canada, but the practice has been altogether more equivocal.[13] Even the comparative successes, like Ottawa, which has come closer than any Canadian city to achieving one of the great model forms of planning theory (Figure 19.1) – the garden city model, complete with green belt and satellite towns – have been plagued by political and jurisdictional conflicts and by limitations on their authority.[14] Some agencies have been able to regulate development fairly effectively, and at least one, Calgary, has been generally successful at containing development within the city limits. But in the larger sense of deliberately structuring the forms of whole metropolitan regions, regional planning has had little effect.

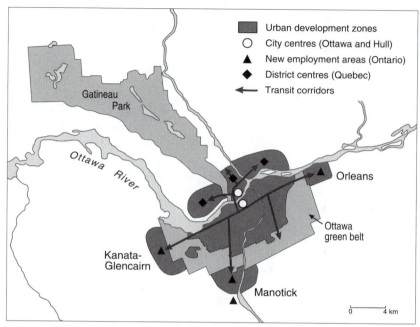

Figure 19.1
Development concept for the Ottawa-Hull metropolitan area, based on the approved plans of
the regional municipality of Ottawa-Carleton (Ontario) and the Outaouais regional
community (Quebec). The growth management strategy on the Ontario side of the Ottawa
River calls for development beyond the green belt to be channelled into three
comprehensively planned new communities.

Planning New Suburbs

Until the late 1940s, urban growth was not planned in any meaningful sense
of the word. Then, in the 1950s, larger houses started to come into favour
and, with that, larger and more varied house lots. Curvilinear street layouts
began to replace the old-style grid, and residential streets were being
planned to fit into a framework of "arterial" through roads. The development industry was changing too. By the 1970s it had come to be dominated
by large, vertically integrated companies that assembled huge tracts of land,
laid out whole subdivisions, installed their own services, and built and mar-
keted the houses. These companies also formed close working relationships
wih their local planning agencies, since adherence to community standards
was essential to their business success. The outcome was the comprehensively planned "corporate suburb," a new form of urban environment for
Canada, although firmly rooted in a long tradition of suburban design. The
most direct influence was the American concept of the neighbourhood unit,

which by the 1950s had become the accepted theoretical base for suburban planning, not just in Canada but throughout the Western world.[15]

Neighbourhood units had two great advantages as far as planners were concerned.[16] First, they were considered to provide a better quality of residential environment than was possible under the grid plan. Their curving streets were attractive, quiet, and safe from traffic; everyday services were close at hand, preferably within walking distance, and it was not necessary to cross busy streets to reach them; traffic-generating activities were located on the major roadways that bounded the neighbourhood, whereas the school and other social facilities so important to community life were grouped at its core; and parks and playgrounds of various kinds were spread liberally around. The neighbourhood unit was not a place of work – indeed, workplaces and living places are strictly segregated in the "planned" city – but it was believed to provide a healthy environment for family life.

Second, neighbourhood units were a practical means of ensuring that new suburbs would develop in an "orderly" manner. This meant two things, most particularly. Neighbourhood unit planning provided a systematic basis for allocating community service facilities to developing residential areas, by ensuring that adequate sites were laid out in advance at appropriate locations; and planners, by treating neighbourhood units as the basic "cells," or building blocks of suburban form, could theoretically control a city's growth pattern. Plans normally envisaged that development would proceed in orderly sequence, from one neighbourhood to the next, following the most efficient extensions of service networks and simultaneously ensuring economical use of land resources – "economical" relative to prevailing density standards, recognizing Canadians' preference for detached houses.

Further elaboration of this approach followed the realization that certain activities and services just cannot be accommodated at the neighbourhood scale.[17] Non-family housing, particularly in large apartment complexes, may be out of place in the family-oriented neighbourhood, and more specialized services, such as high schools, district libraries, and regional shopping centres, require catchment areas larger than a single neighbourhood. The technical problem was to devise a way of organizing suburban space so that extra-neighbourhood elements could be integrated with the system of neighbourhood units. The solution was to plan suburbs as hierarchical structures, by grouping several neighbourhoods into larger units organized around a multi-service centre – the town centre. The units themselves are variously referred to as districts or communities, and even as suburban towns or "cities in the suburbs." They also differ considerably in size and complexity. The smallest may have only three or four neighbourhood units, with perhaps 10–20,000 people; the largest may comprise more than twenty neighbourhoods, plus other facilities and land uses. Whatever the areal scale, however, planning theory requires each unit to be physically distinct, usually separated from its neighbouring units by expressways or other major roadways. The new communities may even be detached from their parent cities, as Kanata and Orleans are from Ottawa (Figure 19.1), although an attached form is more common and provides a more effective basis for managing large-scale expansion.

The prototype for the planned corporate suburb was the Toronto community of Don Mills, built between 1952 and 1962 on a site of some 800 ha assembled and developed by a single corporation.[18] The focal point of the community was fixed by the intersection of two arterial roads, and it is there that the community shopping centre and high school were located. Immediately around this core is a zone of low-rise apartments and row houses, and beyond that are four low-density neighbourhood units, one in each of the quadrants formed by the arterial roads. Several peripheral sites have been developed for industry, as well.

Although Don Mills was a single, privately initiated project, it gave municipal planners a model which they could apply in a more comprehensive way to their planning for suburban growth. The Toronto-era municipality of Scarborough is a good case in point. Scarborough adopted its first official plan under the Ontario Planning Act in 1957, when its population was about 100,000. This plan laid down general policies and land use categories to guide development, and it included a "development plan" to ensure that service infrastructure would be built in phase with houses and factories. The plan also provided for the municipality to be organized into communities (eventually there were thirty-two of them), which were further divided into neighbourhoods on the Don Mills pattern. Each community had its own "secondary plan" which prescribed overall land use, as well as the numbers of single- and multiple-family housing units to be built in its constituent neighbourhoods. The population expected to be generated by this mix of housing was then used to determine the numbers of schools and churches that were required and the amounts of commercial floor space and parkland. By adhering to the framework set by these plans, Scarborough has grown in more orderly fashion than it might otherwise have done, to its present (1991) population of 525,000.

A variation on this procedure is illustrated by Calgary and Edmonton, two other cities that have experienced large-scale suburban development. They were also among the first in Canada to adopt neighbourhood unit planning, and none has used it more consistently or more thoroughly, including the principle of grouping neighbourhoods into larger units, or "sectors," as they became known in Calgary.[19] The suburban units vary in size and formal organization, depending on local site conditions, but a three-level hierarchy is planned for wherever possible. This follows the standard Alberta school hierarchy, under which elementary schools are provided at the neighbourhood level, junior high schools at the community level, and senior high schools at

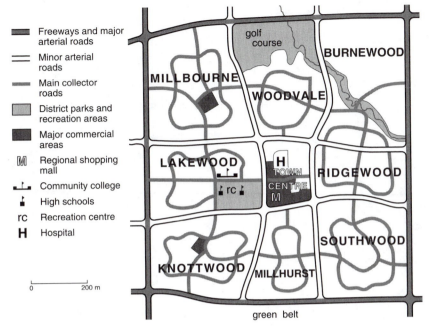

Freeways and major arterial roads
Minor arterial roads
Main collector roads
District parks and recreation areas
Major commercial areas
Ⓜ Regional shopping mall
⌐ Community college
High schools
rc Recreation centre
H Hospital

0   200 m

Figure 19.2
The hierarchical structure of Mill Woods, Edmonton, The "town," as planned and
developed, comprises eight "communities" bounded by arterial roads. Each community is
further divided into two to four neighbourhood units bounded by collector roads. Services
appropriate to each level of the hierarchy are centrally placed in their relevant units.

the sector or suburban town level. A hierarchy of shopping centres commonly parallels this basic structure, and so too does the hierarchy of major
roads. The Edmonton suburb of Mill Woods is the most advanced version of
this general scheme (Figure 19.2). It has been planned for a population of
about 100,000 organized into eight "communities" and twenty-three neighbourhoods, all focused on an elaborate town centre complex which combines commercial, medical, educational, recreational, and public service
facilities with high-density housing.

Another key feature of the town centre concept, as it has come to be applied in the largest Canadian cities, is its strategic integration of land use and
transportation policies. By design, town centres are major traffic destinations
and therefore should be treated as focal points in suburban road networks.
By the same token, however, they are logical nodes on the rapid transit lines
by which more and more suburbs are coming to be linked directly to their
metropolitan cores. Whatever technology is favoured, whether some form of
rail transit (as in Montreal, Toronto, Vancouver, Edmonton, and Calgary) or
a dedicated, high-speed "busway" (such as Ottawa is developing), the com-

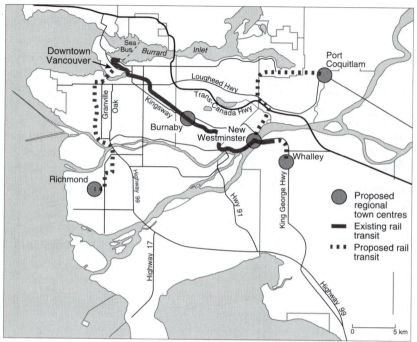

Figure 19.3
Planned regional town centres in the Vancouver metropolitan area in relation to the rail transit network, existing and proposed. The Burnaby and New Westminster centres have experienced major development since the opening of the Skytrain in 1986.

bination of rapid transit and town centres is seen by planners as a powerful growth-management tool.[20] It is, at once, a means of directing suburban growth into desired locations (as laid down in the metropolitan-regional plan) and of giving that growth a well-structured form (by organizing each suburb around its own town centre). The Vancouver plan, for instance, calls for development of five "regional town centres," all of which are to be rail transit stations (Figure 19.3). This ensures that the main suburban growth areas will have good access to the city centre, both for work and for the specialized services that are available there. At the same time, the regional centres are intended to be the major service complexes for their local suburban areas and important employment locations in their own right. It is even hoped, following a general goal of metropolitan decentralization, that they will attract some office development and jobs away from the central business district. This would allow more suburban residents to work close to home and so ease the burden of long journeys to work and congested commuting facilities; by slowing the rate of development in downtown Vancouver, it would minimize the costs of central-area congestion, and growth there might be more effectively managed.

At this stage, it is impossible to predict the long-term consequences of Vancouver's strategy.[21] Still, if Toronto's experience is a reliable guide, huge, multifaceted subcentres are destined to be prominent features in all our largest metropolitan areas, regardless of their effectiveness as instruments of planned decentralization. In the Toronto case, the municipalities of Mississauga, North York, and Scarborough, all of which have come to regard themselves as true "cities in the suburbs" and are anxious to demonstrate their independence from Toronto, are vigorously pressing the development of so-called city centres.[22] These centres are being promoted as alternatives to downtown Toronto as office locations, but their greater purpose is to provide foci for suburban community life by reproducing all the attributes, symbolic as well as material, of traditional city centres. The larger intention is to generate a unifying sense of place in the otherwise "placeless" suburbs.

Beyond these few, somewhat visionary undertakings lies the great bulk of suburban planning practice, which has an altogether more mundane cast to it. There, planning has little or no part in determining whether development will occur, or what kind of development it will be. In this landscape of shopping malls and emerging "megastrips," of giant business "parks" and transportation terminals, all set in a coarse-grained matrix of ring roads and expressways, the planning function is essentially one of administering zoning, building, and subdivision regulations to ensure that development meets the community's standards. By and large, the market prevails, operating within whatever regulatory framework the local planning system imposes. This applies to much residential development, as well. Despite the undoubted influence that neighbourhood unit theory and its associated concepts have had, by no means all post-war development has lived up to the theoretical expectations. In some cities and some suburbs, the imprint of comprehensively planned new communities is bold and clear; in others, it appears only in partial or debased forms.

PLANNING FOR REDEVELOPMENT AND REVITALIZATION

Redevelopment planning is a controversial and problematic field of planning practice, to which it is difficult to do justice in a brief overview. We have therefore chosen to concentrate on two topics: first, we discuss the changing context of redevelopment planning, with particular emphasis on the political context; second, we describe the main approaches that public authorities have taken in their efforts to be a constructive force for change in the urban environment.

The Changing Context

As early as the 1940s, when the first generation of modern city plans was being written, Canadian planners identified three issues that were to loom large in subsequent planning practice. They were the need to accommodate economic growth, especially through expansion of central business districts; the transportation demands of growing cities, and the accessibility of central areas in particular; and the problems of slums and urban blight that were so prominent after decades of depression and neglect. The last issue was seen chiefly as a public responsibility, but the others were directly related to private development actions, both in existing built-up areas and in the suburbs. The future prosperity of the city was thought to require unhindered growth in the downtown office-based economy, so the plans provided for large-scale office construction, for government as well as for business. In turn, this policy implied increasing concentration of employment in the central area, and a continually increasing flow of commuters from the spreading suburbs. To cope with that pressure, expressways were proposed in most cities, complemented by rapid transit (usually rail transit) in the largest of them.[23] Alternatively, some downtown workers could be housed in high-rise apartments located on expressways or transit routes through the older, low-density neighbourhoods that were coming to constitute the "inner city." Very often, in fact, apartment redevelopment was seen as a desirable way of offsetting the "natural" deterioration of these ageing districts.[24]

Expressways were a technical solution to the problem of facilitating employment growth in central office districts; high-rise apartments, though initially a market solution, quickly received technical sanction in city plans and zoning maps. Another market solution was not so well anticipated, though. This was the tendency for white-collar workers to remain living in the older neighbourhoods or to move into them as "white-painters" or gentrifiers, a trend that was well established in some cities by the late 1960s.[25] Here were sown the seeds of planning conflict. These same neighbourhoods were commonly the ones threatened by the planned transport facilities or by apartment redevelopment, and their residents were unwilling to acquiesce in changes that impinged so powerfully on their own interests. They also learned that they could use the political process to protest plans and attempt to have decisions reversed. Often they failed, but sometimes they succeeded, even to the extent of "capturing" city hall when their representatives gained a majority position on city council.[26]

Through the 1960s and 1970s, organized and increasingly effective protests changed the face of Canadian planning. Several affairs were especially celebrated – the Spadina expressway in Toronto, the Milton-Park project in Montreal, renewal of the Strathcona neighbourhood in Vancouver, and the Portage and Main redevelopment in Winnipeg – but they were just the tip of a very large iceberg.[27] In city after city, through fight after fight, it was made plain that a new mood of reform was in the air, signalled by a resurgence of populism in civic politics.[28] The practical achievements did not always live up to the reformers' hopes, but the political nature of redevelopment decisions could no longer be avoided, by either planners or politicians. It was also becoming clear, as it was in the case of Toronto's attempts at regional growth management, that technical plans could not survive without political support and commitment. The problem was not initial lack of support but a change in political climate, such that plans that were viable in the 1960s had lost support a decade or so later. In the upshot, one of the main legacies of the reform era, aside from its contribution to the political education of planners, was that it brought an end to large-scale "master" planning. The dream of being able to chart the course of a city's future development in all its ramifications, a dream that had inspired the planners of the post-war generation, did not stand the test of experience. Today, city plans tend to be loose collections of general (some would say, innocuous or motherhood) policy statements, accompanied by land-use district or zoning maps that are continually being amended to permit redevelopment, and complemented by a battery of more detailed, small-area plans adopted after a long process of community negotiation and debate. It is by means of such highly focused instruments that contemporary planning has its greatest control over redevelopment processes.[29]

Another characteristic of the reform movement of the 1960s and 1970s was that it was largely middle class, or at least was led by middle-class activists. Its voice was essentially conservative: the reformers wished to prevent unwanted changes or to minimize their effects – by limiting growth of central business districts, for example, and by protecting the family-housing character of inner-city neighbourhoods. They also tended to favour public transit as an alternative to expressways, because of its less destructive impact on the residential environment. In some cases governments paid heed, although they learned that rapid transit facilities bring high costs of their own, politically as well as financially.[30] Protecting the inner-city neighbourhoods generated other costs as well, since opponents of technical planning solutions had few if any alternatives to offer to developers of high-rise apartments. To the extent that they prevented apartment redevelopment, then, they shifted the housing supply problem onto others – to tenants who saw their rents rise as supply lagged behind demand or to older suburban districts where developers sought new apartment sites and some homeowners created illegal rental units. As in the central city, however, attempts to grant official sanction to these kinds of changes met insurmountable opposition.

Where planners once attempted to anticipate future development forms in their technically derived plans and zoning maps, proposals for change must now be dealt with by the planning process, case by case and site by site. Some succeed, others fail. The ultimate discretion, of course, is vested in the political decision makers, so the key question always is how they will respond to the many influences that are playing upon them. To the extent that circumstances allow, they tend to respond by using the planning process to facilitate development, especially in central areas where construction of new buildings is still equated with economic and social vitality.[31] Beyond this, however, it is extremely difficult to generalize about the physical outcomes of redevelopment planning. There is no question that the overall impact on Canadian cities has been substantial, since virtually all redevelopment requires planning approval. But although the results are all around us, in every city, they are difficult to read and interpret. Unlike the paramount influence that the neighbourhood unit concept has had on contemporary suburbs, the effects of redevelopment planning decisions assume a great diversity of forms. The real evaluative question is not whether a particular form of built environment has been created but whether that environment, whatever its form may be, meets the community's own standards of acceptable quality.

Public Initiatives

The great majority of redevelopment actions are privately initiated, yet there have been many occasions since 1945 when public authorities have instituted measures of their own. The justifiction has always been the same: the "market" has failed to renew an outmoded part of a city, so public investment must take up the slack. In this section we review the major kinds of public initiatives in chronological sequence, to track the evolution of planning attitudes and public policy. The starting-point is the government of Canada's urban renewal program, which was conducted in two distinct phases, from 1944 to 1964 and from 1964 to 1973.[32]

The primary impetus for a national program came out of concern for the slum housing problem in the 1930s and 1940s. At that time, the solution seemed obvious – tear down the slums and build decent houses in their place – and the government moved in this direction in 1944 when it added provisions for slum clearance to the National Housing Act (NHA). The immediate outcome was a single project, Regent Park North in Toronto,[33] but interest quickened in the 1950s, under the leadership of CMHC, Central (now Canada) Mortgage and Housing Corp. Between 1954 and 1964, fifty redevelopment studies were undertaken in cities that spanned the country from St John's to Victoria, and twenty-one projects were initiated (Table 19.1). Overall, these projects were quite varied in purpose, although the most substantial were of two main kinds. In the first, typified by the Jeanne Mance project in Montreal and Uniacke Square in Halifax, pockets of defective houses were razed and the cleared sites were redeveloped with subsidized public housing for low-income families. In the second, typified by Courtenay Place in Saint John, the targeted sites were in areas of mixed land use

Table 19.1
Planned uses of cleared land in authorized redevelopment projects started before 31 July 1964

Project	Site area (h)	Residential Public (h)	Residential Private (h)	Commercial (h)	Industrial (h)	Open space (h)	Public uses (h)	Year of approval by CMHC
Toronto, Regent Park North	17.0	17.0	–	–	–	–	–	1948
St John's	2.0	1.6	–	–	–	0.4	–	1954
Toronto, Regent Park South	10.6	10.6	–	–	–	–	–	1955
Montreal, Jeanne Mance	7.9	7.9	–	–	–	–	–	1956
Halifax, Jacob Street	6.7	–	1.6	4.3	–	0.8	–	1957
Halifax, Maitland Street	0.6	–	–	0.6	–	–	–	1959
Toronto, Moss Park	7.5	4.1	–	–	–	1.8	1.6	1959
Windsor, Area 1	6.6	5.9	–	–	–	–	0.7	1959
Windsor, Area 2	3.1	–	0.6	0.9	–	–	1.6	1959
Saint John, Courtenay Place	16.8	2.0	2.0	4.0	4.8	2.0	2.0	1960
Sarnia, Bluewater	68.8	–	–	–	68.8	–	–	1960
Vancouver, MacLean Park	14.4	7.6	1.6	–	3.6	1.6	–	1960
Winnipeg, Lord Selkirk Park	19.5	9.9	–	3.2	5.6	0.4	0.4	1961
Hamilton, Van Wagners Beach	28.8	–	–	–	–	28.8	–	1962
Montreal, Dorchester Street	7.0	–	–	–	–	–	7.0	1962
Halifax, Uniacke Square	12.4	11.2	–	–	–	–	1.2	1963
Montreal, Victoriatown	7.6	–	–	–	7.6	–	–	1963
Ottawa, Preston Street	6.7	2.4	1.5	0.4	–	–	2.4	1963
Hamilton, North End	12.0	2.2	–	–	–	1.2	8.6	1964
Kingston, Rideau Heights	6.7	–	5.9	–	–	0.8	–	1964
Toronto, Alexandra Park	10.0	7.6	–	0.4	–	0.8	1.2	1964
Vancouver, Raymur Park	11.4	3.9	–	–	6.5	–	1.0	1964
Totals (hectares)	284.1	93.9	13.2	13.8	96.9	38.6	27.7	
Percentage distribution	100.0	33.0	4.6	5.1	34.0	13.6	9.7	

Sources: Compiled from CMHC records and from Pickett (1968: 236–9).

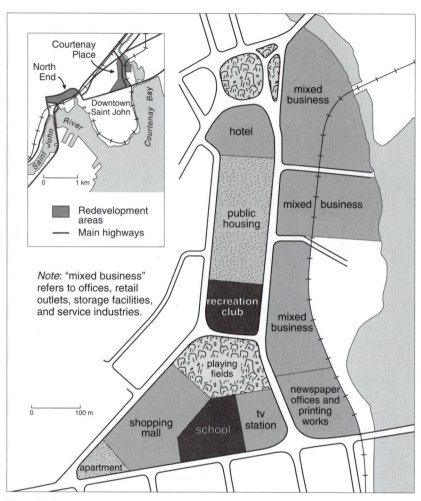

Figure 19.4
Public redevelopment projects in Saint John under the urban renewal program. In the
Courtenay Place project, an area of decrepit houses mixed with outmoded waterfront
industries was replaced by a variety of business, residential, and recreational uses, all neatly
segregated in a new layout plan.

on the blighted margins of industrial or business districts; here, the redevelopment plans provided for several different uses, all neatly separated from
one another (Figure 19.4).

The variety of redevelopment recorded in Table 19.1 was, in itself, a
mark of success for CMHC's planners. From the outset, they argued that

urban renewal should entail much more than residential slum clearance, and they were eager to expand the national program into a truly comprehensive instrument of environmental improvement.[34] Their efforts were rewarded in 1964, when the NHA was broadened to facilitate, first, rehabilitation (rather than demolition) of deteriorating residential areas and, second, redevelopment of non-residential areas. A period of intense planning activity followed. From 1964 through 1972, more than 300 studies were supported by CMHC grants, and about 90 renewal projects were authorized. The great majority of them were aimed at redevelopment for business or public purposes, reflecting the fact that municipal governments had seized upon the urban renewal program as their best available instrument of economic revitalization. Two main approaches were favoured. In the first, exemplified by the North End project in Saint John (Figure 19.4), patches of poorly located houses were demolished to provide new industrial or commercial sites. In the second, which was more common, sites on the fringes of central business districts were redeveloped with such public facilities as city halls, parks, and libraries. The further object was to use public investment as a catalyst to encourage entrepreneurs to reinvest in a part of the city that had ceased to be attractive to them. A particularly successful example was Hamilton's Civic Square project, carried out between 1966 and 1983.[35]

In 1973, national urban renewal policy underwent a second major reorientation as the government retreated from the principle of comprehensive renewal. The program had been in political difficulties for some years, and funds for new planning studies were cut off as early as 1968.[36] The issues were complex, but at their core was a growing belief that the program was not addressing real housing needs, either quantitatively or qualitatively. The economic revitalization thrust, for instance, had no obvious connection with the NHA's mandate of improving housing conditions. It may even have made things worse, since it reduced the supply of low-cost housing and forced people to leave their homes with no guarantee of anything better, except perhaps in a public housing project somewhere.[37] For that matter, the whole concept of large-scale public housing had fallen into disrepute, for reasons that included the social stigma of ghettoization, the institutional environment of the housing projects, and the high cost of rehousing comparatively few families. The prospect of having to move into public housing was especially resented by people who owned their own homes in areas under threat of redevelopment; by 1968 their fear and anger had become a significant political force.[38]

The upshot, in 1973, was refocusing of public aid programs on the specific concept of area rehabilitation. Two programs were provided for: the Residential Rehabilitation Assistance Program (RRAP) and the Neighbourhood Improvement Program (NIP). Henceforward, designated renewal areas – the neighbourhood improvement areas – were to be eligible for two kinds of assistance. Under RRAP, homeowning residents could get grants and low-interest loans to help them renovate their houses; under NIP, municipalities were given grants to upgrade environmental amenities and community infrastructure. The effect was immediate and far-reaching. From 1974 until 1983, when NIP officially ended, 125 cities, large and small, participated in the two programs. All told, 270 neighbourhoods were improved.[39]

NIP's impact was far-reaching in another sense as well. Apart from its obvious physical objectives, it was intended to help local residents gain more control over their community environments. It was therefore expected that plans would be drafted with high levels of citizen involvement and that citizens' groups would be given some powers at the implementation stage. Not all cities lived up to these expectations, but NIP's widespread adoption helped to institutionalize participatory neighbourhood planning in Canada.[40] The need for physical upgrading was not the only concern, and the neighbourhood movement antedated NIP in some cities, but NIP's success had no small part in sensitizing planners and politicians to the value of neighbourhood-based planning.[41]

At the same time, changes to the NHA in 1973 by no means signalled an end to government interest in economic revitalization. Rather, they led to a clearer division of responsibility, since the problems of declining business districts fell within the sphere of agencies other than CMHC, notably the Ministry of State for Urban Affairs (MSUA). In its brief life (1971–79), MSUA was instrumental in securing construction of a number of important projects, including renewal of derelict waterfront areas in Halifax, Saint John, Quebec, Montreal, Toronto, and Vancouver.[42] In the process, it helped to forge a new co-operative approach to redevelopment planning, in the form of public-private partnerships. MSUA was a catalyst, and it fostered close working relations among development agencies at all three levels of government, as well as with the private sector and community interest groups of all kinds. This was a continuation of thinking central to the urban renewal program – that private enterprise would not invest in areas in an advanced state of economic decline and physical deterioration without the stimulus of public investment. There was also the realization, born out of experiences like Hamilton's Civic Square project, that individual development agencies, whether public or private, could not operate on a sufficiently large scale to deal with problem areas in their entirety. Comprehensive treatment demanded effective partnerships between governments and private developers. Finally, the protests and conflicts of the 1960s brought renewed appreciation of one of the planning movement's deepest intellectual roots – the ideal of planning as a co-operative process. The residential revitalization programs, NIP and RRAP, were conceived in that spirit, while MSUA provided an ad-

363 Cities as a Social Responsibility

ministrative framework that extended the co-operative approach to other kinds of renewal actions.

In the event, Canada's largest experiment in co-operative redevelopment, the Core Area Initiative (CAI) in Winnipeg, cannot be attributed directly to MSUA. The ministry was disbanded in 1979, whereas the tri-government agreement launching the CAI was not signed until 1981. Nonetheless, it represented a progression from the publicly initiated renewal projects of the 1960s and 1970s and was the first truly comprehensive one, in all senses: in its areal scale, in the mélange of problems that it was designed to address, and in the package of programs and projects that it spawned across an extraordinarily broad policy spectrum. Economic, social, and physical planning measures were all combined under this one umbrella. Special education programs, job creation and training, and social and cultural services were as vital to the CAI as physical development projects (Figure 19.5). The CAI proved itself as a development catalyst, as well. The initial five-year agreement called for direct expenditure of $96 million assumed equally by the governments of Canada, Manitoba, and Winnipeg. By 1987 that had stimulated further investment in excess of $500 million from other government agencies and the private sector. Both in substantive terms, then, and as a process – that is, as a co-operative partnership among three levels of government, a number of private development agencies, and many different citizens' groups – the CAI is widely regarded as one of the great successes of Canadian planning.[43] Whether this means that it should be copied elsewhere is a moot point, and largely an academic one, since the federal government has shown no inclination to enter into similar agreements with other cities. Its message is nonetheless clear: development partnerships are not only socially acceptable in a modern mixed economy, they may well be indispensable if inner-city decline and obsolescence are to be adequately treated.

Even without the spur of a national revitalization program, several provincial and municipal governments have been prompted to take comparable initiatives, albeit on a smaller scale. To mention just one example, the city of Edmonton has initiated a comprehensive program of revitalization called PRIDE (Program to Improve Downtown Edmonton). This includes public works in the central area, as well as creation of the Downtown Development Association, which is attempting to secure construction of an array of joint-venture projects in accordance with an overall plan. The general goal, as it is in cities all across Canada, is to maintain the centre as the vital "heart" of the city, economically, culturally, and socially. Several common themes stand out, as well: increasing the residential component of downtown development; creating a built environment that is well adapted to pedestrian circulation at all seasons of the year;[44] and capitalizing on the current enthusiasm for heritage conservation, especially in combination with the

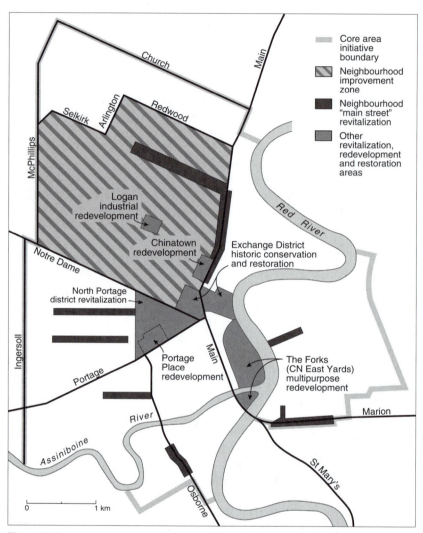

Figure 19.5
Physical concept plan for the Core Area Initiative, Winnipeg. Individual projects include revitalization of inner-city neighbourhoods and "main streets," restoration of historic districts, downtown redevelopment, creation of a new industrial park, and redevelopment of abandoned rail yards.

festival market concept that has been used so successfully in projects such as Market Square in Saint John.[45]

CONCLUSION: HAS PLANNING MADE A DIFFERENCE?

What, then, can be concluded about the effect of planning on Canada's cities? In systematic terms, not a great deal, unfortunately. Although all cities have experienced the effects of planning decisions, it is no straightforward matter to show how their physical forms have changed in consequence, let alone demonstrate that they have been improved. Several factors come into play here: the great variability in planning's effects over space and time; the difficulty of knowing, for sure, that certain effects can be attributed to planning and not to some other cause; and the increased politiciation of planning, which means that physical outcomes depend more on local decision-making contexts than on the technical norms or standards of planners. There is no accepted model form of the well-planned Canadian city against which actual cities can be measured. Rather, the success or effectiveness of planning has to be judged by a community's satisfaction with its consequences, whatever they may be.

On visible evidence alone, planning has had its most obvious effect on suburban residential development. Since the same set of concepts has been applied, with varying degrees of thoroughness, all across Canada, a distinctive pattern has been imposed on huge tracts of our modern suburbs. But have the results been good? Opinions are divided. Most residents of the planned communities seem to be well satisfied; others are more critical.[46] Suburban neighbourhoods provide good physical settings for middle-class, single-income families, but they commonly lack the service and social networks needed by other groups, such as the elderly, the poor, and single parents. Their low density also means that they are not well suited to public transit, and their physical forms cannot easily be adapted to new housing and service needs generated by demographic and social changes. Nonetheless, they continue to be the preferred form for suburban development: developers submit plans based on neighbourhood unit theory, local councils approve them, and people buy the new houses and then defend their "turf" against unwanted intrusions.

In other respects, planning has been less obviously effective, at least in the light of planners' own expectations. The visions of large-scale growth management and comprehensive master planning have largely gone by the board; the urban renewal program, though conceived with the best of intentions, wrought limited changes in most cities, while bringing contumely upon the planning profession; and even the more successful recent initiatives in revitalization have had comparatively modest effects, because of the small areal scale of most projects. Yet we should not read too much into these assessments. They relate to attempts to break out of the restricted role that is generally imposed on planning systems in Canada and so to assume a more positive or forceful direction over development processes. That is notoriously difficult to do. It requires unwavering political commitment, strong community support, and, usually, a great deal of money – and the Canadian experience demonstrates how insecure these bases can be. In reality, the prime function of urban planning systems, as laid down in enabling legislation, is to regulate development, not initiate it. The impulse to development comes primarily from the corporate sector (including public corporations), taking the form of development proposals to which the planning system must react. On the face of it, this is a far remove from the visionary ideals on which the urban planning movement was founded, but it is a vital function of community self-government for all that. If Canadian cities were to escape the worst effects of an unbridled land market, there had to be a public mechanism for co-ordinating private developments and for mitigating their effects on their surroundings. Planning is that mechanism. Its chief social benefit is that it forces development projects, large and small, in suburban locations no less than inner cities, to fit better with their locality and their region and with community values. The consequence, we believe, is that the overall quality of the urban environment is better because of planning. There are exceptions, of course – all too many of them. But in this general sense, planning has indeed made a difference, an important and valuable difference.

Notes and References

NOTES TO CHAPTER ONE

1 Prior to the 1951 census, the urban population was strictly defined as that population living in incorporated cities, towns, and villages. From 1951 on it was
defined to include all cities, towns, and villages, whether incorporated or not,
of 1,000 people or more, as well as the population living in the fringe portions
of the metropolitan areas. All other areas were considered to be rural. Thus the
urban population before 1951 is somewhat underestimated, although probably
not by much.

2 The concept of the (census) metropolitan area (the CMA) was introduced in the
census of 1951 as a statistical means of combining a central city and the adjacent urban and rural fringe areas with which it is closely linked, through social
and economic integration (e.g. commuting to work), into one functional urban
region. See Stone (1967). In the latest census (1991) the CMA was defined as
the labour-market area of a large urbanized core (of 100,000 or more people).
Recognizing that smaller urban places can also have extensive labour catchment
areas, drawing workers from outside their political boundaries, the census also
defines census agglomerations (CAs) for all urban areas over 10,000 population.
Although the concept underlying CMAs and CAs has remained much the same
over time, the criteria for their geographical delimitation have been continually
redefined. These changes introduce some errors or distortions into studies of
urban population growth over time. See Simmons and Bourne (1989), Bunting
and Filion (1991).

3 Goldberg and Mercer (1986).   4 Seeley, Sim, and Loosley (1956).
5 Lipset (1990).   6 Goldberg and Mercer (1986: 152); Newman and Kenworthy
(1989).   7 Charette (1990).   8 Weston (1982).   9 Langlois (1987); Villeneuve and Rose (1988).   10 Rose (1987); Dyck (1989); chapter 18, below.
11 Chapter 14.   12 Hulchanski (1990).   13 Chapters 4, 9, and 15.   14 In
this context, see Chouinard's (1989) discussion of the changing status of housing

co-operatives in Canada.   15 Laws (1988).   16 Riches (1987); Relph (1990).
17 Bellett (1991).   18 Ley and Olds (1988).   19 Dear (1980); Taylor, Elliott,
and Kearns (1989).   20 Bourne (1990); Murdie (1990).   21 The broadly defined service sector employed between 72 and 82 percent of the work-force in the
eight largest CMAs in 1986.   22 Gad (1991).   23 Ley (1988; 1992).   24 Ley
(1981).   25 For a broader discussion of issues of racism, see Kobayashi (1990).
26 Bourne (1989; 1991).   27 For a Canadian example of the range of recent
interdisciplinary work in political theory, see Chorney (1990).   28 Hanson and
Pratt (1988).

REFERENCES FOR CHAPTER ONE

Barlow, M. 1991. *Metropolitan Government.* New York: Routledge.
Beaujot, R. 1991. *Population Change in Canada.* Toronto: McClelland and
Stewart.
Bellett, G. 1991. "Paying the Price." *Vancouver Sun,* 21 March, A9.
Bourne, L.S. 1989. "Are New Urban Forms Emerging? Empirical Tests for Canadian Urban Areas." *Canadian Geographer.* 33: 312–28.
– 1990. *Worlds Apart: The Changing Geography of Income Distributions within
Canadian Metropolitan Areas.* Department of Geography, University of To-
ronto, Discussion Paper No. 36.
– 1991. "Recycling Urban Systems and Metropolitan Areas: A Geographical
Agenda for the 1990s and Beyond." *Economic Geography* 67: 185–209.
– 1992. "Population Turnaround in the Canadian Inner City: Contextual Factors
and Social Consequences." *Canadian Journal of Urban Research* 1 no. 1: 69–
92.
Bunting, T. and Filion, P., eds. 1991. *Canadian Cities in Transition.* Toronto:
Oxford University Press.
Charette, C. 1990. "Inequities between the Inner City and Non-Inner City." Insti-
tute of Urban Studies, University of Winnipeg, *Newsletter* No. 31: 3.
Chorney, H. 1990. *City of Dreams: Social Theory and the Urban Experience.*
Scarborough, Ont: Nelson.
Chouinard, V. 1989. "Explaining Local Experiences of State Formation: The
Case of Cooperative Housing in Toronto." *Society and Space* 7: 51–68.
Dear, M. 1980. "The Public City." In W.A.V. Clark and E. Moore, eds., *Resi-
dential Mobility and Public Policy,* Beverly Hills, Calif.: Sage, 219–41.
Doucet, M., and Weaver, J. 1991. *Housing the North American City.* Montreal:
McGill-Queen's University Press.
Driedger, L. 1991. *The Urban Factor.* Toronto: University of Toronto
Press.
Dyck, I. 1989. "Integrating Home and Wage Workplace: Women's Daily Lives
in a Canadian Suburb." *Canadian Geographer* 33: 329–41.
Gad, G. 1991. "Toronto's Financial District." *Canadian Geographer* 35: 203–7.
Gerecke, K., ed. 1991. *The Canadian City.* Montreal: Black Rose.

Goldberg, M., and Mercer, J. 1986. *The Myth of the North American City*. Vancouver: University of British Columbia Press.

Hanson, S., and Pratt, G. 1988. "Reconceptualizing the Links between Home and Work in Human Geography." *Economic Geography*. 64: 299–321.

Hartshorn, T. 1992. *Interpreting the City*. New York: Wiley.

Hulchanski, D. 1990. "Canada." In W. van Vliet, ed., *International Handbook of Housing Policies and Practices*, Westport, CT: Greenwood Press, 289–325.

Kobayashi, A. 1990. "Racism and Law in Canada: A Geographical Perspective." *Urban Geography* 11: 447–73.

Langlois, S. 1987. "Les familles à un et à deux revenus: changement social et différenciation socio-économique." *Thèmes canadiens* 8: 147–60.

Laws, G. 1988. "Privatisation and the Local Welfare State: The Case of Toronto's Social Services." *Transactions, Institute of British Geographers* 13: 433–48.

Ley, D. 1981. "Inner-City Revitalization in Canada: A Vancouver Case Study." *Canadian Geographer* 25: 124–48.

– 1988. "Attributes of Areas Undergoing Social Upgrading in Six Canadian Cities." *Canadian Geographer* 32: 31–45.

– 1992. "Gentrification in Recession: Social Change in Six Canadian Inner Cities." *Urban Geography* 13: 230–56.

Ley, D., and Olds, K. 1988. "Landscape as Spectacle: World's Fairs and the Culture of Heroic Consumption." *Society and Space* 6: 191–212.

Lipset, S. 1990. *Continental Divide: The Values and Institutions of the United States and Canada*. New York: Routledge.

Miron, J., ed. 1993. *Housing, Home and Community: Progress in Housing Canadians, 1945–86*. Montreal: McGill-Queen's University Press.

Murdie, R.A. 1990. "Economic Restructuring, Changes in Central City Housing and Social Polarization: A Toronto Case Study," unpublished paper, Department of Geography, York University.

Newman, P., and Kenworthy, J. 1989. *Cities and Automobile Dependency*. Aldershot, UK: Gower.

Ram, B., Norris, M., and Skof, K. 1989. *The Inner City in Transition*. Ottawa: Statistics Canada.

Relph, E. 1990. *The Toronto Guide*. Department of Geography, Scarborough College.

Riches, G. 1978. "Feeding Canada's Poor: The Rise of the Food Banks and the Collapse of the Public Safety Net." In J. Ismael, ed., *The Canadian Welfare State*, Edmonton: University of Alberta Press, 126–48.

Rose, D. 1987. "Un aperçu feministe sur la restructuration de l'emploi et sur la gentrification: le cas de Montréal." *Cahiers de géographie du Québec* 31: 205–24.

Seeley, J., Sim, A., and Loosley, E. 1956. *Crestwood Heights*. New York: Basic Books.

Simmons, J., and Bourne, L.S. 1989. *Urban Growth Trends in Canada, 1981–*

86. Centre for Urban and Community Studies, University of Toronto, Major Report No. 25.

Stone, L. 1967. *Urban Development in Canada*. Ottawa: Queen's Printer.

Taylor, S.M., Elliott, S., and Kearns, R. 1989. "The Housing Experience of Chronically Mentally Disabled Clients in Hamilton, Ontario." *Canadian Geographer*. 33: 146–55.

Villeneuve, P., and Rose, D. 1988. "Gender and the Separation of Employment from Home in Metropolitan Montreal, 1971–1981." *Urban Geography* 9: 155–79.

Weston, J. 1982. "Gentrification and Displacement: An Inner City Dilemma." *Habitat* 25 no. 1: 10–19.

Wynn, G., and Oke, T., eds. 1992. *Vancouver and Its Region*. Vancouver: UBC Press.

Yeates, M. 1990. *The North American City*. Fourth edition. New York: Harper and Row.

NOTES TO CHAPTER TWO

1 For the broader arguments of the distinctiveness of Canadian cities in a continental perspective, see Mercer and Goldberg (1986).

2 The legitimacy of these, or any other vignettes, is of course open to debate, and I make no attempt to present models that are reminiscent of the placeless North American urban geography of the 1970s. In pursuing a landscape perspective here, I seek to transcend the idiosyncrasies often associated with traditional cultural geography to explore what Gregory and Ley (1988: 115) regard as landscape: "a 'concept in high tension,' one which contains multiple and competing claims about the constitution of social order."

3 Whether these local variations are significant is a moot point. Relph (1987: especially 166–89) has recently argued that the corporatization of cities makes such identity almost impossible.

4 Semple and Smith (1981).

5 Gad (1991).

6 The financial core is defined as those blocks bounded by Queen, Church, Front, and Simcoe streets. In 1992, the downtown as a whole contained over 5.0 million m² of office space; Royal Lepage, *Toronto Office Leasing Directory* (Spring 1992), 30.

7 Revealing his west coast roots (though he had lived in Toronto for a number of years), Cole Harris juxtaposed Lunenburg draggers with Toronto's "skyscrapers and placeless urban polish" while sitting in a downtown Toronto eatery, and he mused: "How much business for psychiatrists, I wonder, is generated by the several acres surrounding me?; Harris (1981: 143).

8 In the 1984 campaign for the federal Liberal leadership, Jean Chrétien contrasted his Shawinigan roots with John Turner's Bay Street office in the

Royal Bank building, where the windows were tinted with gold. In the 1988 federal election campaign, the NDP's Ed Broadbent used King and Bay to symbolize the corporate power that crushes the "little guy"; Brian Mulroney was Wall Street (because of his free-trade platform), and Turner was Bay Street.

9 The quote continues: "Now, when I see their tops lost in the clouds on a silent, foggy morning, my throat catches. Just like the Rockies, I say. I love them." Robertson (1977: 28).

10 The boom in hardrock mining on the Canadian Shield, financed largely by American interests but managed through Toronto, gave the exchange a new lease on life. That such activity concentrated in Toronto rather than Montreal, the other pole of the Canadian financial system, makes the 1930s a key phase in the shift of metropolitan leadership in Canada. See Kerr (1965).

11 On Toronto's commercial district as a whole and on selected large-scale office buildings, see Gad and Holdsworth (1984; 1985; 1987; 1988).

12 The pole of that shift was the Canadian Bank of Commerce, at 34 stories the tallest in the British Empire; also, west of Bay, the Toronto Star, at 22 stories. At the Bay intersection, a new head office for the Bank of Nova Scotia was planned before the Depression but not completed till 1951, and on the northwest corner, the Bank of Montreal also finished a structure delayed by depression and war, in 1947.

13 The labels First Bank and First Canadian effectively de-regionalize the Montreal origin (and still legal head office) of the Bank of Montreal and support the sense, for real estate purposes at least, that Toronto is the "first" place in the nation.

14 See Lemon (1985).

15 Holdsworth (1987).

16 Rogers (1976).

17 See Collier (1974) and Ley (1983: 302). The same architectural firms were often involved, with minor involvement of local firms.

18 Pacey (1979).

19 Denhez (1978: 184–270).

20 See Freisen (1984: 436–7).

21 A useful summary is by Kalman, "Canada's Main Streets," in Holdsworth (1984: 3–29).

22 Leacock (1912); Weaver (1988); Munro (1974): Laurence (1964). Carla Thomas (1985: 980) asserts that *The Stone Angel* "set the town, Manawaka, firmly in Canada's imaginative landscape."

23 Ennals (1986).

24 McCann (1988b).

25 "Business Activity Flourishing along the River," *Brunswick Business Journal* (June 1988) 23.

26 Robertson (1973); for the roots of the small-town prairie place as the expression of extraregional capital, see Holdsworth and Everitt (1988).

27 Holdsworth (1984).

28 While many of these programs mimic those developed in the United States, especially by the National Trust for Historic Preservation, the two nations diverge in their emphasis on the use of public funds. Typically the American solution appeals to tax incentives and therefore private and individual initiative.

29 For an account of the origins of the Vancouver residential landscape, see Holdsworth (1977; 1986).

30 Residents of the East End neighbourhood called the Beaches might want their claim to special urbanity noted.

31 Ley (1981).

32 Mills (1988).

33 Ley (1987); Cybriwsky, Ley, and Western (1986).

34 Since this land has been sold to the Hong Kong–controlled Concord Pacific Corp. and condominiums sold out in hours in nervous Hong Kong, this development proposal and landscape transformation have taken on enormous social and political meaning in British Columbia. Long-standing ethnic tensions in the province quickly boiled to the surface. Even though some see, or hope to see, Vancouver growing as a financial node for the Pacific Rim economy, the relocation and local benefits of this migration of capital will undoubtedly cause problems for the area's nativists.

35 Sewell (1977); Lemon (1985: 134–6).

36 Clark (1966: 32–4).

37 Planners have attempted to restructure the landscape and the image of the place by inserting a major new community focus (city centre), high-density residential clusters, cultural facilities, and more public transit. Scarborough is now a major centre for luxury condominiums.

38 Relph (1987).

39 One of the most readable accounts of the role of the federal government in this area is provided by Humphrey Carver (1975: esp. 93–166).

40 See Garner (1950).

41 See Carver (1948: 1975); also Rose (1958).

42 See McAfee (1972).

43 Hulchanski (1984).

44 See Holdsworth (forthcoming) for a detailed analysis of the diversity of inner-city residential landscape changes in Toronto.

45 This is most forcefully argued by Lemon (1974; 1978; 1985).

REFERENCES FOR CHAPTER TWO

Carver, Humphrey. 1948. *Houses for Canadians*. Toronto: University of Toronto Press.

– 1975. *Compassionate Landscapes*. Toronto: University of Toronto Press.

Clark. S.D. 1966. *The Suburban Society*. Toronto: University of Toronto Press.

Collier, Robert W. 1974. *Contemporary Cathedrals: Large-Scale Developments in Canadian Cities*. Montreal: Harvest House.

Cybriwsky, Roman A., Ley, D.F., and Western, J. 1986. "The Political and Social Construction of Revitalized Neighborhoods: Society Hill, Philadelphia and False Creek, Vancouver." In N. Smith and P. Williams, eds., *Gentrification of the City*, Boston: Allen and Unwin, 92–120.

Denhez, Mark. 1978. *Heritage Fights Back*. Toronto: Fitzhenry and Whiteside.

Ennals, Peter. 1986. "The Main Streets of Maritime Canada." *Society for the Study of Architecture in Canada* 11: 11–13.

Friesen, Gerald. 1984. *The Canadian Prairies: A History*. Toronto: University of Toronto Press.

Gad, G.H.K. 1986. "The Paper Metropolis: Office Growth in Downtown and Suburban Toronto." *City Planning* Fall: 22–6.

– 1991. "Toronto's Financial District." *Canadian Geographer* 35: 203–7.

Gad, G.H.K., and Holdsworth, Deryck W. 1984. "Building for City, Region, and Nation: Office Development in Toronto, 1834–1984." In V.L. Russell, ed., *Forging a Consensus: Historical Essays on Toronto*. Toronto: University of Toronto Press, 272–322.

– 1985. "Large Office Buildings and Changing Occupancy, Toronto 1880–1950." *Bulletin, Society for the Study of Architecture in Canada* 10: 19–26.

– 1987. "Corporate Capitalism and the Emergence of the High-Rise Office Building." *Urban Geography* 8: 212–31.

– 1988. "Society and Streetscape: The Changing Built Environment of King Street, Toronto." In R. Hall, W. Westfall, and L.S. MacDowell, eds., *Patterns of the Past: Interpreting Ontario's History*, Toronto: Ontario Historical Society, 174–205.

Garner, Hugh, 1950. *Cabbagetown*. Toronto: McGraw-Hill Ryerson.

– 1976. *The Intruders*. Toronto: McGraw-Hill Ryerson.

Gregory, D., and Ley, D. 1988. "Culture's Geographies." *Environment and Planning D: Society and Space* 6: 115–16.

Harris, R.C. 1981. "A Ride on the Rapido." *Acadiensis* 10: 139–43.

Holdsworth, Deryck W. 1977. "House and Home in Vancouver: Images of West Coast Urbanism, 1886–1929." In G.A. Stelter and A.F.J. Artibise, eds., *The Canadian City: Essays in Urban History*, Toronto: McClelland and Stewart, 186–211.

– ed. 1984. *Reviving Main Street*. Toronto: University of Toronto Press.

– 1986. "Cottages and Castles for Vancouver Home-Seekers." In R.A.J. McDonald and J. Barman, eds., *Vancouver Past: Essays in Social History*, Vancouver: University of British Columbia Press, 11–32.

– 1987. *The Parking Authority of Toronto, 1952–87*. Toronto: Parking Authority of Toronto.

– Forthcoming. "Ethnic, Class and Heritage Fabrics: Conflicting Options for Inner-City Revitalization." *Society and Space*.

Holdsworth, Deryck W., and Everitt, John. 1988. "Bank Branches and Country Elevators: Expressions of Big Corporations in Small Prairie Towns." *Prairie Forum* 13: 173–90.

Hulchanski, David. 1984. *St. Lawrence and False Creek: A Review of the Planning and Development of Two Inner-City Neighbourhoods*. UBC Planning Papers, CPI No. 10, Vancouver.

Kerr, D. 1965. "Some Aspects of the Geography of Finance in Canada." *Canadian Geographer* 9: 175–92.

Laurence, Margaret. 1964. *The Stone Angel*. Toronto: McClelland and Stewart.

Leacock, Stephen. 1912. *Sunshine Sketches of a Little Town*. London: John Lane.

Lemon, James T. 1974. "Toronto: Is It a Model for Urban Life and Citizen Participation?" In D. Ley, ed., *Community Participation and the Spatial Order of the City*, Vancouver: Tantalus, 41–58.

– 1978. "The Urban Community Movement: Moving toward Public Households." In D. Ley and M. Samuels, eds., *Humanistic Geography: Prospects and Problems*, Chicago: Maaroufa, 319–37.

– 1985. *Toronto since 1918: An Illustrated History*. Toronto: Lorimer.

Ley, David F. 1981. "Inner-City Revitalization in Canada: A Vancouver Case Study." *Canadian Geographer* 25: 124–48.

– 1983. *A Social Geography of the City*. New York: Harper & Row.

– 1987. "Styles of the Times: Liberal and Neo-Conservative Landscapes in Inner Vancouver, 1968–1986." *Journal of Historical Geography* 14: 40–56.

– 1988. "Attributes of Areas Undergoing Social Uprading in Six Canadian Cities." *Canadian Geographer* 32: 31–45.

McAfee, Ann. 1972. "Evolving Inner-City Residential Environments: The Case of Vancouver's West End." In J.V. Minghi, ed., *Peoples of the Living Land*, Vancouver: Tantalus, 163–82.

McCann, L.D. 1988a. "Living the Double Life: Town and Country in the Industrialization of the Maritimes." In D. Day, ed., *Geographical Perspectives on the Maritime Provinces*, Halifax: St Mary's, 93–103.

– ed. 1988b. *People and Place: Studies of Small Town Life in the Maritimes*. Fredericton: Acadiensis Press.

Mercer, J., and Goldberg, M. 1986. *The Myth of the North American City*. Vancouver: University of British Columbia Press.

Mills, Caroline A. 1988. "Life on the Upslope: The Postmodern Landscape of Gentrification." *Environment and Planning D: Society and Space* 6: 169–90.

Munro, Alice. 1974. *Lives of Girls and Women*. New York: Signet.

Pacey, Elizabeth, 1979. *The Battle of Citadel Hill*. Hantsport, NS: Lancelot.

Relph, Edward. 1987. *Modern Urban Landscapes*. Baltimore: Johns Hopkins.

Robertson, Heather. 1973. *Grass Roots*. Toronto: Lorimer.

– 1977. "Go East, Young Woman, Go East." *Saturday Night* Jan./Feb. 25–8.

Rogers, Stan. 1976. *Fogarty's Cove*. Hannon, Ont.: Fogarty's Cove Music/ Barn Swallow Records.

Rose, Albert. 1958. *Regent Park: A Study in Slum Clearance*. Toronto: University of Toronto Press.

Semple, R. Keith, and Smith, W. Randy. "Metropolitan Dominance and Foreign Ownership in the Canadian Urban System." *Canadian Geographer* 25: 4–26.

Sewell, J. 1977. "Don Mills: E.P. Taylor and Canada's First Corporate Suburb." *City Magazine* 2: 28–38.

Simon, Joan, and Holdsworth, Deryck W. 1989. "Housing Form and Siting." In J. Miron, ed., *Housing Progress in Postwar Canada*, Report to CMHC, Ottawa.

Thomas, Carla. 1985. "Laurence, Margaret." In *The Canadian Encyclopedia*, vol. 2, Edmonton: Hurtig, 980–91.

Weaver, John. 1988. "Society and Culture in Rural and Small-Town Ontario: Alice Munro's Testimony on the Last Forty Years." In R. Hall, W. Westfall, and L.S. MacDowell, eds., *Patterns of the Past: Interpreting Ontario's History*, Toronto: Dundurn Press, 381–402.

NOTES TO CHAPTER THREE

1 See Woodsworth (1911).   2 See Mayhew (1862), Ames (1897), and Woodsworth (1911).   3 See Booth (1902).   4 Davies (1978a) confirmed the existence of a single scale or factor of prosperity from these data.   5 See Marx (1894).   6 For a brief overview of the Chicago school see Knox (1987: 59–63).   7 See Burgess (1925).   8 See Hoyt (1939).   9 See Harris and Ullman (1945).   10 See Firey (1947)   11 See Watson (1957: 495).   12 See Shevky and Bell (1955).   13 See Booth (1893).   14 See Timms (1971: chapter 4) and Davies (1983 and 1984).

15 Factor analysis is a family of related procedures that derives factors, dimensions, or sources of variation from a similarity matrix – usually correlations – calculated between the variables in the data set. Each factor identifies a separate source of variation or dimension measured by different amounts (or loadings) of the individual variables. A factor score measures the importance of each area on a factor.

16 Examples include Peuker and Rase (1971) for Vancouver, Davies (1975 and 1978b) for Calgary, Davies (1978b) for Edmonton, Taylor (1987) for Hamilton, Murdie (1969) for Toronto, Foggin and Polèse (1977) and Le Bourdais and Beaudry (1988) for Montreal, and Cliche (1980) for Quebec City. Comparative analyses include longitudinal studies of Toronto by Murdie (1969); Montreal by Greer-Wootten (1972), Guay (1978), and Le Bourdais and Beaudry (1988); Quebec City by Guay (1981); Hamilton by Taylor (1987); and three Prairie cities (a comparative analysis) by Davies and Barrow (1973).

17 See Davies (1984: chapter 9).

18 See Murdie (1969: chapter 7). Using indices of socioeconomic status, family size, and ethnic diversity, Balakrishnan and Jarvis (1979) confirmed Murdie's findings for most Canadian CMAs, and Pineo (1988) verified the concentric pattern of family status for CMAs using survey research data.

19 See Davies (1984).

20 See Davies (1983; 1984).

21 See Davies (1983) and Harvey's (1985) criticism of studies of residential differentiation.

22 See Firey (1947), Suttles (1972), and Ley (1983).

23 See Davies (1984: chapter 9).

24 See Davies and Murdie (1991) for a review of techniques. All closed number relationships were avoided.

25 Fifty-one tracts were excluded because of incomplete data.

26 This procedure incorporates all the census tracts from all cities into one analysis.

27 Principal-axes component analyses, followed by Direct Oblimin oblique rotation, were run on the 35×35 matrix of correlation values derived from the 2981 tract $\times$ 35 variable data set. Several different solutions were scrutinized in the search for the most suitable solution. The scores were computed using the approximation method of Murdie (1980).

28 Isopleths, based on the scores for each area, are used to highlight the broad social trends, in place of the more typically used choropleth maps based on the actual values.

29 See Hoyt (1939).

30 For a detailed study of the development of one of these suburbs, Kingsway Park in Etobicoke (Toronto), see Paterson (1985).

31 See Ley (1983: 86–7).

32 See, for example, Davies and Lewis (1974) for Britain; Murdie (1969), Davies (1975), Foggin and Polèse (1977), and Le Bourdais and Beaudry (1988) for Canada; and Knox (1987: 128–39) and White (1987: chapter 3), for the United States.

33 See Pratt and Hanson (1988).

34 See Cliche (1980).

35 See, for example, Foggin and Polèse (1977) and Le Bourdais and Beaudry (1988). Emergence of this factor depends on the selection of initial variables. That this factor has not appeared as clearly in other Canadian factorial studies is probably a result of the exclusion of explicit measures related to stage in the life cycle. Two authors have analysed life-cycle stage separately by correlating age groups: Hill (1976a) and Kuz (1984).

36 See Filion (1987) and Ley (1988).

37 These are reported in detail in Davies and Murdie (1991).

38 See Bourne, Baker, and Kalbach (1985).

39 See, for example, Bourne et al. (1986); Balakrishnan (1982); Balakrishnan and Kralt (1987); and Balakrishnan and Selvanathan (1990).

40 For details see McGahan (1986: 109–10).

41 The details of Table 3.4 were drawn primarily from Bourne et al. (1986). Additional indices for blacks were obtained for Ontario cities from Bourne,

Baker, and Kalbach (1985) and for Montreal and Vancouver from
Balakrishnan and Kralt (1987), although data from these sources are not entirely
comparable. Data for blacks and other visible minorities are not easily obtained from the census of Canada. See Boxhill (1984).

42 See, for example, Balakrishnan and Selvanathan (1990).

43 See, for example, Balakrishnan (1982) and Balakrishnan and Selvanathan (1990).

44 The index of ethnic diversity was taken from Bourne et al. (1986: 58–9).

45 See Wolff (1985). For the Asian group see Mercer (1988). For the black
population, housing market differences may also be a factor. In Metropolitan
Toronto, for example, blacks are disproportionately located in public-sector
housing projects which, though locally concentrated, are scattered throughout
the metropolitan area. Evidence on the disproportionate number of blacks
in public housing is given in Henry (1989) and Murdie (1992).

46 Enumeration area data could also be used, but this enormously increases
the detail of the spatial patterns. Davies (1975), as well as Hamm, Currie, and
Forde (1988), has aggregated these data to community and neighbourhood
areas to improve the utility of the spatial areas. The latter showed that the results
are sensitive to different spatial scales.

47 See Taylor (1987), Hamm, Currie, and Forde (1988), Le Bourdais and Beaudry
(1988), and Murdie (1988). Davies and Murdie (1990) completed a parallel
study of the changes for all CMAs in Canada between 1981 and 1986 and found
that most of the axes reported here were similar between the two dates.

48 See Murdie (1969; 1988).

49 The decreased role of female labour-force participation as an indicator of
social variability within the city has also been noted by Pratt and Hanson (1988).

50 See Ley (1988).

51 See Taylor (1987) and Murdie (1988).

52 See comparative studies by Balakrishnan (1982) and, Balakrishnan and
Selvanathan (1990).

REFERENCES FOR CHAPTER THREE

Ames, H.B. 1897. *The City below the Hill*. Reprint 1972. Toronto: University
of Toronto Press.

Balakrishnan, T.R. 1982. "Changing Patterns of Ethnic Residential Segregation in the Metropolitan Areas of Canada." *Canadian Review of Sociology and
Anthropology* 19: 92–110.

Balakrishnan, T.R., and Jarvis, G.K. 1979. "Changing Patterns of Spatial Differentiation in Urban Canada, 1961–1971. "*Canadian Review of Sociology and
Anthropology* 16: 218–27.

Balakrishnan, T.R., and Kralt, J. 1987. "Segregation of Visible Minorities
in Montreal, Toronto and Vancouver." In L. Driedger, ed., *Ethnic Canada*, Toronto: Copp Clark Pitman, 138–57.

Balakrishnan, T.R., and Selvanathan, K. 1990. "Ethnic Residential Segregation in Metropolitan Canada." In S.S. Halli, F. Trovato, and L. Driedger, eds., *Ethnic Demography*, Ottawa: Carleton University Press, 399–413.

Booth, C. 1893. "Life and Labour of the People in London." Presidential Address. *Journal of the Royal Statistical Society* 55: 557–91.

– 1902. *Life and Labour of the People in London*. Reprint 1969. New York: A. Kelley Reprint.

Bourne, L. S., Baker, A.M., and Kalbach, W. 1985. *Ontario's Ethnocultural Population, 1981: Socio-Economic Characteristics and Geographical Distributions*. Ontario Ministry of Citizenship and Culture, Ethnocultural Data Base Series III, Special Report No. 3, Toronto.

Bourne, L.S., Baker, A.M., Kalbach, W., Cressman, R., and Green, D. 1986 *Canada's Ethnic Mosaic: Characteristics and Patterns of Ethnic Origin Groups in Urban Areas*. Centre for Urban and Community Studies, University of Toronto, Major Report No. 24, Toronto.

Boxhill, W.O. 1984. *Limitations to the Use of Ethnic Origin Data to Quantify Visible Minorities in Canada*. Statistics Canada Working Paper (General), Ottawa.

Burgess, E.W. 1925. "The Growth of the City." In R.E. Park, E.W. Burgess, and R.D. McKenzie, *The City*, originally published in 1925, Chicago: University of Chicago Press, 37–44.

Cliche, P. 1980. *Espace social et mobilité résidentielle*. Quebec: Les Presses de l'Université Laval.

Darroch, G., and Marston, W.G 1987. "Patterns of Urban Ethnicity." In L. Driedger, ed., *Ethnic Canada*, Toronto: Copp Clark Pitman, 111–37.

Davies, W.K.D. 1975. "A Multivariate Description of Calgary's Community Areas." In B. Barr, ed., *Calgary*, University of Victoria, Western Geographical Series, vol. 11, Victoria, BC, 231–69.

– 1978a. "Charles Booth and the Measurement of Urban Social Structure." *Area* 290–6.

– 1978b. "A Social Taxonomy of Edmonton's Community Areas in 1971." In P.J. Smith, ed., *Edmonton: The Emerging Metropolitan Pattern*, University of Victoria, Western Geographical Series, Vol. 15, Victoria, BC, 161–97.

– 1983. *Urban Social Structure: A Multivariate Structural Study of Cardiff and Its Region*. University of Wales Press, Social Science Monograph 8, Cardiff.

– 1984. *Factorial Ecology*. Aldershot, England: Gower.

Davies, W.K.D., and Barrow, G.T. 1973. "A Comparative Factorial Ecology of Three Canadian Prairie Cities." *Canadian Geographer* 17: 327–53.

Davies, W.K.D., and Lewis, G.J. 1974. "The Urban Dimensions of Leicester, England." In B. Clark and B. Gleave, eds., *Social Patterns in Cities*, Institute of British Geographers Special Publication No. 5, 71–86.

Davies, W.K.D., and Murdie R.A. 1990 "The Social Complexity of Canadian Metropolitan Areas in 1986." Paper presented at the conference Changing Canadian Metropolis, Urban Studies, York University, Toronto, Oct. 1990.

– 1991. "Consistency and Differential Impact in Urban Social Dimensionality." *Urban Geography* 12 No. 1: 55–79.

Filion, P. 1987. "Concepts of the Inner City and Recent Trends in Canada." *Canadian Geographer* 31: 223–32.

Firey, W. 1947. *Land Use in Central Boston.* Cambridge, Mass.: Harvard University Press.

Foggin, P., and Polèse, M. 1977. *The Social Geography of Montreal in 1971.* Centre for Urban and Community Studies, University of Toronto, Research Paper No. 88, Toronto.

Greer-Wootten, B. 1972. "The Urban Model." In L. Beauregard, ed., *Montreal Field Guide,* Montreal: Les Presses de l'Université de Montréal, 9–31.

Guay, L. 1978. "Les dimensions de l'espace social urbain: Montréal 1951, 1961, 1971." *Recherches sociographiques* 19: 307–48.

– 1981. "Différenciation et ségrégation urbaines: Québec 1951, 1961 et 1971." *Recherches sociographiques* 22: 237–55.

Hamm, B., Currie, R.F., and Forde, D.R. 1988. "A Dynamic Typology of Urban Neighbourhoods: The Case of Winnipeg." *Canadian Review of Sociology and Anthropology* 25: 439–55.

Harris, C.D., and Ullman, E.L. 1945. "The Nature of Cities." *Annals, American Academy of Political and Social Science* 142: 7–17.

Harvey, D. 1985. *The Urbanization of Capital.* Baltimore: Johns Hopkins.

Henry, F. 1989. *Housing and Racial Discrimination in Canada: A Preliminary Assessment of Current Initiatives and Information.* Ottawa: Policy and Research, Multiculturalism and Citizenship.

Hill, F.I. 1976a. "The Family Life Cycle." In D.M. Ray, ed., *Canadian Urban Trends,* vol. 2, *Metropolitan Perspective,* Toronto: Copp Clark, 22–36.

– 1976b. "The Cultural Mosaic of Metropolitan Canada." In D.M. Ray, ed., *Canadian Urban Trends,* vol. 2, *Metropolitan Perspective,* Toronto: Copp Clark, 70–121.

Hoyt, H. 1939. *The Structure and Growth of Residential Neighborhoods in American Cities.* Washington, DC: Federal Housing Administration.

Knox, P. 1987. *Urban Social Geography: An Introduction.* 2nd edition. Essex, England: Longman.

Kuz, T.J. 1984. *Winnipeg Population: Structure and Process, 1951–1981.* Institute of Urban Studies, University of Winnipeg, Working Paper No. 7, Winnipeg.

Le Bourdais, C., and Beaudry, M. 1988. "The Changing Residential Structure of Montreal 1971–1981." *Canadian Geographer* 32: 98–113.

Ley, D. 1983. *A Social Geography of the City.* New York: Harper and Row.

– 1988. "Social Upgrading in Six Canadian Inner Cities." *Canadian Geographer* 32: 31–45.

McGahan, P. 1986. *Urban Sociology in Canada,* 2nd edition. Toronto: Butterworths.

Marx, K. 1894. *Capital.* 3 vols. Reprint 1967. New York: International Publishers.

Mayhew, H. 1862. *London Labour and London Poor*. Reprint 1968. London: Dover Publications.

Mercer, J. 1988. "New Faces on the Block: Asian Canadians." *Canadian Geographer* 32: 360–2.

Murdie, R.A. 1969. *Factorial Ecology of Metropolitan Toronto, 1951–1961.* Department of Geography, University of Chicago, Research Paper No. 116, Chicago.

– 1980. "Factor Scores: A Neglected Element of Factorial Ecology Studies." *Urban Geography* 1: 295–316.

– 1988. "Residential Structure of Metropolitan Toronto, 1951–1981." Paper presented at meetings of the Association of American Geographers, Phoenix, Ariz.

– 1992. *Social Housing in Transition: The Changing Social Composition of Public Sector Housing in Metropolitan Toronto*. Ottawa: CMHC.

Paterson, R. 1985. "The Development of an Interwar Suburb: Kingsway Park, Etobicoke." *Urban History Review* 13: 225–35.

Peuker, T., and Rase, W. 1971. "A Factorial Ecology of Greater Vancouver." In R. Leigh, ed., *Contemporary Geography: Western Viewpoints*, Vancouver: Tantalus Research, 81–96.

Pineo, P.C. 1988. "Socioeconomic Status and the Concentric Zonal Structure of Canadian Cities." *Canadian Review of Sociology and Anthropology* 25: 421–38.

Polèse, M., and Carlos, S. 1978. *L'écologie factorielle d'un système urbain: une analyse globale des facteurs de différenciation spatiale en milieu urbain pour les principales villes du Canada*. Montreal: INRS-Urbanisation.

Pratt, G., and Hanson, S. 1988. "Gender, Class and Space." *Environment and Planning D: Society and Space* 6: 15–35.

Shevky, E., and Bell, W. 1955. *Social Area Analysis*. Stanford, Calif.: Stanford University Press.

Suttles, G.D. 1972. *The Social Structure of Communities*. Chicago: University of Chicago Press.

Taylor, S.M. 1987. "Social Change in Hamilton, 1961–1981." In M.J. Dear, J.J. Drake, and L.G. Reeds, eds., *Steel City: Hamilton and Region*, Toronto: University of Toronto Press, 138–55.

Timms, D.W.G. 1971. *The Urban Mosaic: Towards a Theory of Residential Differentiation*. Cambridge: Cambridge University Press.

Watson, J.W. 1957. "The Sociological Aspects of Geography." In G. Taylor, ed., *Geography in the Twentieth Century*, 3rd edition. London: Methuen, 463–99.

White, M.J. 1987. *American Neighborhoods and Residential Differentiation*. New York: Russell Sage Foundation.

Wolff, P. 1985. "Haitians and Anglophone West Indians in the Ethnic and Socio-Economic Structure of Montreal." In A. Pletsch, ed., *Ethnicity in Canada*, Marburger Geographische Schriften, No. 96, 286–301.

Woodsworth, J.S. 1911. *My Neighbor*. Reprinted 1972. Toronto: University of Toronto Press.

NOTES TO CHAPTER FOUR

1 Dear and Wolch (1987). See also their chapter (16) in this volume.
2 Miron (1988: chapter 3).
3 One measure is the period total fertility rate (PTRF) – the number of live births expected on average of 1,000 women who pass through their childbearing years having the fertility rates specified for that particular year. The PTFR fell from 3,536 births per 1,000 women in 1921 to 2,654 births in 1939.
4 When the PTFR stood at 3,925 births.
5 By the late 1980s, the baby-bust trough hovered near a PTFR of just 1,650 births.
6 A first-marriage rate is the ratio of single persons marrying during a year to the population of single persons enumerated at mid-year.
7 Rodgers and Witney (1981: 729).
8 Adams and Nagnur (1988: 29). In a marital life table, an initial cohort of 100,000 males/females is subjected to the age-sex-marital status–specific rates of marriage, widowhood, divorce, and mortality observed over a period of time, until the last member dies. Thus propensity to ever marry is a function of both first-marriage probabilities and the probability of surviving long enough to marry.
9 In the 1981 census, each person in a household was identified by name, age, sex, marital status, and relationship to "Person 1." In coding of census returns, a common-law union was assumed if a person in the household checked off the box marked "common law partner of Person 1." Common-law relationships went undetected where the relationship was identified otherwise (e.g. "spouse," "partner," or "room-mate") or where neither person was "Person 1." Prior to 1981, since common-law partners might have listed themselves as married or in another category such as "partners," we cannot obtain counts of common-law unions.
10 See Dumas (1987: Table 10).
11 Adams and Nagnur (1988: 29).
12 Miron (1988: 41) has estimated that the increase in longevity between 1945 and 1976 alone would have accounted for an 11-per-cent increase in the steady-state population of Canada.
13 See Dumas (1987: Chart 12).
14 In the 1921 census, individuals were asked to give paternal ethnicity only. In 1986, individuals could report more than one ethnicity, but the 1986 percentage shown is for individuals reporting one ethnic origin only.
15 Even if that person is living with a relative: e.g. grandmother living with a daughter's family.
16 Where a household consists of more than one census family or one non-

family individual living alone, the other families or non-family persons are defined to be secondary: i.e. not maintaining a dwelling. Census forms provide little guidance to respondents as to which person or family is to be designated primary.

17 Miron (1988: 128–45) found that the propensity to maintain a dwelling is sensitive to income but that the propensity of primary families and non-family individuals to live alone is much less sensitive. This suggests that rising post-war affluence may have induced more families and non-family individuals to maintain a dwelling but, among these, not to increase the proportion living alone.

18 There is no single definition of household life. In the present chapter, it is restricted to activities that take place within, or at the site of, the principal residence, such as meal preparation, toilette, house, dish and laundry cleaning, leisure and recreation, reading and education, family upbringing, and home maintenance and repair.

19 Strong-Boag (1986).

20 Spitze (1988: 600–1) reviews studies done worldwide. The amount of time spent on housework by husbands has increased slightly, but not sufficiently to offset the decline in time spent by working wives. See Michelson (1985: 43–59) regarding similar results based on a sample of Toronto women.

21 In 1986, Statistics Canada's family expenditure survey found that food purchased from restaurants accounted for almost 25 per cent of all food expenditures.

22 BRGFCF is private housing investment (exclusive of land costs). It does not include the relatively modest amount of residential investment that originates in the government sector.

23 Throughout this chapter, "real" is used to mean that an expenditure is deflated to correct for price change.

24 There are two principal potential sources of error in doing this: the new construction component of BRGFCF includes investments in conversions as well as new buildings, and investment in a given year may be on new dwellings that have been started but not yet completed.

25 In Canadian censuses, a dwelling is a structurally separate set of living quarters with its own private entrance. A private entrance is a door from outside the building or from a common hall that does not require that residents pass through someone else's living quarters to reach their own.

26 Miron (1988: 183–8).

27 Fallis (1989: Table 2).

28 Miron (1990) discusses the Ontario provisions in detail.

29 Statistics Canada has tracked the growth of CMAs since 1956.

30 1941 datum taken from Bourne (1989: Table 2); 1986 datum calculated from *1986 Census of Canada: Population: Census Metropolitan Areas and Census Agglomerations*, Cat. 92-104, Table 1 and Appendix 1.

31 Construction standards also became more similar as each province wrested control over building codes from the municipalities in the early 1970s. See also chapter 19 in this volume.

32 This is not to understate the contributions made in particular case studies of filtering, racial change, and housing market transition that have used such diverse sources of information as street directories, building and demolition permits, and property tax assessment records.

33 Dansereau (1989: Table 5).

34 See Becker (1981).

35 For example, an evening spent at home playing cards might require a deck of cards, a card table and chairs, electric lighting, drinks and snack food, and the time of the participants. Market-supplied goods that are substitutes, to varying degrees, might include going to a bridge club or a movie.

36 See Ermisch (1981).

37 This is not to deny evidence that wives who enter the paid work-force full time still end up doing much of the household work.

38 Olsen (1987) is a recent review of the theoretical and empirical literature in this area. Modern empirical analysis of the incidence of homeownership using large microdata samples began with David (1962). Ranney (1981) presents an interesting theoretical exposition. Henderson and Ioannides (1987) is a recent example of developments in this field.

39 It is typically assumed that the household has to cover current consumption (i.e. expenditures on goods and services plus mortgage/loan payments) out of current income. The implication is that lenders will not permit households to pay for current consumption by increasing indebtedness even though this is not impossible in practice (e.g. university students who use student loans).

40 Even though change in household composition often causes, precedes, or is at least associated with a change in tenure.

41 See Miron (1988: 223–4). Central to this argument are the ideas that the psychic costs of relocation for a family are greater when there are children who have to change schools and make new friends, and homeownership provides better security of tenure by reducing the probability of eviction.

42 See Alonso (1964) and Muth (1969). Introductory overviews are provided in Miron (1982) and Thrall (1989). Recent reviews of the state of the art can be found in Fujita (1986) and Straszheim (1987).

43 Muth's model assumes that a household consumes q units of housing services and that the market equilibrium price of a unit of housing service is a constant p. The consumer's expenditure on housing is thus pq. However, we can actually observe only expenditure.

44 Sufficiently large to assume a competitive rental market in land or housing as appropriate. The models also assume a large number of landlords for the same reason.

45 I.e. where all jobs are assumed to be located at the centre of the city.

46 Three CMAs are Montreal, Toronto, and Vancouver; consolidated CMA data used. Non-family persons shown as percentage of persons in private households. Calculated by author from 1986 Census of Canada, Cat. 94-217.

47 Boyd (1989: 9–11) points out that elderly immigrant women, many of whom have either not had paid work or worked only in the invisible economy or at low-paid jobs, are sometimes ineligible for CPP or QPP or receive partial benefits. In addition, to be eligible for OAS or GIS, one typically has to have resided in Canada for at least ten years.

48 Overall, the consumer price index rose by 694 per cent.

49 This is instanced in metropolitan areas by the boom in row housing and semi-detached that peaked in the mid-1970s, the link-housing boom that followed, the reduction in lot sizes for detached housing, and even the more-recent "monster homes."

50 Miron (1988: 222–31).

51 Miron (1988: 230).

52 As measured by the shelter component of the CPI.

53 Over the period through the 1970s, shelter expenditure continued to rise more or less proportionately with income: this is something of an enigma given that modern housing policy has typically had a stated goal of making housing more affordable. See Miron (1989).

54 These data were calculated from linked time series for each city of the shelter component (or housing component if the shelter component was unavailable) of the CPI. From the 1974 Survey of Housing Units public-use sample, a hedonic price equation was estimated for each city that relates monthly shelter cost to size, state of repair, neighbouring land use, period of construction, and facilities in the rental dwelling. The average shelter cost of a basket of rental dwellings that corresponded to the national mix of rental dwellings was then estimated for each city. This latter figure was then adjusted by the linked time series for that city to come up with the data shown in Table 4.8. Thus the data in Table 4.8 control for the effect of quality change on monthly shelter cost but not for the effects of either inflation generally or rent regulation in particular.

REFERENCES FOR CHAPTER FOUR

Adams, O.B., and Nagnur, D.N. 1988. *Marriage, Divorce, and Mortality: A Life Table Analysis for Canada and Regions, 1980–1982*. Cat. 84-536E. Ottawa: Statistics Canada.

Alonso, W. 1964. *Location and Land Use*. Cambridge, Mass.: Harvard University Press.

Becker, G.S. 1981. *A Treatise on the Family*. Cambridge, Mass.: Harvard University Press.

Bourne, L.S. 1989. "The Changing Settlement Environment of Housing." In J.R. Miron, ed., *Housing Progress in Canada since 1945*, Report to CMHC, Ottawa.

Boyd, M. 1989. "Immigration and Income Security Policies in Canada: Implications for Elderly Immigrant Women." *Population Research and Policy Review* 8: 5–24.

Dansereau, F. 1989. "Neighbourhood Differentiation and Social Change. In J.R. Miron, ed., *Housing Progress in Canada since 1945*, Report to CMHC, Ottawa.

David, M.H. 1962. *Family Composition and Consumption.* Amsterdam: North Holland.

Dear, M., and Wolch, J. 1987. *Landscapes of Despair: From Deinstitutionalization to Homelessness.* Princeton: Princeton University Press.

Dumas, J. 1987. *Current Demographic Analysis: Report on the Demographic Situation in Canada 1986.* Cat. 91-209E. Ottawa: Statistics Canada.

Ermisch, J.F. 1981. "An Economic Theory of Household Formation." *Scottish Journal of Political Economy* 28 no. 1: 1–19.

Fallis, G. 1989. "The Suppliers of Housing." In J.R. Miron, ed., *Housing Progress in Canada since 1945*, Report to CMHC, Ottawa.

Fujita, M. 1986. "Urban Land Use Theory." In R. Arnott, ed., *Location Theory*, Chur, Switzerland: Harwood Academic Publishers, 73–149.

Henderson, J.V., and Ioannides, Y.M. 1987. "Owner Occupancy: Investment vs Consumption Demand." *Journal of Urban Economics* 21: 228–41.

Michelson, W. 1985. *From Sun to Sun: Daily Obligations and Community Structure in the Lives of Employed Women and Their Families.* Totowa, NJ: Rowman and Allanheld.

Miron, J.R. 1982. "Economic Equilibrium in Urban Land Use." In L.S. Bourne, ed., *Internal Structure of the City: Readings on Urban Form, Growth, and Policy*, 2nd edition, Toronto: Oxford University Press, 124–36.

– 1988. *Housing in Postwar Canada: Demographic Change, Household Formation, and Housing Demand.* Montreal and Kingston: McGill-Queen's University Press.

– 1989. "Household Formation, Affordability, and Housing Policy." *Population Research and Policy Review* 8:55–77.

– 1990. "Security of Tenure, Costly Tenants, and Rent Regulation." *Urban Studies* 27: 167–83.

Muth, R.F. 1969. *Cities and Housing.* Chicago: University of Chicago Press.

Olsen, E.O. 1987. "The Demand and Supply of Housing Service: A Critical Survey of the Empirical Literature." *Handbook of Regional and Urban Economics* 2: 989–1022.

Ranney, S.I. 1981. "The Future Price of Houses, Mortgage Market Conditions, and the Returns to Home Ownership." *American Economic Review* 71: 323–33.

Rodgers, R.H., and Witney, G. 1981. "The Family Life Cycle in Twentieth Century Canada." *Journal of Marriage and the Family* 43: 727–40.

Romaniuc, A. 1984. *Current Demographic Analysis: Fertility in Canada, From Baby-boom to Baby-bust*. Cat. 91-524E. Ottawa: Statistics Canada.

Spitze, G. 1988. "Women's Employment and Family Relations: A Review." *Journal of Marriage and the Family* 50: 595–618.

Straszheim, M. 1987. "The Theory of Urban Residential Location." In E.S. Mills, ed., *Handbook of Regional and Urban Economics* 2, Amsterdam: Elsevier, 717–33 (part).

Strong-Boag, V. 1986. "Keeping House in God's Country: Canadian Women at Work in the Home." C. Heron and R. Storey, eds., *Confronting the Labour Process in Canada*, Kington and Montreal: McGill-Queen's University Press, 124–51.

Thrall, G.I. 1987. *Land Use and Urban Form: The Consumption Theory of Land Rent*. New York: Methuen.

NOTES TO CHAPTER FIVE

I thank Tipu Chowdhury, Joan Ellsworth, and Gordon Shields for assistance in preparing this chapter; Michael Goodchild and Brian Klinkenberg for permission to use examples from our previously published work; and the Social Sciences and Humanities Research Council of Canada and the University of Western Ontario for financial support.

1 Hägerstrand's (1970) time-geography model provides the most explicit view of how temporal and spatial structures influence individual behaviour.

2 For a more complete statement on the timing of space, see Parkes and Thrift (1975).

3 This theme is developed by Lynch (1972).

4 The psychologist Roger Barker (1968) provides a theoretical rationale for the concept of behaviour settings.

5 See chapters 4 and 8 in this volume.

6 An idea expressed by Robert McDaniel (University of Western Ontario) in private conversation.

7 Time pressure comparisons for 1971 and 1981 in Halifax are considered by Harvey and Elliott (1983: 44–5).

8 This discussion is based on a complete recording of all time-related headlines (1 Jan. 1986–31 Dec. 1988) for the *Calgary Herald, Halifax Chronicle Herald, Montreal Gazette, Toronto Globe and Mail, Toronto Star, Vancouver Sun,* and *Winnipeg Free Press*.

9 See documentation in chapter 18 and in Wilson (1982).   10 The Canadian Institute of Public Opinion (10 March 1986).   11 The idea of colonizing time is developed by Melbin (1978).   12 David Harvey (1985) considers the manipulation of space and time in the development of cities.   13 See Elliott, Har-

vey, and Procos (1976) for a description of time-budget research in Halifax; Meis
and Scheu (1973) for Vancouver; and Michelson (1985) for Metro Toronto.   14 Szalai et al. (1972).   15 Reported in volumes of the *Explorations
in Time Use* (1983), edited by M. Catherine Casserly and Brian L. Kinsley.   16 Harvey and Elliott (1983).   17 Michelson (1985).   18 See Osberg
(1988).   19 Canadian Institute of Public Opinion (1986b).   20 Kunin and
Knauf (1988: 27).   21 Yalnizyan (1988: 14).   22 Ibid. 15.   23 Osberg (1988:
7–8).   24 Pronovost (1986: 12).   25 Johnson and Abramovitch (1987:
2–3).   26 Lindsay (1979) reviews these trends from a geographical perspective.
Canadian Business (1987) reports the business view that these are useful employee benefits.   27 Bowman (1982).   28 For similar findings, see Michelson
(1985: 136–41).   29 Burns (1979).   30 Lindsay (1979).   31 Palm and
Pred (1974).   32 Wekerle and Rutherford (1987). Also, see Rutherford and
Wekerle (1988).   33 Villeneuve and Rose (1988).   34 This view is elaborated on in Klodawsky and Mackenzie (1983), Michelson (1983), and Andrews
and Milroy (1988).   35 Dyck (1989).

36 An exception is work by Prior (1986). He used a one-day space-time budget
    survey to identify the activity patterns and social networks of ninety-one
    non-institutionalized retired men between the ages of 57 and 87 in the Vancouver
    area, giving special consideration to mobility constraints.

37 Elliott, Harvey, and Procos (1976) and Harvey and Elliott (1983).   38 Szalai
et al. (1972).   39 Based on straight-line distances between all pairs of origins
and destinations in a respondent's daily space-time path.   40 This procedure follows an example from Taylor and Parkes (1975).   41 See Goodchild and
Janelle (1984).   42 Belloni (1986: 66).   43 C. Rose (1977).

REFERENCES FOR CHAPTER FIVE

Andrew, Caroline, and Moore Milroy, Beth, eds. 1988. *Life Spaces: Gender,
    Household, Employment.* Vancouver: University of British Columbia Press.
Barker, Roger. 1968. *Environmental Psychology.* Stanford, Calif.: Stanford
    University Press.
Belloni, Maria Carmen. 1986. "Social Time Dimensions as Indicators of Class
    Distinction in Italy." *International Social Science Journal* 38 no. 1: 65–76.
Bowman, Dorothy Cindy. 1982. Behavioural Effects of Flexible Working
    Hours: A Time Geography Approach. MA thesis, University of Western Ontario,
    London, Ont.
Burns, Lawrence D. 1979. *Transportation, Temporal, and Spatial Components
    of Accessibility.* Lexington, Mass.: Lexington Books, D.C. Heath.
Canadian Business. 1987. "Family Benefits: Making the Workplace a Good Place
    for Parents." 60 no. 4 (April): 12–13.
Canadian Institute of Public Opinion. 1986a. "4-in-5 Feel Men Should Share Gen-
    eral Housework." *Gallup Report* 10 March: 1–2.

– 1986b. "Opinion Split on Shorter Work Week." *Gallup Report* 14 Aug.: 1–2.

Dyck, Isabel. 1989. "Integrating Home and Wage Workplace: Women's Daily Lives in a Canadian Suburb." *Canadian Geographer* 33 no. 4: 329–41.

Elliott, D.H., Harvey, A.S. and Procos. D. 1976. "An Overview of the Halifax Time Budget Study." *Society and Leisure* no. 3: 145–59.

Goodchild, Michael F., and Janelle, Donald G. 1984. "The City around the Clock: Space-time Patterns of Urban Ecological Structure." *Environment and Planning A* 16: 807–20.

Hägerstrand, Torsten. 1970. "What about People in Regional Science?" *Papers of the Regional Science Association* 24: 7–21.

Harvey, A.S., and Elliott, D.H. 1983. "Time and Time Again." *Explorations in Time Use*, vol. 4. Ottawa: Employment and Immigration Canada.

Harvey, David. 1985. *Consciousness and the Urban Experience*. Baltimore, Md.: Johns Hopkins University Press.

Janelle, Donald G., and Goodchild, Michael F. 1983a. "Diurnal Patterns of Social Group Distributions in a Canadian City." *Economic Geography* 59: 403–25.

– 1983b. "Transportation Indicators of Space-time Autonomy." *Urban Geography* 4: 317–37.

Janelle, D.G., Goodchild, M.F. and Klinkenberg, B. 1988. "Space-Time Diaries and Travel Characteristics for Different Levels of Respondent Aggregation." *Environment and Planning A* 20: 891–906.

Johnson, Laura C., and Abramovitch, Rona. 1987. "Rush Hours: A New Look at Parental Employment Patterns." *Social Infopac* 6 no. 4: 1–4.

Kinsley, Brian L., and O'Donnell, Terry. 1983. "Making Time: Methodology Report of the Canadian Time Use Pilot Study – 1981." *Explorations in Time Use*, vol. 1. Ottawa: Employment and Immigration Canada.

Klodawsky, Fran, and Mackenzie, Suzanne. 1983. "Gender-Sensitive Theory and the Housing Needs of Mother-led Families: Some Concepts and Some Buildings." *Feminist Perspectives* no. 9: 1–30.

Kunin, Roslyn, and Knauf, Joachim. 1988. "Fewer Full-time Jobs." *Canadian Business Review* 15 no. 2: 26–7.

Lindsay, Ian J. 1979. "An Exploratory Time-Space Model of Activity Selection and Travel Behaviour with Special Reference to New Forms of Work Scheduling." PhD thesis, University of Toronto.

Lynch, Kevin. 1972. *What Time Is This Place?* Cambridge, Mass.: M.I.T. Press.

Meis, S., and Scheu, J. 1973. "All in a Day's Work: A Time Budget Analysis of the Daily Activities of Men and Women." Department of Anthropology and Sociology, University of British Columbia.

Melbin, M. 1978. "The Colonisation of Time." In T. Carlstein, D.N. Parkes, and N.J. Thrift, eds., *Human Activity and Time Geography, vol. 2*, London: Edward Arnold, 100–13.

Michelson, William. 1983. *The Impact of Changing Women's Roles on Trans-portation Needs and Usage*. Washington, DC: U.S. Department of Transportation.

– 1985. *From Sun to Sun: Daily Obligations and Community Structure in the Lives of Employed Women and Their Families*. Totowa, NJ: Rowman & Allanheld.

Osberg, Lars. 1988. "The Future of Work in Canada." *Perception* 12 no. 3: 6–9.

Palm, R., and Pred, A. 1974. *A Time-Geography Perspective on Patterns of Inequality for Women*. Working Paper No. 236, Institute of Urban and Regional Development, University of California at Berkeley.

Parkes, D.N., and Thrift, N.J. 1975. "Timing Space and Spacing Time." *Environment and Planning A* 7: 651–70.

Prior, James Bruce. 1986. "The Dwelling as a Node of Daily Activity among Non-Institutionalized Retired Men." PhD thesis, Simon Fraser University.

Pronovost, Gilles. 1986. "Introduction: Time in a Sociological and Historical Perspective." *International Social Science Journal* 38 no. 1: 5–18.

Rose, Courtice. 1977. "Reflection on the Notion of Time Incorporated in Häger-strand's Time-Geographic Model of Society." *Tijdschrift voor Economische en Sociale Geografie* 68: 43–50.

Rutherford, Brent M. and Wekerle, Gerda R. 1988. "Captive Rider, Captive Labor: Spatial Constraints and Women's Employment." *Urban Geography* 9 no. 2: 116–37.

Szalai, Alexander, et al. eds. 1972. *The Use of Time*. The Hague: Mouton.

Taylor, P.J., and Parkes, D.N. 1975. "A Kantian View of the City: A Factor Ecological Experiment in Space in Time." *Environment and Planning A* 7: 671–88.

Villeneuve, Paul, and Rose, Damaris. 1988. "Gender and the Separation of Employment from Home in Metropolitan Montreal, 1971–1981." *Urban Geography* 9 no. 2: 155–79.

Wekerle, Gerda R., and Rutherford, Brent M. 1987. "Employed Women in the Suburbs: Transportation Disadvantage in a Car-Centered Environment." *Alternatives* 14 no. 3: 49–54.

Wilson, S.J. 1982. *Women, the Family and the Economy*. Toronto: McGraw-Hill Ryerson.

Yalnizyan, Armine. 1988. "Economic Recovery and Labour Market Adjustment in Canada." *Perception* 12 no. 2: 14–18.

NOTES TO CHAPTER SIX

1 In the Census of Canada, "usual" is defined as living in a place for three months with no explicit plans to return to the prior place of residence.   2 Roseman (1971).   3 Kopf (1977).   4 Rossi (1980).   5 Speare, Goldstein, and Frey (1975).   6 Rossi (1955).   7 Sjaastad (1962).   8 Herbert and Stevens (1962).   9 Hanushek and Quigley (1978).   10 Michelson (1977).   11 Wolpert (1965).   12 Legaré, Balakrishnan, and Beaujot

(1989).   13 Rose (1984).   14 Clark and Onaka (1983).   15 Dear and
Scott (1981).   16 Gray (1975).   17 Harris (1986).   18 Moore (1982); Rose
(1984).   19 Moore and Clark (1986).   20 Entrikin (1989).   21 Emmi
(1986); Moore and Clark (1986).   22 Winer and Gauthier (1982).   23 Woods
and Rees (1986).   24 Frey (1983); Rees (1986).   25 Sumka (1979); Hartman (1979); Moore (1982).   26 Howell (1986); Ley (1988).   27 Rosenberg,
Moore, and Ball (1989).   28 Burgess (1929).   29 Rees (1986).   30 Liaw
and Ledent (1987).   31 Dahmann and McArthur (1987).   32 Rogerson (1987);
Easterlin (1980).   33 Moore and Clark (1986).   34 See Government of Ontario (1987: Table 4.1).   35 Moore and Clark (1986); Bourne (1987).   36 Simmons (1978).   37 These results were obtained from special tabulations from
the 1986 Census of Canada prepared in conjunction with the research reported in
Moore, Ray, and Rosenberg (1989).   38 Bryant, Russwurm, and McLellan
(1982).   39 Joseph, Keddie, and Smit (1988).   40 Moore and Clark
(1986).   41 Miron (1988).   42 Crossman (1988).   43 Michelson
(1977).   44 Rose (1984).   45 Moore (1982).   46 See Government of Ontario
(1987: Tables 9.1, 9.2).   47 Filion (1987).   48 Filion (1987); Goldberg
and Mercer (1986).   49 Shaw (1985).   50 Clark (1986).   51 McDaniel
(1986); Rosenberg, Moore, and Ball (1989).

REFERENCES FOR CHAPTER SIX

Bourne, L.S. 1987. "Evaluating the Aggregate Spatial Structure of Canadian
    Metropolitan Areas." *Canadian Geographer* 31 no. 3: 194–208.
Bryant, C.R., Russwurm, L.H., and McLellan, A.G. 1982. *The City's Coun-
    tryside: Land and Its Management in the Rural-Urban Fringe*. London:
    Longman.
Burgess, E.W. 1929. Urban Areas. "In T.V. Smith and L.D. White, eds., *Chi-
    cago: An Experiment in Social Science Research*, Chicago: University of Chicago
    Press, 113–38.
Clark, W.A.V. 1986. *Human Migration*. Beverly Hills: Sage.
Clark, W.A.V., and Onaka, J. 1983. "Life Cycle and Housing Adjustment
    as Explanations of Residential Mobility." *Urban Studies* 20: 47–57.
Crossman, S.M. 1988. "Residential Mobility of the Elderly: A Dynamic Per-
    spective." MA thesis, Department of Geography, Queen's University.
Dahmann, D.C., and McArthur, E.K., 1987. *Geographical Mobility and the
    Life Course: Moves Associated with Individual Life Events.*" Working Paper
    No. 8720, US Bureau of the Census, Washington DC.
Dear, M., and Scott, A.J. eds. 1981. *Urbanization and Urban Planning in Capitalist
    Society*. New York: Methuen.
Easterlin, R. 1980. *Birth and Fortune: The Impact of Numbers on Personal Welfare.*
    New York: Basic Books.
Emmi, P.C. 1986. "On the Stability of Housing Sector Interaction: Evidence from
    42 Metropolitan Areas." *Journal of Regional Science* 26: 745–60.

Entrikin, N. 1989. *On Understanding Specificity in the Study of Place and Region.* Discussion Paper No. 100, Department of Geography, Syracuse University, Syracuse, NY.

Filion, P. 1987. "Concepts of the Inner City and Recent Trends in Canada." *Canadian Geographer* 31 no 3: 223–32.

Frey, W.H. 1983. "A Multiregional Projection Framework That Incorporates Both Migration and Residential Mobility Streams: Application to Metropolitan City-Suburb Redistribution." *Environment and Planning A* 15: 1,613–32.

Goldberg, M.A., and Mercer, J. 1986. *The Myth of the North American City.* Vancouver: University of British Columbia Press.

Government of Ontario. 1987. *1985 Housing Market Survey,* vol. 1, *Rental Market.* Toronto: Ministry of Housing.

Gray, F. 1975. "Non-Explanation in Urban Geography." *Area* 7 no. 4: 225–35.

Hanushek, E.A. and J.M. Quigley. 1978. "Housing Market Disequilibrium and Residential Mobility." In W.A.V. Clark and E.G. Moore, eds., *Population Mobility and Residential Change,* Evanston, Ill.: Northwestern University Studies in Geography No. 25, 51–98.

Harris, R. 1986. "Boom and Bust: The Effects of House Price Inflation on Homeownership Patterns in Montreal, Toronto, and Vancouver. *Canadian Geographer* 30 no. 4: 302–15.

Hartman, C. 1979. "Comment on "Neighborhood Revitalization and Displacement: A Review of The Evidence"." *Journal of the American Planning Association* 45 no. 4: 488–90.

Herbert, J.D., and Stevens, B.D. 1962. "A Model for the Distribution of Residential Activity in Urban Areas." *Journal of Regional Science* 2: 137–60.

Howell, L. 1986. "The Affordable Housing Crisis in Toronto." *City Magazine* 9 no. 1: 25–9.

Joseph, A.E., Keddie, P.D., and Smit, B. 1988. "Unravelling the Population Turnaround in Rural Canada." *Canadian Geographer* 32 no. 1: 17–30.

Kopf, E. 1977. "Untarnishing the Dream: Mobility, Opportunity and Order in Modern America." *Journal of Social History* 11: 206–27.

Legaré, J., Balakrishnan, T.R., and Beaujot, R., eds. 1989. *The Family in Crisis: A Population Crisis?* Ottawa: Royal Society of Canada.

Ley, D. 1988. "Social Upgrading in Six Canadian Inner Cities." *Canadian Geographer* 32 no. 1: 131–45.

Liaw, K.-L., and Ledent, J. 1987. "Nested Logit Models and Maximum Quasi-likelihood Methods: A Flexible Methodology for Analyzing Interregional Migration Patterns." *Regional Science and Urban Economics* 17: 67–88.

McDaniel, S.A. 1986. *Canada's Aging Population.* Toronto: Butterworths.

Michelson, W. 1977. *Environmental Choice, Human Behaviour and Residential Satisfaction.* New York: Oxford University Press.

Miron, J. 1988. *Housing in Postwar Canada: Demographic Change, Household Formation, and Housing Demand.* Montreal and Kingston: McGill-Queen's University Press.

Moore, E.G. 1982. "Search Behaviour and Public Policy: The Conflict between Supply and Demand Perspectives." In W.A.V. Clark, ed., *Modelling Housing Market Search*, London: Croom Helm, 224–38.

Moore, E.G., and Clark, W.A.V. 1986. "Stable Structure and Local Variation: A Comparison of Household Flows in Four Metropolitan Areas." *Urban Studies* 23: 185–96.

Moore, E.G., Ray, B.K. and Rosenberg, M.W. 1989. *Redistribution of Immigrants in Canada*. Report prepared for Employment and Immigration Canada, Department of Geography, Queen's University, Kingston, Ont.

Rees, P.H. 1986. "Developments in the Modelling of Spatial Populations." In R. Woods and P. Rees, eds., *Population Structures and Models*. London: Allen and Unwin, 97–125.

Rogerson, P. 1987. "Changes in U.S. National Mobility Levels." *Professional Geographer* 39 no. 3: 344–50.

Rose, D. 1984. "Rethinking Gentrification: Beyond the Uneven Development of Marxist Urban Theory." *Environment and Planning D: Society and Space* 2: 47–74.

Roseman, C. 1971. "Migration as a Spatial and Temporal Process." *Annals of the Association of American Geographers* 61: 589–98.

Rosenberg, M.W., Moore, E.G., and Ball, S. 1989. "Components of Change in the Elderly Population of Ontario, 1976–1986." *Canadian Geographer* 33 no. 3: 218–29.

Rossi, P.H. 1955. *Why Families Move*. Glencoe: Free Press.

– 1980. *Why Families Move*. 2nd edition. Beverly Hills: Sage.

Shaw, R.P. 1985. *Intermetropolitan Migration in Canada: Changing Determinants over Three Decades*. Toronto: NC Press.

Simmons, J.W. 1978. "Migration in the Canadian Urban System." In R.M. Irving, ed., *Readings in Canadian Geography*, 3rd edition, Toronto: Holt, Rinehart and Winston, 47–67.

Sjaastad, L.A. 1962. "The Costs and Returns of Human Migration." *Journal of Political Economy* 70: 80–93.

Speare, A., Goldstein, S. and Frey, W.H. 1975. *Residential Mobility, Migration and Metropolitan Change*. Cambridge, Mass.: Ballinger.

Sumka, H.J. 1979. "Neighborhood Revitalization and Displacement: A Review of the Evidence." *Journal of the American Planning Association* 45 no. 4: 480–8.

Winer, S.L., and Gauthier, D. 1982. *International Migration and Fiscal Structure: An Econometric Study of the Determinants of Interprovincial Migration in Canada*." Ottawa: Economic Council of Canada.

Wolpert, J. 1965. "Behavioral Aspects of the Decision to Migrate." *Papers, Regional Science Association* 15: 159–69.

Woods, R., and Rees, P.H., eds. 1986. *Population Structures and Models*. London: Allen and Unwin.

NOTES TO CHAPTER SEVEN

The authors thank Bernard Thraves (University of Regina) for the photograph of St-Boniface and David Hanna (Université du Québec à Montréal) and Ludger Beauregard (Université de Montréal) for construction statistics.

1 The relationship between immigration and urban construction is debated in an extensive literature, reviewed in Mandel (1975), Easterlin (1968), and Olson (1982).

2 Exceptions are the large number of Filipino immigrants to Winnipeg and South Asian immigrants to Calgary.

3 John Porter (1965) presents a classic and contested statement of the relation between ethnicity and class. For other views see A. Anderson and Frideres (1981) and Breton et al. (1990).

4 Only 2 per cent of couples married across the religious line, and such marriages were often recorded in a "secret" parish register.

5 For example, J.S. Woodsworth in his book *Strangers within Our Gates, or Coming Canadians* (1972). See also Norris (1971) and Burnet and Palmer (1988).

6 Dafoe (1931).

7 Hallett (1972). For a discussion of economic pluralism, see Clarke, Ley, and Peach (1984).

8 The most comprehensive history of Canadian immigration policy is Hawkins (1988). For refugee policy, see Dirks (1977), Abella and Troper (1982), Malarek (1987), and Nash (1988). For specific reference to the Chinese Immigration Act, see McEvoy (1982).

9 For example, Côte des Neiges in Montreal (Lavoie 1989). On residential patterns of other groups in Montreal, see Veltman, Polèse, and Leblanc (1987) and Marois (1988).

10 Canada, Employment and Immigration (1990).

11 Harney (1985: p. 12). "Insider perspectives" on the urban environment are rare. Gans's studies (1962 and 1967) contrast Italian Americans in inner cities and suburbs. For Canadian examples, see Harney and Troper (1975) and Lai (1989). Popular histories (e.g. Takata 1983; Larue 1989) are often more evocative in this respect, as are fictional accounts such as those of Mordecai Richler.

12 Harney (1985: 13).

13 Harney (1985: 15).

14 Zucchi (1985: 126) and Ramirez (1984) describe a system of patronage which also operated in a number of other groups, including Japanese and Chinese immigrants, with their "boss" systems; Nakayama (1921), Ward (1978: 17), Moriyama (1985).

15 The economic success of Italian Canadians has been attributed in part to vertical integration within the work-force. A Toronto study by Kalbach (1981)

shows that Italian Canadians have high average salaries relative to educational levels. They are strongly concentrated residentially and more likely to be surrounded by their own group within the workplace, particularly in the construction industry. The implication is that, through social and kinship networks, job opportunities are provided at all levels of expertise, and social opportunity can in many instances compensate for lack of education.

16 K. Anderson (1987: 594).

17 Ito (1973: 837–44).

18 See Adachi (1976), Sunahara (1981), and Kobayashi (1987).

19 Park, Burgess, and McKenzie (1925: 50–8). This "moral order" is often portrayed as a model of land use, but, as Short (1984: 128) points out, it is simply a crude description of the spatial manifestations of social differentiation, created as different groups are assigned their "place" in the city.

20 Kalbach (1981).

21 Extensive empirical work has been done using social area analysis and the related technique of factorial ecology. For examples see chapter 3. This method retains value as a descriptive device at the early stages of investigation – see Ley (1983: 75–84) for discussion – but it provides no basis for understanding ethnicity itself.

22 Perhaps the two most famous examples of racist violence in Canadian history are the riots against Chinese and Japanese immigrants in Vancouver in 1907 (Ward 1978: 53–76) and the attack by the Swastika Club on Jews in Toronto in 1933, culminating in the "riot at Christie Pits" (Levitt and Shaffir 1987).

23 Porter's (1965) concept of the "vertical mosaic" argues the former view, while Darroch and Marston (1971) and Balakrishnan (1982) provide examples of the latter. Harris (1984) presents a rare theoretical discussion of the complex relationship between class and segregation.

24 Duncan and Duncan (1984) show how ethnicity and status become intertwined, as well as how insider and outsider are defined, in the creation of a valued neighbourhood.

25 See Breton (1964) and Lieberson (1980). This theory maintains that control of institutions (religious, educational, welfare, political, and economic preserves ethnocultural tradition within a geographically bounded area. Leadership combines with propinquity to foster a strong network of interrelationship.

26 Driedger and Church (1974), Thraves (1986). Groups that lose their residential distinctiveness have been termed "ecological assimilators."

27 Ramcharan (1982: 88).

28 George-Etienne Cartier (1814–1873) was co-premier of the Province of Canada and minister of militia and defence under Sir John A. Macdonald after Confederation, which he helped to establish. He was involved in the reform of the Civil Code in Quebec, negotiations to eliminate the seigneurial system, and the machinations of the Grand Trunk Railway.

29 See also Wolfe (1985).

30 On the complex history of the Jewish community in Montreal, see Gutwirth (1973) and Anctil (1988).

31 Notably, the possessive individualism of English liberalism and the French Revolution, evident in Canada's several systems of law: Quebec's civil law is based on the Custom of Paris and Napoleon's codification, while criminal procedure Canada-wide, as well as civil law in most provinces, is derived from the common law of England.

32 The earlier resistance of immigrant communities arose from the fact that they were forced to conform to French as the language of schooling while parents who had received their schooling in English in Quebec were entitled to claim English for their children. The politics of language intersects with the complex issue of Quebec nationalism, which we make no attempt to treat here, although its intricacy is most apparent in Montreal. See Laponce (1984), Langlois (1985), and Veltman (1983).

33 Waddell (1988: 213). In 1986, 16.8 per cent of Canadians could carry on a conversation in both official languages, and 20.5 per cent of the age group 15–24. Half of bilinguals were living in Quebec. From 1970–71 to 1987–88 the numbers in English-language schools in Quebec dropped 56 per cent, and outside Quebec the numbers in French-language schools dropped 22 per cent, despite stability in the number of school-age children of French mother tongue. The number of children in voluntary "immersion" in French as a second language increased 28 per cent. Currently 40 per cent of Quebec children whose first language in English and who are enrolled in English-language schools are, at the choice of their parents, receiving most of their instruction in French.

34 There are constitutional safeguards for certain religions, apparent in the Montreal school commissions. The Constitution Act, 1867, allowed Protestants and Catholics to organize two school systems in Montreal. Anctil (1988) discusses the historical negotiations over the position of the Jewish community in that structure.

35 French Catholic schools were formerly quite homogeneous. Although immigration into Quebec fell in the 1980s (as unemployment rose) to one-third of its 1960s level, children of earlier arrivals continue to enter school, and these communities currently have more youthful age structures and more children per family.

REFERENCES FOR CHAPTER SEVEN

Abella, Irving, and Troper, Harold. 1982. *None Is Too Many: Canada and the Jews of Europe, 1933–1948*. Toronto: Lester and Orpen Dennys.

Adachi, K. 1976. *The Enemy That Never Was: A History of the Japanese Canadians*. Toronto: McClelland and Stewart.

Anctil, Pierre. 1988. *Le rendez-vous manqué: Les juifs de Montréal face au*

Québec de l'entre-deux-guerres. Quebec: Institut québécois de recherche sur la Culture.

Anderson, Alan B., and Frideres, James S. 1981. *Ethnicity in Canada: Theoretical Perspectives.* Toronto: Butterworths.

Anderson, Kay. 1987. "The Idea of Chinatown: The Power of Place and Institutional Practice in the Making of a Racial Category." *Annals of the Association of American Geographers* 77 no. 4: 580–98.

– 1991. *Vancouver's Chinatown: Racial Discourse in Canada, 1875–1980.* Montreal: McGill-Queen's University Press.

Balakrishnan, T.R. 1982. "Changing Patterns of Ethnic Residential Segregation in the Metropolitan Areas of Canada." *Canadian Review of Sociology and Anthropology* 19: 92–110.

– 1988. *Immigration and the Changing Ethnic Mosaic of Canadian Cities.* Report submitted to the Review of Demography and Its Implications for Economic and Social Policy, Health and Welfare Canada, Ottawa.

Breton, Raymond. 1964. "Institutional Completeness of Ethnic Communities and Personal Relations to Immigrants." *American Journal of Sociology* 70: 193–205.

– 1984. "The Production and Allocation of Symbolic Resources: An Analysis of the Linguistic and Ethnocultural Fields in Canada." *Canadian Review of Sociology and Anthropology* 21: 123–44.

Breton, R., Isajiw, W.W., Kalbach, W., and Reitz, G. 1990. *Ethnic Identity and Equality.* Toronto: University of Toronto Press.

Burnet, Jean R., with Palmer, Howard. 1988. *Coming Canadians: An Introduction to a History of Canada's Peoples.* Toronto: McClelland and Stewart.

Canada, Employment and Immigration. 1990. *Annual Report.* Ottawa: Supply and Services.

Clarke, C., Ley, D., and Peach, C., eds. 1984. *Geography and Ethnic Pluralism.* London: Allen and Unwin.

Dafoe, J.W. 1931. *Clifford Sifton in Relation to His Times.* Toronto: Macmillan.

Darroch, G., and Marston, N.G. 1971. "The Social Class Basis of Ethnic Residential Segregation: The Canadian Case. *American Journal of Sociology* 77: 491–510.

Deschamps, Gilles, et Boucher, Marie-Rita. 1983. "Représentation cartographiée de la population allophone de la zone métropolitaine de Montréal selon la langue maternelle en 1981." Quebec: Ministère des communautés culturelles et de l'immigration.

Dirks, Gerald E. 1977. *Canada's Refugee Policy: Indifference or Opportunism?* Montreal: McGill-Queen's University Press.

Dorais, Louis-Jacques, Chan, Kwok B., and Indra, Doreen M., eds. 1988. *Ten Years Later: Indochinese Communities in Canada.* Canadian Asian Studies Association.

Dreidger, L., and Church, Glenn. 1974. "Residential Segregation and Institutional Completeness: A Comparison of Ethnic Groups." *Canadian Review of Sociology and Anthropology* 11 no. 1: 30–52.

Duncan, James, and Duncan, Nancy G. 1984. "A Cultural Analysis of Urban Residential Landscapes in North America: The Case of the Anglophile Elite." In John Agnew, John Mercer, and David Sopher, eds., *The City in Cultural Context*, Boston: Allen and Unwin, 225–76.

Easterlin, Richard A. 1968. *Population, Labor Force and Long Swings in Economic Growth: The American Experience.* New York: Columbia University Press.

Gans, Herbert. 1962. *The Urban Villagers: Group and Class in the Life of Italian-Americans.* New York: Free Press of Glencoe.

– 1967. *The Levittowners: Ways of Life and Politics in a New Suburban Community.* New York: Pantheon.

Gutwirth, Jacques. 1973. "Hassidim et judaicité à Montréal. "*Recherches sociographiques* 14 no. 3: 291–325.

Hallett, M.E. 1972. "A Governor-General's Views on Oriental Immigration to British Columbia, 1904–1911." *B.C. Studies* 14: 51–72.

Harney, Robert ed. 1985. *Gathering Places: Peoples and Neighbourhoods of Toronto, 1834–1945.* Toronto: Multicultural History Society of Ontario.

Harney, Robert, and Troper, Harold. 1975. *Immigrants: A Portrait of the Urban Experience, 1890–1930.* Toronto: Van Nostrand Reinhold.

Harris, Richard. 1984. "Residential Segregation and Class Formation in the Capitalist City: A Review." *Progress in Human Geography* 8: 26–49.

Hawkins, Freda. 1988. *Canada and Immigration: Public Policy and Public Concern.* Montreal: McGill-Queen's University Press.

Isajiw, Wsevolod, ed. 1977. *Identities: The Impact of Ethnicity on Canadian Society.* Toronto: Peter Martin.

Ito, Kazuo. 1973. *Issei: A History of Japanese Immigrants in North America.* Seattle, Wash.: Executive Committee for the Publication of Issei.

Joy, Annamaria. 1988. *Ethnicity in Canada: Social Accommodation and Cultural Persistence among the Sikhs and the Portuguese.* New York: AMS Press.

Kalbach, Warren E. 1981. *Ethnic Residential Segregation And Its Significance for the Individual in an Urban Setting.* University of Toronto, Centre for Urban and Community Studies, Research Paper No. 124, Toronto.

Kobayashi, Audrey. 1987. "From Tyranny to Justice: The Uprooting of Japanese Canadians after 1941." *Tribune juive* 5 no. 1: 28–35.

Kobayashi, Audrey, with Vibert, Dermot, Anderson, Kay J., and Mercer, John. 1988. "Asian Migration to Canada in Historical Context." *Canadian Geographer* 32 no. 4: 351–62.

Lai, David 1989. *Chinatowns: Towns within Cities in Canada.* Vancouver: University of British Columbia Press.

Langlois, André. 1985. "Évolution de la répartition spatiale des groupes ethniques dans l'espace résidentiel montréalais." *Cahiers de géographie du Québec* 19 no. 76: 49–65.

Laponce, J.A. 1984. "The French Language in Canada: Tensions between Geography and Politics." *Political Geography Quarterly* 3: 91–104.

Larue, Monique. 1989. *Promenades littéraires dans Montréal*. Montreal: Editions Québec/Amérique.

Lavoie, Caroline. 1989. "The Residential Mobility of the Vietnamese Canadians: From Dispersal to Concentration." MA thesis, Department of Geography, McGill University.

Levitt, Cyril H., and Shaffir, William. 1987. *The Riot at Christie Pits*. Toronto: Lester and Orpen Dennys.

Ley, David. 1983. *A Social Geography of the City*. New York: Harper and Row.

Lieberson, Stanley. 1980. "A Societal Theory of Race and Ethnic Relations." In J. Goldstein and R. Bienvenue, eds., *Ethnicity and Ethnic Relations in Canada*, Toronto: Butterworths, 67–80.

Linteau, Paul-André. 1981. "La montée du cosmopolitisme montréalais." *Questions de culture* 2: 23–53.

Luciuk, Lubomyr Y., and Kordan, Bohdan S. 1989. *Creating a Landscape: A Geography of Ukrainians in Canada*. Toronto: University of Toronto Press.

McEvoy, F.J. 1982. "A Symbol of Racial Discrimination: The Chinese Immigration Act and Canada's Relations with China, 1942–1947." *Canadian Ethnic Studies* 14 no. 3: 24–42.

McNicholl, Claire. 1986. "L'évolution spatiale des groupes ethniques à Montréal, 1871–1981." Doctoral thesis, École des Hautes Études en Sciences sociales, Paris.

Malarek, Victor 1987. *Heaven's Gate: Canada's Immigration Fiasco*. Toronto: Macmillan.

Mandel, Ernest. 1975. *Late Capitalism*. London: New Left Books.

Marois, Claude. 1988. "Cultural Transformations in Montreal since 1970." *Journal of Cultural Geography* 10: 29–38.

Moriyama, Alan T. 1985. *Imingaisha: Japanese Emigration Companies and Hawaii*. Honolulu: University of Hawaii Press.

Nakayama, J. 1921. *Kanada no Hooko* (The Treasure of Canada). Tokyo: Nakayama.

Nash, Alan, ed., and Humphrey, John P., rapp. 1988. *Human Rights and the Protection of Refugees under International Law*. Ottawa: Canadian Human Rights Foundation.

Norris, John. 1971. *Strangers Entertained: A History of the Ethnic Groups of British Columbia*. Vancouver: British Columbia Centennial '71 Committee.

Olson, Sherry. 1982. "Urban Metabolism and Morphogenesis." *Urban Geography* 3 no. 2: 87–109.

Park, Robert, Burgess, E.W. and McKenzie, R.D. 1925. *The City*. Chicago: University of Chicago Press.

Polèse, Mario, Hamel, Charles, and Bailly, Antoine. 1981. *La géographie résidentielle des immigrants et des groupes ethniques: Montréal, 1971.* Montreal: INRS-Urbanisation.

Porter, John. 1965. *The Vertical Mosaic: An Analysis of Social Class and Power in Canada.* Toronto: University of Toronto Press.

Proulx, Normand. 1979. *La répartition sectorielle des travailleurs immigrants au Québec.* Québec Ministère de l'immigration, Direction de la recherche, Études et documents no. 8.

Ramcharan, Subhas. 1982. *Racism: Nonwhites in Canada.* Toronto: Butterworths.

Ramirez, Bruno. 1984. *Les premiers Italiens de Montréal.* Montreal: Boréal Express.

Robinson, J. Lewis. 1988. "Vancouver: Changing Geographical Aspects of a Multicultural City." *B.C. Studies* 79: 59–80.

Short, John R. 1984. *An Introduction to Urban Geography.* London: Routledge and Kegan Paul.

Sunahara, Ann Gomer. 1981. *The Politics of Racism: The Uprooting of Japanese Canadians during the Second World War.* Toronto: Lorimer.

Takata, Toyo. 1983. *Nikkei Legacy: The Story of Japanese Canadians from Settlement to Today.* Toronto: NC Press.

Thraves, Bernard D. 1986. "An Analysis of Ethnic Intra-Urban Migration: The Case of Winnipeg." PhD thesis, University of Manitoba.

Veltman, Calvin. 1983. "L'évolution de la ségrégation linguistique à Montréal, 1961–1984." *Recherches sociographiques* 24 no. 3: 379–90.

Veltman, Clavin, Polèse, Mario, and Leblanc, Marc. 1987. "Évolution de la localisation résidentielle des principaux groupes ethniques et immigrants, Montréal 1971–1981." *Actualité immobilière* 10 no. 4: 20–33.

Waddell, Eric. 1988. "The Influence of External Relations on the Promotion of French in Canada." *Language Culture and Curriculum* 1 no. 3: 203–14.

Ward, Peter. 1978. *White Canada Forever.* Montreal: McGill-Queen's University Press.

Wolfe, Peter. 1985. "Haitians and Anglophone West Indians in the Ethnic and Socio-economic Structure of Montreal." In Alfred Pletsch, ed., *Ethnicity in Canada: International Examples and Perspectives,* Marburg, 286–301.

Woodsworth, J.S. 1972. *Strangers within Our Gates, or Coming Canadians.* First published 1909. Toronto: University of Toronto Press.

Zucchi, John. 1985. "Italian Hometown Settlements and the Development of an Italian Community in Toronto, 1875–1935." In Harney (1985: 121–46).

NOTES TO CHAPTER EIGHT

We are grateful to the editors, and also to Clifford Hastings, for comments on earlier drafts of this chapter. The financial assistance of the Social Sciences and Humanities Research Council of Canada and of the Fonds pour la formation de chercheurs et l'aide à la recherche (Quebec Ministry of Higher Education and Science) is also gratefully acknowledged. The authors share equal responsibility for this chapter.

1 Mackenzie and Rose (1983).

2 See Andrew and Milroy (1988) and Pratt (1990).

3 See Goldberg and Mercer (1986).

4 Rose (1987).

5 Tertiarization means a process by which various services – producer services, consumer services, and public and parapublic services – occupy an increasing share of the labour force. Tertiarization is studied using data on industry divisions, as in Table 8.1, or on occupations, as in Table 8.2. See Polèse (1988b) on the structural transformations entailed by tertiarization.

6 See Galois and Mabin (1987).

7 See Semple (1985) for an extension of central place theory which includes high-order services, sometimes called quaternary activities. See also Bunting and Filion (1988), Filion (1987), and Polèse (1988b) for recent statements on the transformation of the central districts of Canadian metropolitan areas.

8 Although the region still lives partly off staples produced all over the country and its industrialization remains largely tied to the American economy; see Simmons (1991).

9 See Norcliffe and Stevens (1979). This helps us to understand how Toronto replaced Montreal as the pivot of the Canadian economy between the wars (McCann and Smith 1991: 84–9).

10 Bradbury (1987: 401); see also Ley and Hutton (1987).

11 See Anderson (1985).

12 Statistics Canada (1986).

13 Green (1987).

14 Also, Parent (1983) has shown how state contracts in Quebec have propelled forward very large engineering firms such as Lavalin and S.N.C., which have their headquarters in Montreal and have diversified their activities into manufacturing in other parts of Canada and abroad.

15 Goldberg and Mercer (1986: 82ff); see also Leo and Fenton (1990).

16 Lamonde and Polèse (1984); Polèse (1990); McCann and Smith (1991). Since its re-election in 1988, however, the Conservative federal government has reprivatized major crown corporations in communications (Telesat), air transportation (Air Canada), and natural resources (Petro-Canada), while slashing subsidies to the remaining crown corporations.

17 Lamonde and Polèse (1984).

18 See Black and Myles (1986).

19 Coffey and Polèse (1987a); Polèse (1988b). The link between growth of these services and production of goods is underlined by Polèse (1988b: 21), who comments that rather than simply talking about the economy "ter-

tiarizing," we should think of the tertiary sector as industrializing, given that the growth in producer services is in part a result of manufacturing firms purchasing from other firms the services they used to produce internally.

20 See Theilheimer (1988); Saarinen (1990).

21 In 1985, women counted for almost 40 per cent of 25–34-year-old managers and administrators, up from only one-fifth in 1976; calculated from data in Dumas (1986: 102).

22 See Moore (1989).

23 Rose and Wexler (1989).

24 See Andrew (1984), Fortin (1986), Séguin and Villeneuve (1987), and Séguin (1988). For a review of recent work on women and urban social movements see Fincher and McQuillen (1989).

25 Rose (1984); Lauria and Knopp (1985); Mills (1989).

26 Langlois (1987).

27 Burch and Madan (1986).

28 Statistics Canada, Census of 1986, Cat. 93-106, Table 4. See also Le Bourdais with Rose (1986), Pool and Moore (1986), and Klodawsky and Spector (1988).

29 See also Rose (1990).

30 Statistics Canada, Labour Force Survey, unpublished data (available on microfiche).

31 See Mackenzie and Rose (1983) for an overview; see also Lowe (1982).

32 Villeneuve and Rose (1988).

33 Dufour (1988: 67).

34 Myles, Picot, and Wannell (1988).

35 Source as for Table 8.1: in 1986, 67 per cent of workers in finance and insurance in Montreal were female as against 64 per cent in Toronto. Although small, this difference runs counter to the trend in other sectors.

36 Notably, in Montreal between 1971 and 1981, out-migration of finance and insurance head offices was reflected in an absolute decline in the number of male professionals in the finance, insurance, and real estate sector; see Rose (1987).

37 See Rose (1987). Significantly, however, and probably reflecting recent changes in the political-economic climate in Quebec, the proportion of all Quebec women who work in the government sector fell behind that of Ontario between 1986 and 1988.

38 Statistics Canada (1987: 8).

39 See Saarinen (1990). In Sudbury, 2,000 new federal jobs were created from 1981 to 1986, almost all of them going to women (Statistics Canada, Census of 1986, Cat. 93-156, Table 17).

40 Statistics Canada, Census of 1986, Cat. 93-156, Table 17.

41 Saarinen (1990); Theilheimer (1988).

42 Mackenzie (1987).

43 Flexibility refers to a set of strategies for restructuring the labour process and labour markets in ways that permit more rapid response to changing market conditions. A major debate has developed around this question; see, for example, Harvey (1987), Gertler (1988), Sayer (1989), and Storper and Scott (1989). For a critical evaluation of the political and economic dynamics of this process in Canada, see Myles (1988a).

44 See Economic Council of Canada (1990).

45 See Dumas (1986), Villeneuve and Rose (1988), and Chicoine, Germain, and Rose (1992).

46 Holmes (1986).

47 For instance, the numbers employed in bookkeeping – an overwhelmingly "female" occupation – fell from 1981 to 1986.

48 Robinson (1986).

49 Christopherson (1989); Nielsen (1991); see also Walsh (1990).

50 See in particular Myles (1988b).

51 See Maroney (1983); Myles (1988b). Notably – and contrary to trends in the United States and Britain, for example – unionization rates among women were still increasing in the mid-1980s; see Clemenson (1989).

52 Economic Council of Canada (1990: 14–15).

53 Myles (1988b).

54 Myles, Picot, and Wannell (1988); Marcoux, Morin, and Rose (1990).

55 Rose and Villeneuve (1988; 1990).

56 See Sassen-Koob (1984).

57 Especially if the share of public- and parapublic-sector employment declines significantly and if the state allows wages and working conditions to deteriorate further in consumer services; see Marcoux, Morin, and Rose (1990).

58 Social Planning Council of Metro Toronto (1989).

59 Dufour (1988: 68–9).   60 Rose and Villeneuve (1992).   61 Gad and Holdsworth (1987).   62 See, for example, Nelson (1986).   63 Huang (1989) shows that in Toronto the differences between the downtown office core and the suburban office subcentres are much less pronounced than current theory suggests.   64 Villeneuve and Rose (1986).   65 Ley (1985a; 1985b).   66 Gad (1985: 346).   67 Polèse (1988a).   68 (Polèse (1988a).   69 See Beauregard (1986).   70 See Ley (1988).   71 Huang (1989); Gad (1986).   72 Rose (1987); Rose and Le Bourdais (1986).   73 See, for example, Norcliffe (1984).   74 Chicoine, Germain, and Rose (1992).   75 Séguin and Villeneuve (1987).   76 Villeneuve and Viaud (1987); Dansereau (1988).   77 Cloutier, Careau, and Drolet (1988).   78 Rose (1984).   79 See Hanson and Pratt (1990) and the special issue of *Urban Geography*, 9 no. 2 (1988).   80 Michelson (1986).   81 Klodawsky and Spector (1988).   82 Rose and Le Bourdais (1986).   83 Villeneuve and Rose (1988).   84 Chicoine and Rose (1989).   85 Wekerle and Rutherford (1988).   86 Ley (1985a).   87 Ley (1985a: 36).   88 Ley (1985b).   89 Simpson

(1987).   90 See Godbout and Blais (1983), Villeneuve and Morency (1990);
Villeneuve and Viaud (1987), Morency (1988), and Viaud
(1988).   91 Villeneuve and Viaud (1987).   92 Pratt and Hanson (1988); cf.
Wallman (1984).   93 Scott (1981).   94 See Hanson and Pratt
(1990).   95 See Villeneuve and Rose (1988) for a more detailed analysis.
96 Unfortunately, because of data limitations, length is measured here in terms
   of distance rather than time. Since fewer women own automobiles, the same
   increment of distance may entail proportionately more travel time for
   women than men, except for those workers located within easy access of rapid
   transit.
97 See, for example, Le Bourdais, Hamel, and Bernard (1987); and see Rouffignat
   and Vallée (1990).
98 See, for example, Morris (1990); and see, for example, Wallman (1984), Fortin
   (1988), and Finch (1989).
99 Le Bourdais and Beaudry (1988); Economic Council of Canada (1990); Séguin
   and Villeneuve (1987); Dansereau avec Lacroix (1988).

REFERENCES FOR CHAPTER EIGHT

Anderson, F.J. 1985. *Natural Resources in Canada: Economic Theory and
   Policy.* Toronto: Methuen.
Andrew, Caroline. 1984. "Women and the Welfare State." *Canadian Journal
   of Political Science* 17 no. 4: 667–83.
Andrew, Caroline, and Milroy, Beth M. 1988. "Introduction." In C. Andrew
   and B.M. Milroy, eds., *Life Spaces: Gender, Household, Employment*, Van-
   couver: University of British Columbia Press, 1–12.
Beauregard, Robert. 1986. "The Chaos and Complexity of Gentrification." In N.
   Smith and P. Williams, eds., *Gentrification of the City*, Boston: Allen and
   Unwin, 35–55.
Black, Don, and Myles, John. 1986. "Dependent Industrialization and the Ca-
   nadian Class Structure: A Comparative Analysis of Canada, the United States
   and Sweden." *Canadian Review of Sociology and Anthropology* 23 no. 2:
   157–81.
Bradbury, John. 1987. "British Columbia: Metropolis and Hinterland in Mi-
   crocosm." In L.D. McCann, ed., *Heartland and Hinterland: A Geography of
   Canada*, 2nd edition, Scarborough, Ont.: Prentice-Hall, 400–40.
Bunting, Trudi, and Filion, Pierre, eds. 1988. *The Changing Canadian Inner City.*
   University of Waterloo, Department of Geography, Publication Series No. 31.
Burch, Thomas K., and Madan, Ashok K. 1986. *Union Formation and
   Dissolution/Formations et ruptures d'unions.* Statistics Canada, Results from
   the 1984 Family History Survey, Cat. no. 99-963, Ottawa.
Chicoine, Nathalie, Germain, Annick, and Rose, Damaris. 1992. "From Economic
   Restructuring to the Fabric of Everyday Life: Families' Use of Childcare

Services in Multiethnic Neighbourhoods in Transition." In F. Remiggi, ed., *The Changing Geography of Montréal*. University of Victoria Press (in press).

Chicoine, Nathalie, and Rose, Damaris. 1989. "Restructuration économique, division sexuelle du travail et répartition spatiale de l'emploi dans la région métropolitaine de Montréal." *Espaces, populations, sociétés* no. 1: 53–64.

Christopherson, Susan. 1989. "Flexibility in the United States Service Economy and the Emerging Spatial Division of Labour." *Transactions of the Institute of British Geographers* new series, 14 no. 2: 131–43.

Clemenson, Heather. 1989. "Unionization and Women in the Service Sector." Statistics Canada, Cat. no. 75-001E, *Perspectives on Labour and Income* 1 no. 2: 30–44.

Cloutier, Richard, Careau, Louise and Drolet, Jacques. 1988. "La garde partagée: implications psychologiques. "Paper presented at the annual meeting of the Corporation des Psychologues du Québec, Montreal, June.

Coffey, William J., and Polèse, Mario. 1987a. "Trade and the Location of Producer Services: A Canadian Perspective." *Environment and Planning A* 19 no. 5: 597–611.

– 1987b. "Intrafirm Trade in Business Services: Implications for the Location of Office-Based Activities." *Papers of the Regional Science Association* 62: 71–80.

Dansereau, Francine. 1988. "Les transformations de l'habitat et des quartiers centraux: singularités et contrastes des villes canadiennes." *Cahiers de recherche sociologique* 6 no. 2: 95–115.

Dansereau, Francine, avec la collaboration de Benoit Lacroix. 1988. *Habiter au centre: tendances et perspectives socio-économiques de l'habitation dans l'arrondissement Centre*. Dossier Montréal 3. Montreal: Ville de Montréal and INRS-Urbanisation.

Dufour, Alain. 1988. "Les marchés du travail de Montréal et de Toronto: rétrospective 1975–1987. *Le marché du travail* Dec.: 66–72.

Dumas, Cécile. 1986. Occupational Trends Among Women in Canada: 1976–1985/L'évolution professionnelle des femmes au Canada, 1976 à 1985." Statistics Canada, Cat. no. 71-001, *The Labour Force* Oct.: 83–127.

Economic Council of Canada. 1990. *Good Jobs, Bad Jobs: Employment in the Service Economy*. Cat. no. EC22-164/1990E. Ottawa: Supply and Services Canada.

Filion, Pierre. 1987. "Concepts of the Inner City and Recent Trends in Canada." *Canadian Geographer* 31 no. 3: 223–32.

Finch, Janet. 1989 *Family Obligations and Social Change*. Oxford: Blackwell.

Fincher, Ruth, and McQuillen, Jacinta. 1989. "Women in Urban Social Movements." *Urban Geography* 10 no. 6: 604–13.

Fortin, Andrée. 1986. "Familles, réseaux et stratégies de sociabilité." In S. Langlois and S. Trudel, eds., *La morphologie sociale en mutation au Québec*. Actes du colloque annuel de l'Association des sociologues et anthropologues de langue française, *Cahiers de l'Acfas* 41: 159–69.

Gad, Gunter. 1985. "Office Location Dynamics in Toronto: Suburbanization and Central District Specialization." *Urban Geography* 6 no. 4: 331–51.

– 1986. "The Paper Metropolis." *City Planning* 4 no. 1: 22–6.

Gad, Gunter, and Holdsworth, Deryck W. 1987. "Corporate Capitalism and the Emergence of the High-rise Office Building. "*Urban Geography* 8 no. 3: 212–31.

Galois, Robert M., and Mabin, Alan. 1987. "Canada, the United States, and the World-System: The Metropolis-Hinterland Paradox. In L.D. McCann, ed., *Heartland and Hinterland: A Geography of Canada*, Scarborough, Ont.: Prentice Hall, 38–67.

Gertler, Meric. 1988. "The Limits to Flexibility: Comments on the Post-Fordist Vision of Production and Its Geography." *Transactions of the Institute of British Geographers* new series, 13 no. 4: 419–32.

Godbout, Jacques, and Blais, Serge. 1983. *L'accessibilité financière au logement neuf.* INRS-Urbanisation, Rapports de recherches no. 8, Montreal.

Goldberg, Michael A., and Mercer, John. 1986. *The Myth of the North American City: Continentalism Challenged.* Vancouver: University of British Columbia Press.

Green, Milford B. 1987. *A Geography of Canadian Merger Activity, 1962–1984.* University of Western Ontario, Department of Geography, Paper no. 54, London, Ont.

Hanson, Susan, and Pratt, Geraldine. 1990. "Geographic Perspectives on the Occupational Segregation of Women." *National Geographic Research.* 6 no 4: 376–99.

Harvey, David. 1987. "Flexible Accumulation through Urbanisation: Reflections on 'Post-Modernism' in the American City." *Antipode* 19 no. 3: 260–86.

Holmes, John. 1986. "The Organization and Locational Structure of Production Subcontracting." In A.J. Scott and M. Storper, eds., *Production, Work, and Territory: The Geographical Anatomy of Industrial Capitalism*, Boston: Allen and Unwin, 80–106.

Huang, Shirlena. 1989. "Office Suburbanization in Toronto: Fragmentation, Workforce Composition and Laboursheds." PhD thesis, Department of Geography, University of Toronto.

Klodawsky, Fran, and Spector, Aron. 1988. "New Families, New Housing Needs, New Urban Environments: The Case of Single-Parent Families." In C. Andrew and B.M. Milroy, eds., *Life Spaces: Gender, Household, Employment*, Vancouver: University of British Columbia Press, 141–52.

Lamonde, Pierre, and Polèse, Mario. 1984. "L'évolution de la structure économique de Montréal, 1971–1981: désindustrialisation ou reconversion?" *L'actualité économique* 60 no. 4: 471–94.

Langlois, Simon. 1987. "Les familles à un et à deux revenus: changement social et différenciation socio-économique." *Thèmes canadiens* 8: 147–60.

Lauria, Mickey, and Knopp, Larry. 1985. "Toward an Analysis of the Role of Gay Communities in the Urban Renaissance." *Urban Geography* 6 no. 2: 152–69.

Le Bourdais, Céline, and Beaudry, Michel. 1988. "The Changing Residential Structure of Montreal, 1971–1981." *Canadian Geographer* 32 no. 2: 98–113.

Le Bourdais, Céline, Hamel, Pierre J., and Bernard, Paul. 1987. "Le travail et l'ouvrage, charge et partage des tâches domestiques chez les couples québécois." *Sociologie et sociétés* 19 no. 1: 37–55.

Le Bourdais, Céline, with Rose, Darmais. 1986. "Vers une caractérisation des familles monoparentales québécoises à chef féminin." In S. Langlois and S. Trudel, eds., *La morphologie sociale en mutation au Québec*, Montreal: Actes du colloque annuel de l'Association des sociologues et anthropologues de langue française, *Cahiers de l'Acfas*, 41: 141–58.

Leo, Christopher, and Fenton, Robert. 1990. "Mediated Enforcement and the Evolution of the State: Development Corporations in Canadian City Centres." *International Journal of Urban and Regional Research* 14 no. 2: 185–206.

Ley, David. 1985a. "Downtown or the Suburbs? A Comparative Study of Two Vancouver Head Offices." *Canadian Geographer* 29 no. 1: 30–43.

– 1985b. "Work-Residence Relations for Head Office Employees in an Inflating Housing Market." *Urban Studies* 22 no. 1: 21–38.

– 1988. "Social Upgrading in Six Canadian Inner Cities." *Canadian Geographer* 32 no. 1: 31–45.

Ley, David, and Hutton Thomas. 1987. "Vancouver's Corporate Complex and Producer Services Sector: Linkages and Divergence within a Provincial Staple Economy." *Regional Studies* 21 no. 5: 413–24.

Lowe, Graham. 1982. "Class, Job and Gender in the Canadian Office." *Labour/Le travailleur* 10 (autumn): 11–37.

McCann, Larry, and Smith, Peter J. 1991. "Canada Becomes Urban: Cities and Urbanization in Historical Perspective." In T.E. Bunting and Pierre Filion, eds., *Canadian Cities in Transition*, Toronto: Oxford University Press, 69–99.

Mackenzie, Suzanne. 1987. "Neglected Spaces in Peripheral Places: Home-workers and the Creation of a New Economic Centre." *Cahiers de géographie du Québec* 31 no. 83: 247–60.

Mackenzie, Suzanne, and Rose, Damaris. 1983. "Indutrial Change, the Domestic Economy and Home Life." In J. Anderson, S. Duncan, and R. Hudson, eds., *Redundant Spaces in Cities and Regions?: Studies in Industrial Decline and Social Change*, London: Academic Press, 155–200.

Marcoux, Richard, Morin, Richard, and Rose, Damaris. 1990. "Jeunes et précarisation économique: analyse de la situation des couples." *Cahiers québécois de démographie* 19 no. 2: 273–307.

Maroney, Heather. 1983. "Feminism at Work." *New Left Review* 141: 51–71.

Michelson, William. 1986. *From Sun to Sun: Daily Obligations and Community Structure in the Lives of Employed Women and Their Families*. Totowa, NJ: Rowman and Allanheld.

Mills, Caroline. 1989. "Interpreting Gentrification: Post Industrial, Post Patriarchal, Post Modern?" PhD thesis, Department of Geography, University of British Columbia.

Moore, Maureen. 1989. "Dual Earner Families: The New Norm." Statistics Canada, Cat. no. 11-008E, *Canadian Social Trends* no. 12: 24–6.

Morency, René. 1988. "Travail, espace et familles dans la région de Québec." MA thesis, Département de géographie, Université Laval.

Morris, Lydia. 1990 *The Workings of the Household*. Cambridge, UK: Polity.

Myles, John. 1988a "Decline or Impasse? The Current State of the Welfare State." *Studies in Political Economy* 26: 73–107.

– 1988b. "The Expanding Middle: Some Canadian Evidence on the Deskilling Debate." *Canadian Review of Sociology and Anthropology* 25 no. 3: 335–64.

Myles, John, Picot, Garnett, and Wannell, Ted. 1988. "The Changing Wage Distribution of Jobs, 1981–1986/La répartition salariale des emplois: variations de 1981 à 1986." Statistics Canada, Cat. no. 71-001, *The Labour Force* Oct.: 85–129.

Nelson, Kristin. 1986. "Labor Demand, Labor Supply and the Suburbanization of Low-Wage Office Work." In A.J. Scott and M. Storper, eds., *Production, Work and Territory: The Geographical Anatomy of Industrial Capitalism*, Boston: Allen and Unwin, 149–71.

Nielsen, Lise Drewes. 1991. "Flexibility, Gender and Local Labour Markets – Some Examples from Denmark." *International Journal of Urban and Regional Research* 15 no. 1: 42–54.

Norcliffe, Glen B. 1984. "Nonmetropolitan Industrialization and the Theory of Production." *Urban Geography* 5 no. 1: 25–42.

Norcliffe, Glen, and Stevens, J.H. 1979. "The Heckscher-Ohlin Hypothesis and Structural Divergence in Québec and Ontario, 1961–1969." *Canadian Geographer* 23 no. 3: 239–54.

Parent, Robert. 1983. "Les multinationales québécoises de l'ingénierie." *Recherches sociographiques* 24 no. 1: 75–94.

Polèse, Mario. 1988a. *Les activités de bureau à Montréal: structure, evolution et perspectives d'avenir*. Dossier Montréal 1. Montreal: Ville de Montréal and INRS-Urbanisation.

– 1988b. "La transformation des économies urbaines: tertiarisation, délocalisation et croissance économique." *Cahiers de recherche sociologique* 6 no. 2: 13–27.

– 1990. "La thèse du déclin économique de Montréal, revue et corrigée." *L'actualité économique* 66 no. 2: 133–46.

Pool, I., and Moore, M. 1986. *Lone Parenthood: Characteristics and Determinants/L'état de parent seul: caractéristiques et déterminants*. Results from the 1984 Family History Survey, Statistics Canada, Cat. no. 99-961.

Pratt, Geraldine. 1990. "Feminist Analyses of the Restructuring of Urban Life." *Urban Geography* 11 no. 6: 594–605.

Pratt, Geraldine, and Hanson, Susan. 1988. "Gender, Class and Space." *Environment and Planning D: Society and Space* 6: 15–35.

Robinson, Patricia. 1986. *Women's Work Interruptions/Interruptions de travail chez les femmes*. Results from the 1984 Family History Survey, Statistics Canada, Cat. no. 99-962.

Rose, Damaris. 1984. "Rethinking Gentrification: Beyond the Uneven Development of Marxist Urban Theory." *Environment and Planning D: Society and Space* 2 no. 1: 47–74.

– 1987. "Un aperçu féministe sur la restructuration de l'emploi et sur la gentrification." *Cahiers de géographie du Québec* 31 no. 83: 205–24 (English version "A Feminist Perspective on Employment Restructuring and Gentrification: The Case of Montréal," in J. Wolch and M. Dear, eds., *The Power of Geography: How Territory Shapes Social Life*, Boston: Unwin Hyman, 1989, 118–38).

– 1990. "'Collective Consumption' Revisited: Analysing Modes of Provision and Access to Childcare Services in Montréal, Québec." *Political Geography Quarterly* 9 no. 4: 353–80.

Rose, Damaris, and Le Bourdais, Céline. 1986. "The Changing Conditions of Female Single Parenthood in Montréal's Inner City and Suburban Neighbourhoods." *Urban Resources* 3 no. 2: 45–52.

Rose, Damaris, and Villeneuve, Paul. 1988. "Women Workers and the Inner City: Some Implications of Labour Force Restructuring in Montréal, 1971–81." In C. Andrew and B.M. Milroy, eds., *Life Spaces: Gender, Household, Employment*, Vancouver: University of British Columbia Press, 31–64.

– 1992. "Gender and Occupational Restructuring in Montréal in the 1970s." In A. Kobayashi, ed., *Women, Work and Place*, Montreal: McGill-Queen's University Press (in press).

Rose, Damaris, and Wexler Martin. 1989. "Post-war Social and Economic Change and Housing Adequacy." In J. Miron, ed., *Housing Progress in Canada*, Report to CMHC, Ottawa.

Rouffignat, Joël, and Vallée, Anne. 1990. "Acculturation alimentaire et mutation urbaine: le développement de l'industrie de la restauration dans l'agglomération urbaine de Québec." *Le géographe canadien* 34 no. 3: 194–208.

Saarinen, Oiva. 1990. "Sudbury: A Historical Case Study of Multiple Urban Economic Transformation." *Ontario History* 82 no. 1: 53–81.

Sassen-Koob, Saskia. 1984. "The New Labour Demand in Global Cities." In M.P. Smith, ed., *Cities in Transformation: Class, Capital and the State*, Beverly Hills: Sage, 139–71.

Sayer, Andrew. 1989. "Postfordism in Question." *International Journal of Urban and Regional Research* 13 no. 4: 666–95.

Scott, Allen J. 1981. "The Spatial Structure of Metropolitan Labor Markets and the Theory of Intra-urban Plant Location." *Urban Geography* 2 no. 1: 1–30.

Séguin, Anne-Marie. 1988. "Formes domestiques marginales et État local: le cas du quartier Saint-Jean-Baptiste de Québec. "In G. Fontaine, L. Landry, A.-M. Séguin, and P. Villeneuve, *Marginalité et territorialité à Québec*, Québec: Université Laval, *Les cahiers du Centre de recherches en aménagement et développement* 11 no. 2: 22–40.

Séguin, Anne-Marie, and Villeneuve, Paul. 1987. "Du rapport hommes-femmes au centre de la Haute-Ville de Québec." *Cahiers de géographie du Québec* 31 no. 83: 189–204.

Semple, R. Keith. 1985. "Toward a Quaternary Place Theory." *Urban Geography* 6 no. 4: 285–96.

Simmons, James W. 1991. "The Urban System." In T.E. Bunting and Pierre Filion, eds., *Canadian Cities in Transition*, Toronto: Oxford University Press, 100–25.

Simpson, Wayne. 1987. "Workplace Location, Residential Location and Urban Commuting." *Urban Studies* 24 no. 2: 119–28.

Social Planning Council of Metro Toronto. 1989. *A Social Report for Metro 1988, no. 2*. Toronto: Council.

Statistics Canada. 1986. *Canada's International Trade in Services, 1969 to 1984/Le commerce international des services du Canada, 1969 à 1984*. Cat. no. 67-510 (occasional), Ottawa.

– 1987. *Corporations and Labour Unions Returns Act/Loi sur les déclarations des corporations et des syndicats ouvriers, 1985 Report*. Cat. no. 61-210, Ottawa.

– 1989. *Labour Force Annual Averages, 1981–1988/Moyennes annuelles de la population active, 1981–1988*. Cat. no. 71-529 (occasional), Ottawa.

Storper, Michael, and Scott, Allen J. 1989. "The Geographical Foundations and Social Regulation of Flexible Production Complexes. "In J. Wolch and M. Dear, eds., *The Power of Geography: How Territory Shapes Social Life*, Boston: Unwin Hyman, 21–40.

Theilheimer, Ish. 1988. "Commmunity-Driven Development Planning in Sudbury." *Transition* Dec.: 11–13.

Viaud, Gilles. 1988. "Une géographie sociale du double revenu familial à Québec." MA thesis, Département de géographie, Université Laval.

Villeneuve, Paul. 1989. "Gender, Employment and Territory in Metropolitan Environments." In G.J.R. Linge and G.A. van der Knaap, eds., *Labour, Environment and Industrial Change*, London: Routledge, 67–86.

Villeneuve, Paul, and Morency, René. 1990. "Couples à double emploi et hétérogénéité social dans les quartiers de Montréal." *Le Géographe canadien* 34 no. 3: 239–50.

Villeneuve, Paul, and Rose, Damaris. 1986. "De la place des femmes dans la division spatiale du travail: le cas de Québec entre 1971 et 1981." In R. De Koninck and L. Landry, eds., *Les genres de vie urbains: essais exploratoires*, Université Laval, Département de géographie, *Notes et documents de recherche* 26: 71–92.

– 1988. "Gender and the Separation of Employment from Home in Metropolitan Montréal, 1971–1981." *Urban Geography* 9 no. 2: 155–79.

Villeneuve, Paul, and Viaud, Gilles. 1987. "Asymétrie occupationnelle et localisation résidentielle des familles à double revenu à Montréal." *Recherches sociographiques* 28 nos. 2–3: 371–91.

Wallman, Sandra. 1984. *Eight London Households*. London: Tavistock.

Walsh, Tim. 1990. "Flexible Labour Utilisation in the Private Sector." *Work, Employment and Society* 4 no 4: 517–30.

Wekerle, Gerda, and Rutherford, Brent. 1988. "Captive Rider, Captive Labour: Spatial Constraints and Women's Employment Equity." *Urban Geography* 9 no. 2: 116–37.

NOTES TO CHAPTER NINE

1 In this chapter, following standard practice, we shall use the term *housing* in two principal ways: first, a stock of dwellings that changes relatively slowly, and second, a flow of services, which may change much more rapidly. The latter concept incorporates the full range of benefits (e.g. equity, security, privacy) and costs (e.g. rents, taxes, user charges) that flow from ownership and occupancy of the stock; see Bourne (1981), Fallis (1985), and Sayegh (1987).

2 See Michelson (1977) and Logan and Molotch (1987) for a general discussion of housing and neighbourhood satisfaction.

3 From the 1941 and 1986 censuses of Canada.

4 See Baer (1986), CMHC (1986; 1987a; 1987b), Grebler and Burns (1987), and Fallis (1989).

5 Many argue that this decline has come about, at least in part, as a result of the construction industry's response to rent controls in some provinces (L. Smith and Tomlinson 1981). Other factors are rising management and maintenance costs, inflation, the relative shortage of serviced land for rental housing in many municipalities, and the tendency of tenants to have lower incomes than owner-occupiers, thus depressing rents (see also Hulchanski 1985).

6 In 1986, condominiums totalled only 3.5 per cent of the nation's owner-occupied housing stock, but in a few urban areas in which they are concentrated – Vancouver (14 per cent), Toronto (11 per cent), and Edmonton (8 per cent) – they have affected the housing market substantially both in the inner city and the suburbs (see Skaburskis 1988). Rapid tenure conversion from rental to condominium within the existing stock also has led to recent legislation to control such conversions (e.g. in Ontario).

7 Technical terms used in this table are defined in CMHC (1986).

8 The remainder of the stock was divided (in 1989) into single, attached (8.5 per cent), apartments and flats (32.4 per cent), and mobile homes (2.1 per cent).

9 The unique tenure composition of Montreal's housing developed in the latter half of the nineteenth century. See Hertzog and Lewis (1986), Choko (1987), and Harris and Choko (1988).

10 For elaboration of the effects of inflation on homeownership see Harris (1986).

11 The corresponding new-housing price index for the Toronto CMA is relatively low as a result of pre-1981 price increases. Dramatic rises in residential real estate prices in Toronto and subsequently Vancouver post-date the 1986 figures shown in Table 9.1.

12 See Pratt (1986), Badcock (1989), and chapter 15 for an extended discussion.

13 See Miron (1989) and chapter 4.

14 Owner-occupiers both produce and consume housing services.

15 This role is dealt with in some detail by Goldberg and Mark (1985).

16 See Clayton Research Associates (1986; 1988) and McKellar (1989).

17 Goldberg (1980) debates the issue of whether insistence 'on high servicing standards for new residential developments reflects municipal arrogance or economic rationality.

18 A detailed empirical study of residential mortgage lending in Toronto is provided by Murdie (1991). Although institutional lenders generally avoided inner-city areas (in the 1970s), there is little evidence for Canada of the negative effects of redlining documented in American cities (Adams 1987).

19 This point is taken up by Hulchanski (1989).

20 Discussions of this process are found in Whitehead (1987) and Clayton Research Associates (1988).

21 See Rowe (1989).

22 See also Sewell (1977), Lorimer (1978), and Greenspan (1978) for critical appraisals of the development process.

23 The flavour of this debate, and the contrasting views, are evident in Spurr (1976), Markusen and Scheffman (1977), Gunton (1978), Roweis and Scott (1981), Fallis (1985), Bourne (1992), and most issues of City Magazine.

24 The short-run supply of serviced land available for development is relatively fixed.

25 Land costs, which typically represent 20 to 30 per cent of new housing prices, now account for 50 per cent or more in rapidly growing areas.

26 Submarkets can be defined as relatively autonomous subdivisions of an urban housing market, between which there are few household movements and within which price (rent) changes may behave independently; see Bourne (1981).

27 Although newly built units comprise a relatively small portion of the short-run stock of dwellings (1 to 3 per cent annually), they may constitute from 40 to 60 per cent of all units on the market at any given time.

28 Definitions of neighbourhoods vary widely, depending largely on whether physical (territorial) or social criteria are employed. See Logan and Molotch (1987) and Dansereau (1989) for recent interpretations.

29 For discussion with reference to US housing conditions, see Adams (1984; 1987).

30 For further commentary and empirical evidence, see Ley (1988) and chapter 11.

31 Recent work on residential expansion in the rural-urban fringe is found in Coppack and Russwurm (1988).

32 See, for example, Bunting (1987).

33 As initially suggested by McLemore et al. (1975), and modified by the authors.

34 For further details, refer to Bunting and Filion (1988) and Dansereau (1989).
35 P. Smith and McCann (1981).
36 Logan and Molotch (1987).
37 See Ley (1986; 1988) and Filion and Bunting (1990).
38 See Cullingworth (1987) and Clayton Research Associates (1988).
39 For examples see Coppack and Russwurm (1988).
40 These processes are described in detail in Downs (1981) and Adams (1987).
41 See Hulchanski (1984) for a detailed review of both developments.
42 See, for example, Patterson (1989).
43 For a further and more detailed commentary, see Bunting and Filion (1988).
44 See chapter 16.
45 Provincial programs to accelerate growth of the social housing sector, such as Ontario's Homes Now Program, have had a positive effect, but not immediately.
46 See Bourne (1990).
47 The affordability issue is dealt with in detail in chapter 4 and in Miron (1989).
48 There is evidence that new rental housing supply is forthcoming, even under rent control, but indirectly through rental of single-family houses and condominiums; see Steele (1990).
49 See Fallis and Murray (1990) and Ray and Moore (1991).

REFERENCES FOR CHAPTER NINE

– Adams, J. 1984. "The Meaning of Housing in America." *Annals, Association of American Geographers* 74: 515–26.
– 1987. *Housing America in the 1980s.* New York: Russell Sage Foundation.
Badcock, B. 1984. *Unfairly Structured Cities.* Oxford: Blackwell.
– 1989. "Homeownership and the Accumulation of Real Wealth." *Society and Space* 7 no. 1: 69–92.
Baer, W. 1986. "The Shadow Market in Housing." *Scientific American* 255: 29–35.
Ball, M. 1986. "Housing Analysis: Time for a Theoretical Refocus." *Housing Studies* 1 no. 3: 147–65.
Bourne, L.S. 1981. *The Geography of Housing.* London: Arnold.
– 1986. "Recent Housing Policy Issues in Canada: A Retreat from Social Housing?" *Housing Studies* 1 no. 2: 122–8.
– 1989. "The Changing Settlement Environment of Housing." In J. Miron, ed., *Housing Progress in Canada since 1945*, Report to CMHC, Ottawa.
– 1990. *Worlds Apart: The Changing Geography of Income Distributions in Canadian Metropolitan Areas.* Department of Geography, University of Toronto, Discussion Paper 36.
– 1992. "Choose Your Villain: Five Ways to Oversimply the Price of Housing

and Urban Land." In G. Arbuckle, and H. Bartel, eds., *Readings in Canadian Real Estate*, Downsview, Ont.: Captus University Publications, 19–32.

Bunting, T. 1987. "The Invisible Upgrading of Inner-City Housing: A Study of Homeowners' Reinvestment Behaviour in Central Kitchener." *Canadian Geographer* 31: 209–22.

Bunting, T., and Filion, P., eds. 1988. *The Changing Canadian Inner City*. University of Waterloo, Department of Geography Publication Series.

Canada Mortgage and Housing Corporation (CMHC). 1986. *Canadian Residential Construction Industry: Perspective and Prospective*. Ottawa: CMHC.

– 1987a. *Housing in Canada 1945 to 1986: An Overview and Lessons Learned*. Ottawa: CMHC.

– 1987b. *A Consultation Paper on Housing Renovation*. Ottawa: CMHC.

Clayton Research Associates. 1986. *The Outlook for Residential Construction in Canada*. Institute for Research in Construction, National Research Council of Canada.

– 1988. *The Evolution of the Housing Industry in Canada*. Report to CMHC, mimeo.

Choko, M. 1987. *The Characteristics of Housing Tenure in Montreal*. Centre for Urban and Community Studies, University of Toronto, Research Paper 164.

Coppack, P., and Russwurm, L., eds. 1988. *Essays on Canadian Urban Process and Form III: The Urban Field*. University of Waterloo, Department of Geography Publications.

Cullingworth, J.B. 1987. *Urban and Regional Planning in Canada*. New Brunswick, NJ: Transaction Books.

Dansereau, F. 1989. "Intraurban Differentiation and Social Change." In J.R. Miron, ed., *Housing Progress in Canada since 1945*, Report to CMHC, Ottawa.

Downs, A. 1981. *Neighborhoods and Urban Development*. Washington, DC: Brookings.

Fallis, G. 1985. *Housing Economics*. Toronto: Butterworths.

– 1989. "Post-war Changes in the Supply Side of Housing." In J. Miron, ed., *Housing Progress in Canada since 1945*, Report to CMHC, Ottawa.

Fallis, G., and Murray, A., eds. 1990. *Housing the Homeless and Poor*. Toronto: University of Toronto Press.

Filion, P. 1991. "The Gentrification–Social Structure Dialectic: A Toronto Case Study." *International Journal of Urban and Regional Research* 15 no. 4: 553–74.

Filion, P., and Bunting, T. 1990. "Socio-Economic Change within the Older Housing Stock of Canadian Cities." *Housing Studies* 5 no. 2: 79–91.

Goldberg, M.A. 1980. "Municipal Arrogance or Economic Rationality: The Case of High Servicing Standards." *Canadian Public Policy* 6: 78–88.

Goldberg, M.A., and Mark, J. 1985. "The Role of Government in Housing Policy." *Journal, American Planning Association* 51: 34–42.

Goldberg, M.A., and Mercer, J. 1986. *The Myth of the North American City: Continentalism Challenged*. Vancouver: University of British Columbia Press.

Grebler, L., and Burns, L. 1987. "Long Term Prospects for U.S. Housing Markets: Fewers Units, Greater Investment." *Housing Studies* 2: 143–56.

Greenspan, D. 1978. *Down to Earth.* 2 vols. Report of the Federal Task Force on the Supply and Price of Serviced Residential Land, Toronto.

Gunton, T. 1978. "The Urban Land Question: Who Is Right?" *City Magazine* 3 no. 3: 39–45.

Harris, R. 1986. "Boom and Bust: The Effects of House Price Inflation on Home Ownership Patterns in Montreal, Toronto and Vancouver." *Canadian Geographer* 30 no. 4: 302–15.

Harris, R., and Choko, M. 1988. *The Evolution of Housing Tenure in Montreal and Toronto since the Mid-19th Century.* Centre for Urban and Community Studies, University of Toronto, Research Paper No. 166.

Hertzog, S., and Lewis, R.D. 1986. "A City of Tenants: Homeownership and Social Class in Montreal: 1846–1881." *Canadian Geographer* 30 no. 4: 316–23.

Hulchanski, J.D. 1984. *St. Lawrence and False Creek.* School of Community and Regional Planning, University of British Columbia, CPI no. 10.

– 1985. *Market Imperfections and the Role of Rent Regulations in the Residential Rental Market.* Commission of Enquiry into Residential Tenancies, Toronto, Research Study No. 6.

– 1989. "New Forms of Owning and Renting." In J.R. Miron, ed., *Housing Progress in Canada since 1945*, Report to CMHC, Ottawa.

Ley, D. 1986. "Alternative Explanations for Inner-City Gentrification: A Canadian Assessment." *Annals, Association of American Geographers* 76: 521–35.

– 1988. "Social Upgrading in Six Canadian Inner Cities." *Canadian Geographer* 32 no. 1: 31–45.

Logan, J., and Molotch, H. 1987. *Urban Fortunes: The Political Economy of Place.* Berkeley: University of California Press.

Lorimer, J. 1978. *The Developers.* Toronto: James Lorimer.

McKellar, J. 1989. "Building Technology and the Production Process." In J.R. Miron, ed., *Housing Progress in Canada since 1945.* Report to CMHC Ottawa.

McLemore, R., Aass, C., and Keilhofer, P. 1975. *The Inner City: Problems, Trends and Federal Policy.* Ottawa: Ministry of State for Urban Affairs.

Markusen, J., and Scheffman, D. 1977. *Speculation and Monopoly in Urban Development.* Toronto: Ontario Economic Council.

Michelson, W. 1977. *Environmental Choice, Human Behaviour, and Residential Satisfaction.* New York: Oxford University Press.

Miron, J.R. 1988. *Housing in Postwar Canada: Demographic Change, Household Formation, and Housing Demand.* Montreal: McGill-Queen's University Press.

– ed. 1989. *Housing Progress in Canada since 1945.* Report to CMHC, Ottawa.

Murdie, R. 1991. "Local Strategies in Resale Home Financing in the Toronto Housing Market." *Urban Studies* 28 no. 3: 465–83.

Ontario Ministry of Housing. 1991. *Rent Control: Issues and Options*. Toronto:
Ministry.

Patterson, J. 1989. "Housing and Community Development Policies." In J.R.
Miron, ed., *Housing Progress in Canada since 1945*, Report to CMHC Ottawa.

Pratt, G. 1986. "Housing Tenure and Social Cleavages in Urban Canada." *Annals, Association of American Geographers* 76: 366–80.

Ray, B., and Moore, E. 1991. "Access to Homeownership among Immigrant
Groups in Canada." *Canadian Review of Sociology and Anthropology* 28: 1–29.

Rowe, A. 1989. "Self-Help Housing Provision: Production, Consumption, Accumulation and Policy in Atlantic Canada." *Housing Studies* 4 no. 2: 75–91.

Roweis, S., and Scott, A. 1981. "The Urban Land Question." In M. Dear and
A. Scott, eds., *Urbanization and Planning in Capitalist Society*, New York:
Methuen, 123–58.

Sayegh, K.S. 1987. *Housing: A Canadian Perspective*. Ottawa: ABCD-Academy
Books.

Sewell, J. 1977. "Where the Suburbs Came from." In J. Lorimer and E. Ross,
eds., *The Second City Book*, Toronto: Lorimer, 10–12.

Skaburskis, A. 1988. "The Nature of Canadian Condominium Submarkets and Their
Effects on Urban Spatial Structure." *Urban Studies* 25 no. 2: 109–23.

Smith, L., and Tomlinson, P. 1981. "Rent Controls in Ontario: Roofs or Ceilings?"
Journal of the American Real Estate and Urban Economics Association 9:
93–114.

Smith, P., and McCann, L. 1981. "Residential Land Use Change in Inner Edmonton." *Annals, Association of American Geographers* 71 no. 4: 536–51.

Spurr, P. 1976. *Land and Urban Development*. Toronto: Lorimer.

Steele, M. 1990. *Report on Aspects of the Structure of the Rental Housing Market*.
Toronto: Ontario Ministry of Housing.

van Vliet, W., ed. 1990. *International Handbook of Housing Policy and Practice*.
New York: Greenwood.

Whitehead, J. 1987. "Decision Making in the Development Industry During a
Boom-Bust Cycle." PhD thesis, Department of Geography, University of
British Columbia.

NOTES TO CHAPTER TEN

1 Statistics Canada (1983).   2 Relph (1987: chapter 8).   3 Mackenzie and
Rose (1983: 156).   4 Park (1936).   5 Park (1967).   6 Burgess (1925: 50–6;
also see Wirth (1928).   7 Alonso (1960).   8 Shevky and Bell (1955).
9 Davies (1984). See Williams (1986) for a critique of this approach, and see chapter 3 in this volume.
10 Note, however, that Shevky and Bell's original formulation of factorial ecology
was predicated on the theory that social change leads to increasing social
differentiation; see Davies (1984: 24).

11 See Duncan and Duncan (1955) for an example of an early geographical
   study of socioeconomic status. Harvey (1973) was an important catalyst in re-
   orienting the conception of class within social geography.

12 See, for example, Gordon (1978) and Walker (1981; 1985).   13 Harris
(1984).   14 See, for example, Harris (1988: 60–1).   15 Some of these factors are addressed, from a Marxist viewpoint, in Harvey (1985:
112–16).   16 Mackenzie and Rose (1983), Mackenzie (1986), Klausner
(1986), Pahl (1984), and Williams (1987).   17 Mackenzie (1986:
91–4).   18 Pahl (1984: 78–81); Pratt and Hanson (1988).   19 For reviews
of this literature, see Peach (1983) and Jackson (1985).   20 Yancey, Ericksen,
and Juliani (1976) and Anderson (1988).   21 Smith (1984: 366).   22 On
ethnicity as a resource used in social struggle, see Cohen (1974) and Wallman
(1979). On the socio-spatial aspects of this struggle, see Boal (1976) and Jackson and Smith (1981).   23 Schreuder (1989) and Ward (1989: chapter 6).   24 Bodnar (1985: chapter 5).   25 Sociologists have long been interested in this question in Canada, especially since the pioneering work of Porter
(1965). Also see Reitz (1980; 1982) and Li (1988).   26 Li (1988: chapter 5).   27 The need for this type of research is recognized in Scott (1986). Research linking class and gender is the most advanced; see, for example,
Mackenzie (1986) and Pratt and Hanson (1988).   28 On the early development
of this area, see Spelt (1973: 41–5) and Careless (1984: 96).   29 Canada,
Dominion Bureau of Statistics (DBS), 1901 Census, Vol. I, Table XI.   30 Thirty
per cent of the residents of ward 4 in 1901 were Irish, and slightly less than
half of this group were Catholic. DBS, 1901 Census, Vol. I, Tables X and XI.
31 Rosenberg (1954).   32 Speisman (1979: chapter 5).

33 Toronto's clothing work-force doubled in size between 1891 and 1901; DBS
   1891 Census, Vol. III, Table I; 1901 Census, Vol. III, Table VIII. On lack
   of organization among Toronto's clothing workers, see Roberts (1978).
   Speisman (1979: 71–4) provides a discussion of the intersection between
   the demands of orthodoxy and the nature of clothing production.

34 Data on the ethnicity of Toronto's work-force were drawn from city tax
   assessment rolls. Although the rolls specify the religious affiliation of the household "head" and therefore provide a reasonably accurate portrayal of the
   ethnicity of male workers (certainly Jews are easily identified), they provide
   virtually no information on other people, including women, children, and
   boarders.

35 The proportion of garment workers among Jews was calculated using assessment data and therefore refers to males only. The ratio of Jews to non-Jews
   in Toronto's garment industry includes males and females and is based on
   data reported in DBS, 1931 Census, Vol. V, Tables 41 and 49, and DBS (1931a;
   1931b).

36 The DBS collected information only for firms with five or more workers. Therefore the exact ratio of custom v. factory clothing production is impossible
   to determine. However, the census, together with the advertisement section of

Might's Directory of Toronto, suggests that approximately 55 per cent of clothing produced in Toronto in 1901 was factory made, while the remainder was produced by individual artisan tailors or in "custom shops". Also see Steed (1976).

37 "The Scribe" (1919).

38 Information on the Eaton's factory complex was drawn from a combination of tax rolls and various newsletters written for Eaton's employees. These are housed in the Eaton's Archives, Toronto.

39 The organization of Eaton's factory was described in Anon (1908).

40 Hiebert (1987: chapter 7).

41 Most of these shops eluded the factory inspection process in Toronto, since inspectors could not keep track of their constant movement from one location to another. Inspectors argued for more exacting laws prohibiting home-based shops as well as for regulations concerning use of outside workers (i.e. individual home-workers, usually women, who contracted directly with manufacturers for specific jobs): Ontario, Department of Agriculture (various years). On use of home work in modern-day clothing production, see Johnson and Johnson (1982).

42 The shop-floor organization of these firms was described by former workers in oral history accounts. Interviews were conducted by Karen Levine and are housed at the Multicultural Historical Society of Ontario (MHSO) in Toronto.

43 Many researchers argue that introduction of machine-controlled methods of production, along with new styles of shop-floor organization, has ushered in a new era of "post-Fordist" manufacturing. See Scott and Storper (1986). On technological change in the contemporary garment industry, see Mather (1988).

44 *Might's Directory of Toronto* (1901), Business Advertisements; *The Toronto Jewish Directory* (1931), Business Section. The Jewish "takeover" of the garment industry was but one example of the close correspondence between ethnicity and economy in Toronto. Italians gained almost complete control over fruit peddling during the same period, while Macedonians specialized in low-priced restaurants, and the Chinese in laundering. Zucchi (1988) and Nipp (1985).

45 However, Jewish-owned firms were still less than half the size of non-Jewish firms and therefore accounted for only between one-quarter and one-third of all production in the city. Still, this was remarkably higher than the corresponding figure would have been in 1901 (less than 1 per cent). Dun and Bradstreet financial records were used to estimate the relative size of firms.

46 I have discussed these changes at length in Hiebert (1987: chapter 8).

47 On the success of Jewish- and Italian-owned garment firms in American cities during this period, see Passero (1978: chapter 2).

48 Tax assessment rolls were used as the principal source for this comparison.

49 Hiebert (1987: Table 8.8).

50 Detailed descriptions of Jewish residential patterns are provided in Hiebert (1987: section 8.2.3).

51 It appears that Italian entrepreneurs in Toronto also continued to reside in their ethnic enclave but that other groups, such as the Finns and Chinese, were less spatially concentrated. Zucchi (1988) and Nipp (1985).

52 The fact that Jews were moving to new residential areas may have been a factor underlying the Christie Pits riot between Jews and anti-Semites in 1932: Levitt and Shaffir (1987: 182). Also see Lemon (1985: chapter 2).

53 Speisman (1985).

54 Tax rolls indicate that the rate of homeownership among Jewish factory owners in 1931 was 76 per cent, 10 per cent higher than that for blue-collar Jews.

55 Speisman (1979: 325–6). A similar struggle for control over Jewish community institutions occurred in many other North American cities. See, for example, Rischin (1962).

56 This was a pejorative phrase used by radical Jews to refer to upwardly mobile Jews: Kayfetz (1959: 24).

57 See Toll (1982).

58 Speisman (1979: chapter 7).

59 According to oral testimony (interview with Al Hershkovitz, MHSO), radical Jews frequently held events on the Sabbath specifically to elicit such a response.

60 On the goals of Kehilloth, see Goren (1970: chapter 1).

61 Speisman (1979: 278).

62 Hiebert (1987: chapter 6).

63 Stevens and Kennedy (1937: 111).

64 Strike activity among garment workers peaked three times in the period 1901–31: in 1912, during the First World War, and just after the crash of 1929: Canada (1901–31). Initially, union activity among Jews was organized by men; over time, however, Jewish women also became central actors. See MacLeod (1974).

65 Paris (1980: 137). This was undoubtedly the case during industry-wide strikes, when workers picketed even the smallest shops.

66 Oral history testimony of Sam Galinsky and Jack Isacoff (MHSO).

67 Billy Newton Davis and Elias Zaron, "Spadina China syndrome," composed for a performance of the Second City comedy troupe, 1980.

68 As of 1981, Jews comprised fewer than 3 per cent of residents in the Spadina area: Bourne and Cressman (1985).   69 Hiebert (1991: Table 2).   70 Statistics Canada (1983).   71 On the viability of a dualist structure of garment production, see Fraser (1983).   72 Mann (1986: chapter 1).

REFERENCES FOR CHAPTER TEN

Alonso, William. 1960. "A Theory of the Urban Land Market." *Papers and Proceedings of the Regional Science Association* 6: 149–57.

Anderson, K.J. 1988. "Cultural Hegemony and the Race-Definition Process in Chinatown, Vancouver: 1880–1980." *Environment and Planning D: Society and Space* 6: 127–50.

Anon. 1908, "A Women's Clothing Factory: How the Clothes You Wear Are Made." *Ladies' Home Journal* 25: 43–9.

Boal, Fred W. 1976. "Ethnic Residential Segregation." In D.T. Herbert and R.J. Johnston, eds., *Social Areas in Cities,* vol. 1: *Spatial Process and Form,* London: John Wiley, 41–79.

Bodnar, John. 1985. *The Transplanted: A History of Immigrants in Urban America.* Bloomington: Indiana University Press.

Bourne, L.S., and Cressman, R., eds. 1985. *Ethnic Origin Groups, 1981, for Census Metropolitan Areas in Canada.* Centre for Urban and Community Studies, University of Toronto.

Burgess, Ernest. 1925. "The Growth of the City: An Introduction to a Research Project." In Robert E. Park, Ernest Burgess, and Roderick D. McKenzie, *The City,* Chicago: University of Chicago Press, 47–62.

Canada, Department of Labour. 1901–31. *The Labour Gazette: the Journal of Labour.*

Canada, Dominion Bureau of Statistics. 1931a. *Report of the Men's Factory Clothing Industry in Canada.* Ottawa: King's Printer.

– 1931b. *Report of the Women's Factory Clothing Industry in Canada.* Ottawa: King's Printer.

Careless, J.M.S. 1984. *Toronto to 1918: An Illustrated History.* Toronto: Lorimer.

Cohen, Abner. 1974. "The Lesson of Ethnicity." In Abner Cohen, ed., *Urban Ethnicity,* London: Tavistock, ix–xxiv.

Davies, Wayne K.D. 1984. *Factorial Ecology.* Aldershot, England: Gower.

Duncan, O.D., and Duncan, B. 1955. "Residential Distribution and Occupational Segregation." *American Journal of Sociology* 60: 493–503.

Fraser, Steven. 1983. "Combined and Uneven Development in the Men's Clothing Industry." *Business History Review* 57: 522–47.

Gordon, D.M. 1978. "Capitalist Development and the History of American Cities." In W.K. Tabb and L. Sawers, eds., *Marxism and the Metropolis: New Perspectives in Urban Political Economy,* New York: Oxford University Press, 25–63.

Goren, Arthur A. 1970. *New York Jews and the Quest for Community: The Kehillah Experiment, 1908–1922.* New York: Columbia University Press.

Harris, Richard. 1984. "Residential Segregation and Class Formation in Canadian Cities." *Canadian Geographer* 28 no. 2: 187–96.

– 1988. *Democracy in Kingston: A Social Movement in Urban Politics, 1965–1970.* Kingston and Montreal: McGill-Queen's University Press.

Harvey, David. 1973. *Social Justice and the City.* Baltimore: Johns Hopkins University Press.

– 1985. *The Urbanization of Capital: Studies in the History and Theory of Capitalist Urbanization.* Baltimore: Johns Hopkins University Press 1985.

Hiebert, Daniel J. 1987. "The Geography of Jewish Immigrants and the Garment Industry in Toronto, 1901–1931: A Study of Ethnic and Class Relations." PhD thesis, University of Toronto.

– 1991. "Ethnicity, Work and the Urban Economy." Paper presented at the Canadian Association of Geographers Conference, Kingston, Ont.

Jackson, Peter. 1985. "Urban Ethnography." Progress in Human Geography 9: 157–76.

Jackson, Peter, and Smith, Susan J. 1981. "Introduction." In Peter Jackson and Susan J. Smith, eds., Social Interaction and Ethnic Segregation, London: Academic Press, 1–17.

Johnson, L.C., and Johnson, R.E. 1982. The Seam Allowance: Industrial Home Sewing in Canada. Toronto: Women's Press.

Kayfetz, Ben G. 1959. "The Evolution of the Jewish Community in Toronto." In Albert Rose, ed., A People and Its Faith: Essays on Jews and Reform Judaism in a Changing Canada, Toronto: University of Toronto Press, 14–29.

Klausner, D. 1986. "Beyond Separate Spheres: Linking Production with Social Reproduction and Consumption." Environment and Planning D: Society and Space 4: 29–40.

Lemon, James. 1985. Toronto since 1918: An Illustrated History. Toronto: Lorimer.

Levitt, Cyril H., and Shaffir, William. 1987. The Riot at Christie Pits. Toronto: Lester and Orpen Dennys.

Li, Peter S. 1988. Ethnic Inequality in a Class Society. Toronto: Wall and Thompson.

Mackenzie, Suzanne. 1986. "Women's Responses to Restructuring: Changing Gender, Changing Space." In Roberta Hamilton and Michele Barrett, eds., The Politics of Diversity: Questions for Feminism, Montreal: Book Centre, 81–100.

Mackenzie, Suzanne, and Rose, Damaris. 1983. "Industrial Change, the Domestic Economy and Home Life." In J. Anderson, S. Duncan, and R. Hudson, eds., Redundant Spaces in Cities and Regions: Studies in Industrial Decline and Social Change, London: Academic Press, 155–200.

Macleod, Catherine. 1974. "Women in Production: The Toronto Dressmakers' Strike of 1931." In J. Acton, P. Goldsmith, and B. Shepard, eds., Women at Work: Ontario, 1850–1930, Toronto: Canadian Women's Educational Press, 309–25.

Mann, Michael. 1986. The Sources of Social Power, vol. 1, A History of Power from the Beginning to A.D. 1760." Cambridge: Cambridge University Press.

Mather, Charles. 1988. "Flexible Manufacturing in Vancouver's Clothing Industry." MA thesis, University of British Columbia.

Nipp, Dora. 1985. "The Chinese in Toronto." In Robert F. Harney, ed., Gathering Places: Peoples and Neighbourhoods of Toronto, 1834–1945, Toronto: Multicultural Historical Society of Ontario, 147–75.

Ontario, Department of Agriculture. Various years. *Report of the Inspectors of Factories for the Province of Ontario*. Ontario Legislative Assembly, *Sessional Papers*, Toronto.

Pahl, R.E. 1984. *Divisions of Labour*. Oxford: Blackwell.

Paris, Erna. 1980. *Jews: An Account of Their Experience in Canada*. Toronto: Macmillan.

Park, Robert E. 1936. "Human Ecology." *American Journal of Sociology* 42: 1–26.

– 1967. "The Urban Community as a Spatial Pattern and a Moral Order." In Robert H. Turner, ed., *Robert E. Park: Selected Papers on Social Control and Collective Behavior*, Chicago: Phoenix Books, 55–68.

Passero, Rosara Lucy. 1978. "Ethnicity in the Men's Ready-made Clothing Industry, 1880–1950: The Italian Experience in Phildelphia." PhD thesis, University of Pennsylvania.

Peach, G.C.K. 1983. "Ethnicity." In Michael Pacione, ed., *Progress in Urban Geography*, London: Croom Helm, 103–27.

Porter, John. 1965. *The Vertical Mosaic: An Analysis of Social Class and Power in Canada*. Toronto: University of Toronto Press.

Pratt, G., and Hanson, S. 1988. "Gender, Class, and Space." *Environment and Planning D: Society and Space* 6: 15–35.

Reitz, Jeffrey J. 1980. *The Survival of Ethnic Groups*. Toronto: McGraw-Hill Ryerson.

– 1982. "Ethnic Group Control of Jobs." Centre for Urban and Community Studies, Research Paper No. 133, University of Toronto.

Relph, Edward. 1987. *The Modern Urban Landscape*. Baltimore: Johns Hopkins University Press.

Rischin, Jacob. 1962. *The Promised City: New York's Jews, 1870–1914*. Cambridge, Mass.: Harvard University Press.

Roberts, W.D. 1978. "Studies in the Toronto Labour Movement, 1896–1914." PhD thesis, University of Toronto.

Rosenberg, Louis. 1954. *A Study of the Changes in the Geographic Distribution of the Jewish Population in the Metropolitan Area of Toronto, 1851–1951*. Canadian Jewish Population Studies, Jewish Community Series No. 2.

Schreuder, Yda. 1989. "Labour Segmentation, Ethnic Division of Labor, and Residential Segregation in American Cities in the Early Twentieth Century." *Professional Geographer* 41: 131–43.

Scott, J. 1986. "Industrialization and Urbanization: A Geographical Agenda." *Annals of the Association of American Geographers* 76: 25–37.

Scott, A.J., and Storper, Michael, eds. 1986. *Production, Work, Territory: The Geographical Anatomy of Industrial Capitalism*. London: Allen and Unwin.

"The Scribe." 1919. In *Golden Jubilee, 1869–1919: A Book to Commemorate the Fiftieth Anniversary of the T. Eaton Co. Limited*. Toronto: T. Eaton Co., 141–9.

Shevky, E., and Bell, W. 1955. *Social Area Analysis: Theory, Illustrative Applications and Computational Procedures*. Stanford, Calif.: Stanford University Press 1955.

Smith, Susan J. 1984. "Negotiating Ethnicity in an Uncertain Environment." *Ethnic and Racial Studies* 7: 360–73.

Speisman, Stephen A. 1979. *The Jews of Toronto: A History to 1937*. Toronto: McClelland and Stewart.

– 1985. "St. John's Shtetl: The Ward in 1911." In Robert F. Harney, ed., *Gathering Places: People and Neighbourhoods of Toronto, 1834–1945*, Toronto: Multicultural History Society of Ontario, 107–20.

Spelt, Jacob. 1973. *Toronto*. Toronto: Collier-Macmillan.

Statistics Canada. 1983. *Census Tracts, Toronto*. Cat. 95–976, Vol. 3, Profile Series B. Ottawa: Minister of Supply and Services Canada.

Steed, Guy P.F. 1976. *An Historical Geography of the Canadian Clothing Industries: 1800–1930s*. University of Ottawa, Department of Geography, Research Note 11.

Stevens, H.H., and Kennedy, W.E. 1937. *Report of the Royal Commission on Price Spreads*. Ottawa: King's Printer.

Toll, William. 1982. *The Making of an American Middle Class: Portland Jewry over Four Generations*. Albany: State University of New York Press.

Walker, Richard A. 1981. "A Theory of Suburbanization: Capitalism and the Construction of Urban Space in the United States." In M. Dear and A.J. Scott, eds., *Urbanization and Planning in Capitalist Society*, London: Methuen, 383–429.

– 1985. "Class, Division of Labor and Employment in Space." In Derek Gregory and John Urry, eds., *Social Relations and Spatial Structures*, New York: St. Martin's Press, 164–89.

Wallman, Sandra, ed. 1979. *Ethnicity at Work*. London: Macmillan.

Ward, David. 1989. *Poverty, Ethnicity, and the American City, 1840–1925: Changing Conceptions of the Slum and the Ghetto*. Cambridge: Cambridge University Press.

Williams, Peter. 1986. "Social Relations, Residential Segregation and the Home." In Keith Hoggart and Eleonore Kofman, eds., *Politics, Geography and Social Segregation*, London: Croom Helm, 247–73.

– 1987. "Constituting Class and Gender: A Social History of the Home, 1700–1901." In Nigel Thrift and P. Williams, eds., *Class and Space: The Making of Urban Society*, London: Routledge and Kegan Paul, 154–204.

Wirth, Louis. 1928. *The Ghetto*. Chicago: University of Chicago Press.

Yancey, W.L., Ericksen, E.P., and Juliani, R.N. 1976. "Emergent Ethnicity: A Review and Reformulation." *American Sociological Review* 41: 391–403.

Zucchi, John E. 1988. *Italians in Toronto: Development of a National Identity, 1875–1935*. Kingston and Montreal: McGill-Queen's University Press.

NOTES TO CHAPTER ELEVEN

1 Oziewicz (1986).   2 City of Toronto (1984).   3 City of Toronto (1984). Additional data from the Metro Toronto Real Estate Board.   4 See chapter 3 in this volume.   5 Goldberg and Mercer (1986). For elite patterns to 1940 in a range of Canadian cities, see Beaudet (1987; 1988a; 1988b).   6 See Ley (1985) for the compilation of these scores; also chapter 3 in this volume.   7 From a large literature, Canadian examples might include Moore (1982a), Anderson (1988), Mills (1988), and the case studies in Hasson and Ley (1993).   8 Duncan and Duncan (1984).   9 Bryce (1990).   10 Robertson (1977).   11 MacKay (1987: 157). For other sources on Montreal's elite, see Rémillard and Merritt (1986) and Beaudet (1988b).   12 MacKay (1987: 185).   13 MacKay (1987: 90).   14 MacKay (1987: 182).   15 MacKay (1987: 7-8).   16 MacKay (1987: 71). See also note 34 below.

17 Gowans (1966: 120). Typical were the château-style hotels built by the railways across Canada. Symptomatically, the heaviest concentration of this architecture seems to have been in the Square Mile, home of the elite that controlled the CPR, where it survives in such buildings as the Royal Victoria Hospital. At the other end of the continent, in Victoria, the CPR's château-style Empress Hotel is part of an ensemble of late Victorian and Edwardian buildings constructed in a picturesque eclecticism, chief of which is the idiosyncratic Parliament Buildings (1897) designed, like the Empress, by the flamboyant English immigrant Francis Rattenbury (Barrett and Liscombe 1983). The staged aestheticism of the Parliament Buildings is more than a little reminiscent of the film-set design of some elements of post-modernism. For further discussion on this convergence, see Ley (1989).

18 Careless (1984: 179).   19 Robertson (1977: 26).   20 The suggestion belongs to Gowans (1966: 126). The more sensational examples would include Casa Loma in Toronto and Craigdarroch "Castle," home of the Dunsmuir family in Victoria.   21 Kalman and Roaf (1974: 136); also Holdsworth (1979: 201).   22 Holdsworth (1979: 203).   23 Duncan and Duncan (1984); also Duncan (1993).   24 Pratt (1981).   25 Seeley, Sim, and Loosley (1956: 49).   26 MacKay (1987: 204).   27 MacKay (1987: 170).   28 MacKay (1987: 203).   29 Robertson (1977).   30 Seeley, Sim, and Loosley (1957: 224, 306).   31 Cooper (1971).   32 Clement (1975: 244, 338); Porter (1965).   33 Clement (1975: 248).

34 Robertson (1987: 32). The design of the club, for which English immigrant architects were selected, has been described as "a convincing copy ... [that] emulates the aristocratic, 19th century gentlemen's clubs of St. James" (Ward 1989).

35 Cooper (1971); Pratt (1981).   36 MacKay (1987: 113).   37 Robertson (1977: 46); MacKay (1987: 122).   38 Robertson (1977: 48); Wilson

(1960).   39 Cooper (1971: 67). This was equally true of the elite two generations earlier in its sponsorship of Vancouver General Hospital: Robertson (1977:
33–5).   40 MacKay (1987: 64, 192).   41 Seeley, Sim, and Loosley (1956:
292).   42 Pratt (1981: 157).   43 Seeley, Sim, and Loosley (1956:
396).   44 Seeley, Sim, and Loosley (1956: 307).   45 MacKay (1987: 60,
72, 158, 172, 175).   46 Beshers (1962).   47 MacKay (1987:
33–4).   48 Robertson (1977: 270); McAfee (1973).49 Hoyt (1939); Johnston (1971).   50 Firey (1945); Johnston (1971: 89–93); Ley (1983: 84–7,
159–62).   51 Johnston (1966).   52 Goheen (1970: 184,
218).   53 Beaudet (1988a).   54 Davies (1975); Beaudet (1987); Davies and
Murdie, chapter 3 in this volume.   55 Fairbairn (1978: 206).   56 Millward
(1981).   57 Forward (1973).   58 Beaudet (1987).   59 Lemon (1985: 33);
Duncan (1993).   60 Forward (1973).   61 Winsor (1981); Duncan
(1993).   62 Winsor (1981).   63 City of Toronto (1984).   64 City of Toronto
(1986); Howell (1986). Deconversion occurs when a house, converted into several units, reverts back to occupation by a single household.   65 For the rapidly
expanding Canadian literature on this topic, refer to Ley (1985; 1988; 1992)
and Bunting and Filion (1988).   66 Corral (1986); Beaudet (1988b).   67 Bourne
(1967); Smith and McCann (1981).   68 For a review of studies reaching
these conclusions, see Ley (1985).   69 Ley (1992).   70 Ley (1981).   71 Gee
(1978).   72 Martin (1989). There is not space here to treat such important
consequences of gentrification as housing displacement and social polarization: see
Ley (1985); Howell (1986); and also Bourne (1990).   73 Moore (1982b).
Product credentialism (designer labels) is pervasive in "yuppie" shopping districts
such as Yorkville or Vancouver's Robson Street.   74 Mills (1988).   75 Mills
(1988); Hanson and Pratt (1988).   76 Majury (1990); Lamphier (1992).

REFERENCES FOR CHAPTER ELEVEN

Anderson, Kay. 1988. "Cultural Hegemony and the Race-Definition Process
in Chinatown, Vancouver." *Society and Space* 6: 127–49.
Barrett, Anthony, and Liscombe, Rhodri. 1983. *Francis Rattenbury and British
Columbia: Architecture and Challenge in the Imperial Age.* Vancouver: Uni-
versity of British Columbia Press.
Beaudet, Paul. 1987. *Historical Spatial Dimensions of the Upper Status Population
of Halifax, Hamilton, Winnipeg and Calgary.* Department of Geography and
Planning, State University College at Buffalo, mimeo.
– 1988a. "Elite Residential Areas in Toronto and Buffalo." In T. Bunting and P.
Filion, eds., *The Changing Canadian Inner City*, Department of Geography,
University of Waterloo, Publication Series No. 31, 25–35.
– 1988b. *Historical Spatial Dimensions of the Elite Population of Montreal and
Quebec City.* Department of Geography and Planning, State University College at Buffalo, mimeo.

Beshers, James. 1962. *Urban Social Structure*. New York: Free Press.

Bourne, L.S. 1967. *Private Redevelopment of the Central City*. Department of Geography, University of Chicago, Research Paper No. 112.

– 1990. *Worlds Apart: The Changing Geography of Income Distributions within Canadian Metropolitan Areas*. Department of Geography, University of Toronto, Discussion Paper No. 36.

Bryce, J. Stephen. 1990. The Construction of an Elite Landscape in Westmount, Quebec, 1870–1940. MA thesis, Department of Geography, McGill University.

Bunting, Trudi, and Filion, Pierre, eds. 1988. *The Changing Canadian Inner City*. Department of Geography, University of Waterloo, Publication Series No. 31.

Careless, J.M.S. 1984. *Toronto to 1918: An Illustrated History*. Toronto: James Lorimer.

City of Toronto Planning and Development Department. 1984. *Toronto Region Incomes*. Research Bulletin 24.

– 1986. *Trends in Housing Occupancy*. Research Bulletin 26.

Clement, Wallace. 1975. *The Canadian Corporate Elite*. Toronto: McClelland and Stewart.

Cooper, Marion. 1971. "Residential Segregation of Elite Groups in Vancouver, B.C." MA thesis, Department of Geography, University of British Columbia.

Corral, Isabel. 1986. *Inner City Gentrifications: The Case of Shaughnessy Village, Montreal*. School of Urban Planning, McGill University.

Davies, Wayne K.D. 1975. "A Multivariate Description of Calgary's Community Areas." In B. Barr, ed., *Calgary: Metropolitan Structure and Influence*, Department of Geography, University of Victoria, Western Geographical Series No. 11, 231–68.

Duncan, James. 1993. "Shaughnessy Heights: The Protection of Privilege." In S. Hasson and D. Ley, eds., *Neighbourhood Organizations and the Welfare State*, Toronto: University of Toronto Press.

Duncan, James, and Duncan, Nancy. 1984. "A Cultural Analysis of Urban Residential Landscapes in North America: The Case of the Anglophone Elite." In J. Agnew, J. Mercer, and D. Sopher, eds., *The City in Cultural Context*, Boston: Allen and Unwin, 255–76.

Fairbairn, Kenneth J. 1978. "Location Changes of Edmonton's High Status Residents, 1937–1972." In P.J. Smith, ed., *Edmonton: The Emerging Metropolitan Pattern*, Department of Geography, University of Victoria, Western Geographical Series No. 15, 199–231.

Firey, Walter. 1945. "Sentiment and Symbolism as Ecological Variables." *American Sociological Review* 10: 140–8.

Forward, Charles. 1973. "The Immortality of a Fashionable Residential District: The Uplands. "In C. Forward, ed., *Residential and Neighbourhood Studies in Victoria*, Department of Geography, University of Victoria, Western Geographical Series No. 5, 1–39.

Gee, Marcus. 1978. "The Avenue Recultured." *Vancouver Province* 25 May.

Goheen, Peter. 1970. *Victorian Toronto, 1850 to 1900*. Department of Geography, University of Chicago, Research Paper No. 127.

Goldberg, Michael, and Mercer, John. 1986. *The Myth of the North American City*. Vancouver: University of British Columbia Press.

Gowans, Alan. 1966. *Building Canada: An Architectural History of Canadian Life*. Toronto: Oxford University Press.

Hanson, Susan, and Pratt, Geraldine. 1988. "Reconceptualizing the Links between Home and Work in Urban Geography." *Economic Geography* 64: 299–321.

Hasson, Shlomo, and Ley, David, eds., 1993. *Neighbourhood Organizations and the Welfare State*. Toronto: University of Toronto Press.

Holdsworth, Deryck. 1979. "House and Home in Vancouver: Images of West Coast Urbanism, 1886–1929." In G. Stelter and A. Artibise, eds., *The Canadian City: Essays in Urban History*, Toronto: Macmillan 186–211.

Howell, Leigh. 1986. "Gentrification and the Loss of Affordable Rental Housing: The Toronto Experience." *City Magazine* 9 no. 1: 25–9.

Hoyt, Homer. 1939. *The Structure and Growth of Residential Neighborhoods in American Cities*. Washington, DC: Federal Housing Administration.

Johnston, R.J. 1966. "The Location of High Status Residential Areas." *Geografiska Annaler* 48B: 23–35.

– 1971. *Urban Residential Patterns*. New York: Praeger.

Kalman, Harold, and Roaf, John. 1974. *Exploring Vancouver*. Vancouver: University of British Columbia Press.

Lamphier, Gary. 1992. "Vancouver's New Power Elite." *Financial Times of Canada* 1 June.

Lemon, James. 1985. *Toronto since 1918: An Illustrated History*. Toronto: James Lorimer.

Ley, David. 1981. "Inner City Revitalization in Canada: A Vancouver Case Study." *Canadian Geographer* 25 no. 2: 124–48.

– 1983. *A Social Geography of the City*. New York: Harper and Row.

– 1985. *Gentrification in Canadian Inner Cities: Patterns, Analysis, Impacts and Policy*. Ottawa: Canada Mortgage and Housing Corporation.

– 1988. "Social Upgrading in Six Canadian Inner Cities." *Canadian Geographer* 32 no. 1: 31–45.

– 1989. "Modernism, Post-modernism and the Struggle for Place." In J. Agnew and J. Duncan, eds. *The Power of Place*, Boston: Unwin Hyman, 44–65.

– 1992. "Gentrification in Recession: Social Change in Six Canadian Inner Cities, 1981–1986." *Urban Geography* 13 no. 3: 230–56.

– Forthcoming. *Labour, Lifestyle and Landscape: The New Middle Class and the Inner City*. Oxford: Oxford University Press.

McAfee, Ann. 1973. "Evolving Inner City Residential Environments: The Case of Vancouver's West End." In J. Minghi, ed., *Peoples of the Living Land*. BC Geographical Series No. 15, Vancouver: Tantalus, 163–81.

MacKay, Donald. 1987. *The Square Mile: Merchant Princes of Montreal*. Vancouver: Douglas and McIntyre.

Majury, Niall. 1990. "Identity, Place, Power and the 'Text': Kerry's Dale and the 'Monster House'." MA thesis, Department of Geography, University of British Columbia.

Martin, Bruce. 1989. "Faith without Focus: Neighbourhood Transition and Religious Change in Inner-City Vancouver." MA thesis, Department of Geography, University of British Columbia.

Mills, Caroline. 1988. "Life on the Upslope: The Postmodern Landscape of Gentrification." *Society and Space* 6: 169–90.

– 1989. "Interpreting Gentrification: Postindustrial, Postpatriarchal, Postmodern?" PhD thesis, Department of Geography, University of British Columbia.

Millward, Hugh. 1981. *The Geography of Housing in Metropolitan Halifax, Nova Scotia*. Department of Geography, Saint Mary's University, Atlantic Region Geographical Studies No. 3.

Moore, Peter. 1982a. "Zoning and Neighbourhood Change: The Annex in Toronto, 1900–1970." *Canadian Geographer* 26 no. 1: 21–36.

– 1982b. "Gentrification and the Residential Geography of the New Class." Unpublished paper, Department of Geography, University of Toronto.

Oziewicz, Stanley. 1986. "Toronto the Rich." *Globe and Mail* 8 Nov.

Porter, John. 1965. *The Vertical Mosaic*. Toronto: University of Toronto Press.

Pratt, Geraldine. 1981. "The House as an Expression of Social Worlds." In J. Duncan, ed., *Housing and Identity*, London: Croom Helm, 135–80.

Rémillard, François, and Merritt, Brian. 1986. *Les Demeures Bourgeoises de Montréal: Le Mille Carré Doré, 1850–1930*. Montreal: Méridien.

Robertson, Angus. 1977. "The Pursuit of Power, Profit and Privacy: A Study of Vancouver's West End Elite, 1886–1914." MA thesis, Department of Geography, University of British Columbia.

Seeley, J., Sim, R.A., and Loosley, E. 1956. *Crestwood Heights*. New York: Basic Books.

Smith, P.J., and McCann, L. 1981. "Residential Land Use Change in Inner Edmonton." *Annals, Association of American Geographers* 71: 536–51.

Ward, Robin. 1989. "Vancouver Club, West Hastings." *Vancouver Sun* 3 June.

Wilson, Ethel. 1960. *The Innocent Traveller*. Toronto: Macmillan.

Winsor, John. 1981. "The First Shaughnessy Plan." *Quarterly Review* (City of Vancouver Planning Department) 8 (Oct.: 3–4).

NOTES TO CHAPTER TWELVE

We would like to thank R.C. Brown, dean of arts, Simon Fraser University; M. Gaudreau, INRS-Urbanisation, Montreal; E. Hacke, Simon Fraser University; J.T. Lemon, University of Toronto; T. Mallinson, Simon Fraser University;

E. Noah, Simon Fraser University; D. Rose, INRS-Urbanisation, Montreal; and P. Villeneuve, Université Laval.

1 R. Walker (1981).   2 Stone (1967).   3 Bourne (1989).   4 *City Magazine*, various issues, but particularly the special issue on suburbs, 6 no. 2 (1977).   5 Lacoste (1958).   6 Seeley, Sim, and Loosley (1956).   7 Park Forest, Ill., was the setting for *The Organization Man*, by W.H. Whyte Jr. For Reisman's remark see Seeley, Sim, and Loosley (1956: xi).   8 Goldberg and Mercer (1986).   9 These comments reflect a large body of Clark's work.   10 Lemon (1989: 24) draws attention to the historical existence of "shack towns ... on the margins of Toronto."   11 See Dawson (1926) for perhaps the earliest use of such an ecological model in Canada, as applied to Montreal.   12 Clark (1978: 141–63).   13 Bourne et al. (1986: 170–4).   14 Lai (1988); Mercer (1988).   15 Clark (1978: 140).   16 Foggin and Polèse (1976: 23).   17 Brunet (1980: 20).   18 Pearson (1957).   19 Bryant, Russwurm, and McLellan (1982).   20 British Columbia, LMRPB (1956; 1963).   21 Belec (1984).   22 Choko (1989).   23 Collin (1986).   24 Fortin (1987).   25 Lorimer and Ross (1977).   26 Hancock (1961).   27 Sewell (1977: 37).   28 Sewell (1977: 35).   29 Weaver (1978); Paterson (1985).   30 Harasym and Smith (1975).   31 Smith (1972; 1991).   32 Boal and Johnson (1965).   33 Harasym and Smith (1975: 158).   34 Divay and Gaudreau (1984).   35 Hodge (1986: 70).   36 Hancock (1961).   37 Hodge (1986: 70).   38 Hutton and Davis (1985); Relph (1991).   39 Carver (1975: 157).   40 Ashcroft (1980).   41 Miron (1982).   42 This is largely undocumented except in confidential municipal reports and in newspaper articles. See also Harris (1986: 307).   43 Evenden (1991).   44 Langlois (1984).   45 Villeneuve and Rose (1988: 176).   46 Rutherford and Wekerle (1988: 130).   47 Villeneuve and Rose (1988: 176).   48 Rutherford and Wekerle (1988: 130).   49 Villeneuve and Rose (1988: 165–6).   50 Dyck (1989).   51 Villeneuve et Viaud (1987).   52 Villeneuve and Rose (1988: 167).   53 Metropolitan Toronto Social Planning Council (1979: 132).   54 Metropolitan Toronto Social Planning Council (1979: 132).   55 Metropolitan Toronto Social Planning Council (1979: 132); Lemon (1978).   56 Vischer (1987).   57 Priest (1984).   58 Prior (1986).   59 Carlyle (1991).   60 The neighbourhood of Maillardville retains a distinctively French-Canadian image within the district municipality of Coquitlam. Under pressure of rapid development, however, the image is fading; Villeneuve (1972).   61 Lai (1988).   62 Mercer (1988).   63 Gibson (1974).   64 Harris (1984).   65 Bridger and Greer-Wootten (1965).   66 Harris (1986: 304).   67 Murdie (1986).   68 Turcotte (1980).   69 Pratt (1986).   70 Cromwell (1972).   71 G. Walker (1987: 91).

72. Relph (1991). During the 1950s, Putnam, in studying the urban sprawl extending west from Toronto around the end of Lake Ontario, speculated on the inevitability of local government reform there, proposing the name

Mississauga to refer to the emerging conurbation of this large area; Putnam
(1953); Pearson 1960). But a planning designation of a large, inchoate urban
region is not necessarily meaningful to local communities. With creation
first of the town (1968) and then the city of Mississauga (1974), a politically
legitimate if more geographically restricted place was brought into existence. The task that followed included the "struggle to build a civic identity
in the shadow of a giant metropolitan neighbour and in the face of a tradition of local village identification"; Riendeau (1988: 1,367).

73 McIlwraith (1988: 43).   74 Watson (1979: 42); Williams
(1987).   75 Currie and Thacker (1986).   76 Weightman (1976); Kuitenbrouwer
(1988).   77 Ley (1984).   78 Fortin (1987).

REFERENCES FOR
CHAPTER TWELVE

Ashcroft, R.J. 1980. "Intrametropolitan Branch Banking: Ottawa, 1930–79." MA
thesis, Department of Geography, Carleton University.

Belec, J. 1984. "Recent Work on Local Government Reorganization: Toward a
Theory of the Local State in Canada." *Canadian Geographer* 28 no. 1: 90–6.

Boal, F.W., and Johnson, D.B. 1965. "The Functions of Retail and Service Establishments on Commercial Ribbons." *Canadian Geographer* 9 no. 3:
154–69.

Bourne, L.S. 1987. "Evaluating the Aggregate Spatial Structure of Canadian
Metropolitan Areas." *Canadian Geographer* 31 no. 3: 194–208.

– 1989. "Are New Urban Forms Emerging? Empirical Tests for Canadian Urban
Areas." *Canadian Geographer* 33 no. 4: 312–28.

Bourne, L.S., Baker, A.M., Kalbach, W., Cressman, R., and Green, D. 1986.
*Canada's Ethnic Mosaic: Characteristics and Patterns of Ethnic Origin
Groups in Urban Areas.* University of Toronto, Centre for Urban and Community
Studies, Major Report No. 24.

Bridger, M.K., and Greer-Wootten, B. 1965. "Landscape Components and Residential Growth in Western Montreal Island." *Revue de géographie de Montréal* 19 nos. 1 and 2: 75–90.

British Columbia, Lower Mainland Regional Planning Board. 1956. *Urban
Sprawl.* New Westminster: The Board.

– 1963. *The Urban Frontier.* New Westminster: The Board.

Brunet. Y. 1980, "L'exode urbain: essai de classification de la population exur-
baine des Cantons de l'Est." *Le Géographe canadien* 24 no. 4: 385–405.

Bryant, C.R, Russwurm, L.H., and McLellan, A.G. 1982. *The City's Countryside: Land and Its Management in the Rural-Urban Fringe.* London:
Longman.

Bussière, Y. 1989. "L'automobile et l'expansion des banlieues: le cas de Montréal, 1901–2001." *Revue d'histoire urbaine* 18 no. 2: 159–65.

Carlyle, I.P. 1991. "Ethnicity and Social Areas within Winnipeg." In G.M. Robinson, ed., *A Social Geography of Canada: Essays in Honour of J. Wreford Watson*. Toronto: Dundurn Press, 195–219.

Carver, H. 1962. *Cities in the Suburbs*. Toronto: University of Toronto Press.

– 1975. *Compassionate Landscape*. Toronto: University of Toronto Press.

Choko, M.H. 1989. "De la cité idéale à la maison de banlieue familiale: l'expérience de Cité-jardin du tricentenaire, Montréal, 1940–1947." *Plan Canada* 29 no. 3: 38–50.

Clark, S.D. 1966. *The Suburban Society*. Toronto: University of Toronto Press.

– 1978. *The New Urban Poor*. Toronto: McGraw-Hill Ryerson.

Collett, C.W. 1982. "The Congregation of Italians in Vancouver." MA thesis, Department of Geography, Simon Fraser University.

Collin, J.-P. 1986. *La cité coopérative canadienne-française: Saint-Léonard-de-Port-Maurice, 1955–1963*. Sainte-Foy: Presses de l'Université du Québec à Montréal, INRS-Urbanisation.

Cromwell, J. 1972. "Perceptual Differences between Established and New Residents in the Urban-Rural Fringe: Surrey, B.C." In J.V. Minghi, ed., *Peoples of the Living Land: Geography of Cultural Diversity in British Columbia*, BC Geographical Series No. 15, Vancouver: Tantalus, 229–42.

Currie, R.F., and Thacker, C. 1986. "Quality of the Urban Environment as Perceived by Residents of Slow and Fast Growth Cities." *Social Indicators Research* 18: 95–118.

Dawson, C.A. 1926. *The City as an Organism*. McGill University Publications Series 13 (Art and Architecture), No. 10, Montreal.

Divay, G., and Gaudreau, M. 1984. *La formation des espaces résidentiels: le système de production de l'habitat urbain dans les années soixante-dix au Québec*. Sainte-Foy: Presses de l'Université du Québec à Montréal, INRS-Urbanisation.

Dyck, I.J. 1989. "Integrating Home and Wage Workplace: Women's Daily Lives in a Canadian Suburb." *Canadian Geographer* 33 no. 4: 329–41.

Evenden, L.J. 1978. "Shaping the Vancouver Suburbs." In L.J. Evenden, ed., *Vancouver: Western Metropolis*, Western Geographical Series No. 16, University of Victoria, Department of Geography, 179–200.

– 1991. "The Expansion of Domestic Space on Vancouver's North Shore." In G.M. Robinson, ed., *A Social Geography of Canada: Essays in Honour of J. Wreford Watson*, Toronto: Dundurn Press, 220–44.

Foggin, P. and Polèse, M. 1976. *La géographie sociale de Montréal en 1971*. Études et documents 1. Montreal: INRS-Urbanisation.

Fortin, A. 1987. "Des banlieues ... pas si plates!" *Histoire de familles et de réseaux, la sociabilité au Québec d'hier à demain*. Montreal: Éditions Saint-Martin, 121–45.

Gibson, E.M.G. "Lotus Eaters, Loggers and the Vancouver Landscape." In L.J. Evenden and F.C. Cunningham, eds., *Cultural Discord in the Modern*

World: Geographical Themes, BC Geographical Series No. 20, Vancouver: Tantalus, 57–74.

Goldberg, M.A., and Mercer, J. 1986. *The Myth of the North American City: Continentalism Challenged*. Vancouver: University of British Columbia Press.

Goliger, G. 1983. "The Changing Canadian Suburb." *Habitat* 26 no. 2: 20–33.

Hancock, M.L. 1961. "Flemingdon Park: A New Urban Community." *Plan Canada* 2 no. 1: 4–24.

Harasym, D.G., and Smith, P.J. 1975. "Planning for Retail Services in New Residential Areas since 1944." In B.M. Barr, ed., *Calgary: Metropolitan Structure and Influence*, Western Geographical Series No. 11, University of Victoria Department of Geography, 157–92.

Hardwick, W.G. 1974. *Vancouver*. Don Mills: Collier-Macmillan.

Harris, R. 1984. "Residential Segregation and Class Formation in Canadian Cities." *Canadian Geographer* 28 no. 2: 186–96.

– 1986. "Boom and Bust: The Effects of House Price Inflation on Home-Ownership Patterns in Montreal, Toronto and Vancouver." *Canadian Geographer* 30 no. 4: 302–15.

Hodge, G. 1986. *Planning Canadian Communities: An Introduction to the Principles, Practices and Participants*. Toronto: Methuen.

Hutton, T.A., and Davis, H.C. 1985. "Multinucleation and Regional Town Centre Planning: The Case of Vancouver." *Canadian Journal of Regional Science* 8 no. 1: 17–34.

Kuitenbrouwer, P. 1988. "Mohawks Claim to Own 8 Suburbs." *Montreal Gazette* 4 June.

Lacoste, N. 1958. *Les caractéristiques sociales de la population du grand Montréal*. Montreal: Université de Montréal.

Lai, D.C. 1988. *Chinatowns: Towns within Cities in Canada*. Vancouver: University of British Columbia Press.

Langlois, S. 1984. "L'impact de double revenue sur la structure des besoins dans les ménages." *Recherches sociographiques* 25 no. 2: 211–65.

Lemon, J.T. 1978. "The Urban Community Movement: Moving towards Public Households." In D. Ley and M. Samuels, eds., *Humanistic Geography: Prospects and Problems*, Chicago: Maaroufa Press, 319–37.

– 1989. "Plans for Early 20th Century Toronto: Lost in Management." *Urban History Review* 18 no. 1: 11–31.

Ley, D. 1984. "Pluralism and the Canadian State." In C. Clarke, D. Ley, and C. Peach, eds., *Geography and Ethnic Pluralism* London: George Allen and Unwin.

– 1985. "Downtown or the Suburbs? A Comparative Study of Two Vancouver Head Offices." *Canadian Geographer* 29 no. 1: 30–43.

Linteau, P.-A. 1987. "Canadian Suburbanization in a North American Context: Does the Border Make a Difference?" *Journal of Urban History* 13 no. 3: 252–74.

Lorimer, J., and Ross, E., eds. 1977. *The Second City Book: Studies of Urban and Suburban Canada*. Toronto: James Lorimer.

McCarthy, M.P. 1986. "The Politics of Suburban Growth: A Comparative Approach." In G.A. Stelter and A.F.J. Artibise, eds., *Power and Place: Canadian Urban Development in the North American Context*, Vancouver: University of British Columbia Press, 323–39.

McIlwraith, T.F. 1988. "The Nazar House: A Demonstration of Intergovernmental Cooperation in Ontario." *Urban History Review* 17 no. 1: 41–3.

Mercer, J. 1988. "New Faces on the Block: Asian Canadians." *Canadian Geographer* 32 no. 4: 360–2.

Metropolitan Toronto Social Planning Council. 1979. *Metro's Suburbs in Transition*. 2 vols. Toronto: The Council.

Miron, J.R. 1982. *The Two-Person Household: Formation and Housing Demand*. University of Toronto, Centre for Urban and Community Studies, Research Paper no. 131.

Murdie, R.A. 1986. "Residential Mortgage Lending in Metropolitan Toronto: A Case Study of the Resale Market." *Canadian Geographer* 30 no. 2: 98–110.

Paterson, R. 1985. "The Development of an Interwar Suburb: Kingsway Park, Etobicoke." *Urban History Review* 13 no. 3: 225–35.

Pearson, N. 1957. "Hell Is a Suburb: What Kind of Neighbourhoods Do We Want?" *Community Planning Review* 7 no. 3: 124–8.

– 1960. "Regional Planning." *Plan Canada* 1 no. 3: 189–94.

Pratt, G.J. 1986. "Housing Tenure and Social Cleavages in Urban Canada." *Annals, Association of American Geographers* 76 no. 3: 366–80.

Priest, G.E. 1984. "The Family Life Cycle and Housing Consumption in Canada: A Review Based on 1981 Census Data." *Canadian Statistical Review* No. 11-003E (Sept.) vi–xxi.

Prior, J.B. 1986. "The Dwelling as a Node of Daily Activity among Non-institutionalized Retired Men." PhD thesis, Department of Geography, Simon Fraser University.

Putnam, D.F. 1953. "Geographic Factors in Canadian Planning." *Revue canadienne de géographie* 7 nos. 3–4: 64–71.

Relph, E. 1991. "Suburban Downtowns of the Greater Toronto Area." *Canadian Geographer* 35 no. 4: 421–5.

Riendeau, R.E. 1988. "Mississauga." In J.H. Marsh, ed., *The Canadian Encyclopedia*, 2nd edition, Edmonton: Hurtig, vol. 3: 1366–7.

Ross, G. 1984. *Census Metropolitan Area/Census Agglomeration Program: A Review 1941–1981*. Statistics Canada, Geography Division, Working Paper No. 8–GEO 24, Ottawa.

Rutherford, B.M., and Wekerle, G.R. 1988. "Captive Rider, Captive Labor: Spatial Constraints and Women's Employment." *Urban Geography* 9 no. 2: 116–37.

Seeley, J.R., Sim, R.A., and Loosley, E.W. 1956. *Crestwood Heights: A Study of the Culture of Suburban Life*. Introduction by David Reisman. Toronto: University of Toronto Press.

Séguin, A.-M. 1989. "Madame Ford et l'espace: lecture féministe de la suburbanisation." *Recherches féministes* 2 no. 1: 51–68.

Sewell, J. 1977. "The Suburbs." *City Magazine* 2 no. 6: 19–55.

Simmons, J.W. 1974. *Patterns of Residential Movement in Metropolitan Toronto*. Department of Geography, Research Publication No. 13, Toronto: University of Toronto Press.

Smith, P.J. 1972. "Changing Forms and Patterns in the Cities." In P.J. Smith, ed., *The Prairie Provinces*, Toronto: University of Toronto Press, 99–117.

– 1991. "Community Aspirations, Territorial Justice and the Metropolitan Form of Edmonton and Calgary." In G.M. Robinson, ed., *A Social Geography of Canada: Essays in Honour of J. Wreford Watson*, Toronto: Dundurn Press, 245–66.

Stone, L.O. 1967. *Urban Development in Canada: An Introduction to the Demographic Aspects*. Ottawa: Dominion Bureau of Statistics.

Turcotte, G. 1980. *Différenciation sociale et accessibilité à l'espace neuf. Études et documents* 18. Montreal: INRS-Urbanisation.

Villeneuve, P.Y. 1972. "Residential Location Problems in the French Canadian Community of Maillardville: A Linear Programming Approach." In J.V. Minghi, ed., *Peoples of the Living Land: Geography of Cultural Diversity in British Columbia*, B.C. Geographical Series No. 15, Vancouver: Tantalus, 85–106.

Villeneuve, P.Y., and Rose, D. 1988. "Gender and the Separation of Employment from Home in Metropolitan Montreal, 1971–81." *Urban Geography* 9 no. 2: 155–79.

Villeneuve, P.Y., and Viaud, G. 1987. "Asymétrie occupationnelle et localisation résidentielle des familles à double revenu à Montréal." *Recherches sociographiques* 28 nos. 2–3: 371–91.

Vischer, J.C. 1987. "The Changing Canadian Suburb." *Plan Canada* 27 no. 5: 130–40.

Walker, G. 1987. *An Invaded Countryside: Structures of Life on the Toronto Fringe*. York University, Atkinson College, Department of Geography, Geographical Monographs No. 17.

Walker, R. 1981. "A Theory of Suburbanization: Capitalism and the Construction of Urban Space in the United States." In M. Dear, and A.J. Scott, eds., *Urbanization and Urban Planning in Capitalist Society*, New York: Methuen.

Watson, J.W. 1979. "Mental Distance in Geography: Its Identification and Representation." In J.K. Fraser, ed., *Congress Proceedings, 22nd International Geographical Congress, Montréal, 1972*, Ottawa: Canadian Committee for Geography, 38–50.

Weaver, J.C. 1978. "From Land Assembly to Social Maturity: The Suburban Life of Westdale (Hamilton) Ontario, 1911–51." *Histoire sociale/Social History* 11: 411–40.

Weightman, B.A. 1976. "Indian Social Space: A Case Study of the Musqueam Band of Vancouver, British Columbia." *Canadian Geographer* 20 no. 2: 171–86.

Whyte, W.H., Jr. 1956. *The Organization Man*. New York: Simon and Schuster.

Williams, D. 1987. "Perception of Neighbourhood: Gender Differences in the Mental Mapping of Urban and Suburban Communities." MA thesis, Department of Geography, Carleton University.

Yeates, M. 1987. "The Extent of Urban Development in the Windsor–Quebec City Axis." *Canadian Geographer* 31 no. 1: 64–9.

NOTES TO CHAPTER THIRTEEN

1 We would like to acknowledge the receipt of an SSHRC grant that helped fund this research. We also wish to thank the many Canadian geographers who aided us in assembling much of the bibliography for this research, and in particular Fredric Dahms and Gerald Hodge, who commented on an earlier draft of this chapter.

2 Smith (1970); Robertson (1973); Rees (1974; 1988).

3 Hodge and Qadeer (1976; 1983). Qadeer and Chinnery (1986) identified 2,039 communities from the 1981 census with a population of under 10,000; over half (57.9 per cent) had fewer than 1,000 inhabitants.

4 Hodge (1983: 20).   5 Fuller (1985: 2). Break (1988); Dahms (1988).   6 Bowles and Johnston (1987).   7 Dahms (1986b: 174).

8 Everitt and Sturko (1980); Hodge and Qadeer (1983); Dahms (1986a; 1986b); De Benedetti and Price (1987).

9 Native communities are not discussed in this chapter because of their distinctive cultural and social characteristics and their political autonomy.

10 Williams (1987: 65).

11 Robertson (1973); Dahms (1980a); Todd, (1981; 1982; 1983); Hodge and Qadeer (1983).

12 Dahms (1980b; 1986a); G. Walker (1988).

13 G. Walker (1988: 1).

14 Everitt and Sturko (1980).

15 Sometimes the process of official definition was simply serendipitous. When Brandon, Manitoba, was incorporated as a city in 1882, "it had been discovered that incorporation as a city would cost the same as that for a town, and therefore [it was concluded that] it would be wiser to ask incorporation as a city"; Everitt and Stadel (1988: 63).

16 While these are often identified as "single industry communities," there
seems little consensus on the number of such communities. A report by DREE
(Canada, Department of Regional Economic Expansion) in 1979, based on
1971 census data, identified 811 such places (excluding those in Northwest Territories and Yukon), which in total accounted for 25.5 per cent of the non-
metropolitan population of Canada. Alternatively, the Canadian Association
of Single Industry towns claims that there are 1,500 such communities,
about 600 of them in Newfoundland, in the guise of small fishing villages and
towns; Canada, Employment and Immigration Advisory Council (1987).

17 Hodge (1988); Russwurm and Bryant (1984).

18 Coffey and Polèse (1988). Post office data have been sometimes used to
define small settlement centrality (Ironside et al 1981: 230), but there are numerous problems in using these data (Everitt, Kempthorne, and Schafer,
1989). Similarly, data on the extent of small-town services, including the classic
example of newspaper circulation, have been investigated, but without definitive results.

19 Brown and Wardwell (1980).   20 Dahms (1986b: 172).   21 Dahms
(1980a, 1984); Fuller (1985: 2); Break (1988).   22 Jankunis (1972) gives an example for southern Alberta.   23 Meredith (1975). For similar trends in Manitoba and Alberta, see Jankunis (1972: 84–5) and Everitt (1986).

24 J. Walker (1988); Everitt, Annis, and McGuinness (1990). The classic
study by Berry (1967: chapter 1) of southwest Iowa finds many of the same
processes at work. See also Murdie's (1965) research on the behaviour patterns of Old Order Mennonites and "modern" Canadians in rural southwestern
Ontario.

25 Jankunis (1972: 84–5); Mulligan and Ryder (1985); Coffey and Polèse (1988);
Hart (1988).

26 De Benedetti and Price (1987: 207); J. Walker (1988). Hodge (1988) suggests
a classification of different types of rural regions.

27 Compare, for example, studies of Newfoundland (Mannion 1974), the prairies
(Holdsworth and Everitt, 1988; Davies 1990), and the eastern Canadian
"heartland" (Dahms 1988).

28 Jankunis (1972); Stabler (1987).

29 For instance, Rees found that the larger settlements (of over 1,000 population)
may have been growing, whereas the smaller ones were declining or had
already disappeared (1974: 14). Qadeer and Chinnery (1986) found that communities with fewer than 2,500 people were vulnerable to population decline, as they often fell beneath the threshold size for provision of critical services
such as water and other utilities.

30 Hodge (1987) estimates that 3,000 of approximately 9,000 towns/villages/
hamlets in Canada already have 20 per cent of their population over 65 years
old.

31 As one resident put it, "If the people over 65 stopped doing things, this town would dry up like a pretzel"; Turner (1988: 1).

32 J. Walker (1988).   33 Bekkering (1987).   34 Bowles and Johnston (1987: 432).   35 Paul (1977).   36 Todd (1983: 905).   37 Dahms (1988).   38 *Macleans* (1988).   39 Bradbury (1984).   40 Bradbury and St Martin (1983); Canada, Employment and Immigration Advisory Council (1987).   41 Bradbury and St Martin (1983).   42 Ennals and Holdsworth (1988).   43 Rees (1988).   44 Gill (1989).   45 Hart (1988: 272).   46 Holdsworth (1985); Dahms (1988).   47 Ennals and Holdsworth (1988: 5).   48 Mannion (1974: chapter 3).   49 Rees (1988: 60).   50 Warkentin (1959); Evans (1973).   51 Tyman (1972).   52 Rees (1988: 43).   53 Robertson (1973).   54 Biays (1968); Spelt (1968).   55 Bradbury (1980); McCann (1982).

56 McCann (1982) notes the conformity of land use in five new towns (Gold River, British Columbia; Fort McMurray, Alberta; Pinawa, Manitoba; Lebel sur Quevillon, Quebec; and Wabush, Newfoundland).

57 Pressman (1976).

58 Gill (1989).

59 Canada, Employment and Immigration Advisory Council (1987).

60 Rowe (1988) has demonstrated the importance of self-help housing to home-ownership in the non-metropolitan housing stock, particularly in Atlantic Canada.

61 Hodge and Qadeer (1983: 149).   62 Wellar (1978) details some of the government programs put into place in the prairie provinces.   63 Holdsworth (1985).   64 Carter (1987).   65 Jess (1987: 50).

66 Rowe (1988). Mortgages, for instance, usually are not available in a rural/Native/northern context in part because of the lack of a normal market for housing, but also because irregular income is more characteristic of a rural context and does not dovetail well with a system of regular payments.

67 Pinfield and Etherington (1985).

68 Bradbury and St Martin (1983); Bradbury (1985).

69 Bradbury (1985: 2).

70 Gill (1986); Hodge (1989). Lucas (1971) identifies the stages of development associated with resource town evolution, each characterized by distinctive demographic features.

71 Kinnear, Stadel, and Everitt (1989).   72 Dahms (1988): G. Walker (1988).   73 Roberts and Fisher (1984); Gill and Smith (1985); Corbett (1988).   74 Everitt (1986); Pierce (1986: 7).   75 Everitt (1986); J. Walker (1988).   76 Everitt (1986). Similar results were found by Jankunis (1972).   77 Porteous (1979); Bowles (1982); Gill (1986).   78 Gill (1989).

79 Mackenzie and Menzies (1988) demonstrate in their study of Trail, British Columbia, that inadequate daycare facilities often result in the setting up of informal arrangements.

80 "Professional rookies": recently trained professionals who obtain jobs in small communities to get experience until they can gain employment in a large centre and thus often remain for a relatively short time; Redekop (1983).

81 Canada, Immigration and Employment Advisory Council (1987).

82 Hodge and Qadeer (1983).

83 Social problems may be quite common in small towns, even though they are often praised for their smallness and their "close-knit" environment. For instance, a recent study in Manitoba has showed that "women in small towns are reluctant to seek shelter from family violence because of the lack of confidentiality." A task force that discovered this problem also unearthed difficulties with daycare, literacy, and affordable housing; Martens (1988).

84 Dahms (1988).   85 Todd (1982; 1983); Rees (1988).   86 Hart (1988: 286).   87 Jankunis (1972); Paul (1977); Bradbury and St Martin (1983); Rees (1988).   88 Burns (1982).   89 Todd (1982: 110).

REFERENCES FOR
CHAPTER THIRTEEN

Bekkering, M. 1987. "Residential Migration and Spatial Mobility of the Elderly in Brandon, Manitoba." BA thesis, Department of Geography, Brandon University.

Berry, J.L. 1967. *Geography of Market Centers and Retail Distribution.* Englewood Cliffs, NJ: Prentice Hall.

Biays, P. 1968. "Southern Quebec." In John Warkentin, ed., *Canada: A Geographical Interpretation*, Toronto: Methuen, 281–333.

Bowles, Roy T. 1982. *Little Communities and Big Industries.* Toronto: Butterworths.

Bowles, Roy T., and Johnston, Cynthia. 1987. "Communities in a Nonmetropolitan Region: An Ontario Case." *Rural Sociologist* 7 no. 5: 421–33.

Bradbury, J.H. 1980. "Instant Resource Town Policy in British Columbia: 1965–72." *Plan Canada* 20: 19–38.

– 1984. "The Impact of Industrial Cycles in the Mining Sector: The case of the Quebec-Labrador Region of Canada." *International Journal of Urban and Regional Research* 8 no. 3: 311–31.

– 1985. "Housing Policy and Home Ownership in Mining Towns: Quebec, Canada." *International Journal of Urban and Regional Research* 9 no. 1: 14–28.

Bradbury, J.H., and I. St Martin. 1983. "Winding-down in a Quebec Mining Town: A Case Study of Schefferville." *Canadian Geographer* 17 no. 2: 128–44.

Break, Helen. 1988. "Change Impacts Ontario's Small Rural Service Centres: Implications for Planning." *Small Town* 18: 4–9.

Brown, D.L., and Wardwell, J.M., eds., 1980. *New Directions in Urban-Rural Migration.* New York: Academic Press.

Burns, Nancy. 1982. "The Collapse of Small Towns on the Great Plains: A Bibliography." *Emporia State Research Studies* 31 no. 1, Emporia, Kan.: Emporia State University.

Canada, Department of Regional Economic Expansion. 1979. *Single Sector Communities*. Ottawa: Ministry of Supply and Services.

Canada, Employment and Immigration Advisory Council. 1987. *Canada's Single-Industry Communities: A Proud Determination to Survive*. Ottawa: Ministry of Supply and Services.

Carter, Tom. 1987. "Housing Requirements and Public Policy in Small Towns in Saskatchewan." In F.W. Dykeman, ed., *Rural and Small Town Housing: Issues and Approaches*, Mount Allison: Rural and Small Town Research and Studies Programme, 29–48.

Coffey, William J., and Polèse, Mario. 1988. "Locational Shifts in Canadian Employment, 1971–1981: Decentralization vs. Decongestion." *Canadian Geographer* 32 no. 3: 248–56.

Corbett, Ron. 1988. "Between the Devil and the Deep Blue Atlantic: The Dilemma of Rural and Small Town Development in Atlantic Canada." Paper presented to the Canadian Urban and Housing Studies Conference, Winnipeg, Man., Feb.

Dahms, Fred A. 1980a. "The Evolving Spatial Organization of Small Settlements in the Countryside – an Ontario Example." *Tijdschrift voor Economische en Sociale Geografie* 71 no. 5: 295–306.

– 1980b. "Small Town and Village Ontario." *Ontario Geography* 7/8 no. 16: 19–32.

– 1984. "Wroxeter, Ontario: The Anatomy of a "Dying" Village." *Small Town* 14: 17–23.

– 1986. "Diversity, Complexity, and Change: Characteristics of Some Ontario Towns and Villages." *Canadian Geographer* 30 no. 2: 158–66.

– 1986b. "'Regional Urban History: Small Towns and Their Hinterlands, 1820–1985. *Urban History Review* 15 no. 2: 172–4.

– 1986c. "Residential and Commercial Renaissance: Another Look at Small Town Ontario." *Small Town* 17: 10–15.

– 1988. *The Heart of the Country*. Toronto: Deneau.

Davies, W.K.D. 1990. "What Population Turnaround?: Some Canadian Prairie Settlement Perspectives, 1971–1986." *Geo Forum* 21 no. 3: 303–20.

De Benedetti, George, and Price, Richard J. 1987. "Population Growth and the Industrial Structure of Maritime Small Towns, 1971–1981." In Larry McCann, ed., *People and Place: Studies of Small Town Life in the Maritimes*, Victoria, BC: Morriss Printing, 191–208.

Ennals, Peter, and Holdsworth, Deryck. 1988. "The Cultural Landscape of the Maritime Provinces", In D. Day, ed., *Geographical Perspectives on the Maritime Provinces*, Halifax: Saint Mary's University, 1–14.

Evans, S.M. 1973. "The Dispersal of Hutterite Colonies in Alberta, 1918–1971: The Spatial Expression of Cultural Identity." MA thesis, University of Calgary.

Everitt, John. 1986. *Central Plains Lead Planning Phase*. 5 vols. Brandon, Man.: Westarc Group Inc.

Everitt, John, Annis, R.A., and McGuinness, F. 1990. "The Responsibility of Urban Dwellers to Foster Sustainable Rural Communities." In M.A. Beavis, ed., *Ethical Dimensions of Sustainable Development and Urbanization: Seminar Papers*, Winnipeg: Institute of Urban Studies, 118–48.

Everitt, John, Kempthorne, Roberta, and Schafer, Charles. 1989. "Controlled Aggression: James J. Hill and the Brandon, Saskatchewan and Hudson's Bay Railway." *North Dakota History* 56 (spring): 3–19.

Everitt, John, and Stadel, Christoph. 1988. "Spatial Growth of Brandon." In J. Welsted et al. eds., *Brandon: Geographical Perspectives on the Wheat City*, Regina: Canadian Plains Research Centre, 61–88.

Everitt, John, and Sturko, Derek. 1980. "When a City Is a Town Is a City: The 'Place' of Brandon, Manitoba, in the Urban Hierarchy." *Great Plains – Rocky Mountains Geographical Journal* 9: 31–9.

Fuller, A.M. 1985. "The Development of Farm Life and Farming in Ontario." In A.M. Fuller, ed., *Farming and the Rural Community in Ontario: An Introduction*, Toronto: Foundation for Rural Living, 1–46.

Gill, Alison. 1986. "New Resource Communities: The Challenge of Meeting the Needs of Canada's Modern Frontierspersons." *Environments* 18 no. 3: 21–34.

– 1989. "Experimenting with Environmental Design in Canada's Newest Mining Town." *Applied Geography* 9: 177–95.

Gill, Alison, and Smith, G.C. 1985. "Residents' Evaluative Structures of Northern Manitoba Mining Communities." *Canadian Geographer* 24: 15–27.

Hart, John Fraser. 1988. "Small Towns and Manufacturing." *Geographical Review* 78 no. 3: 272–87.

Hodge, Gerald. 1983. "Canadian Small Town Renaissance: Implications for Settlement System Concepts." *Regional Studies* 17 no. 1: 19–28.

– 1987. *The Elderly in Canada's Small Towns: Recent Trends and Their Implications*. Vancouver, BC: Centre for Human Settlements.

– 1988. "Canada." In Paul Cloke, ed., *Policies and Plans for Rural People: An International Perspective*, London: Unwin Hyman, 166–91.

– 1989. *Seniors in Small Towns in British Columbia: Demographic Tendencies and Trends 1961–86*. Vancouver, BC: Centre for Human Settlements.

Hodge, G., and Qadeer, M. 1976. *Towns and Villages in Urban Canada*. Ottawa: Ministry of State for Urban Affairs.

– 1983. *Towns and Villages in Canada: The Importance of Being Unimportant*. Toronto: Butterworths.

Holdsworth, Deryck, ed. 1985. *Reviving Main Street*. Toronto: University of Toronto Press.

Holdsworth, Deryck, and Everitt, John. 1988. "Bank Branches and Elevators: Expressions of Big Corporations in Small Prairie Towns." *Prairie Forum* 13 no. 2: 173–90.

Ironside, R.G., Preston, R.E., Fairbairn, K.J., Johnson, D.B., Savitt, R., and Brown, S. 1981. "The Definition of Trade Areas and Their Boundaries: A Literature Review and Annotated Bibliography." Report to the Department of Tourism and Small Business, Government of Alberta, March.

Jankunis, Frank J. 1972. "Urban Development in Southern Alberta." In F.J. Jankunis, ed., *Southern Alberta: A Regional Perspective*, Lethbridge: University of Lethbridge, 74–85.

Jess, C.R. 1987. "Using Housing to Develop Human Resources in Rural Areas." In F.W. Dykeman, ed., *Rural and Small Town Housing: Issues and Approaches*. Mount Allison: Rural and Small Town Research and Studies Programme, 49–60.

Kinnear, M., Stadel, C. and Everitt, J. 1989. "Recreational Homes and Hinterlands in Southwest Manitoba: The Example of Minnedosa Lake Developments." In R. Keith Semple and Lawrence Martz, eds., *Prairie Geography*, Saskatchewan Geography No. 2, Department of Geography, University of Saskatchewan, 23–39.

Lucas, R. 1971. *Minetown, Milltown, Railtown: Life in Canadian Communities of Single Industry*. Toronto: University of Toronto Press.

McCann, L.D. 1982. "The Changing Internal Structure of Canadian Resource Towns." In R.T. Bowles, ed., *Little Communities and Big Industries*, Toronto: Butterworths, 61–81.

Mackenzie, S., and Menzies, H. 1988. "Transforming Communities, Transforming Economy: Issues and Strategies in the Politics of Restructuring." Paper presented to the Canadian Urban and Housing Studies Conference, Winnipeg, Man., Feb.

Macleans. 1988. "Life after Death: Dinosaur Shows Have Become Blockbuster Hits as Scientists Heatedly Debate New Theories." 5 Dec.: 51–6.

Mannion, John J. 1974. *Irish Settlements in Eastern Canada: A Study of Cultural Transfer and Adaptation*. University of Toronto, Department of Geography, Research Publication No. 12.

Martens, Kathleen. 1988. "Family Violence 'Hidden' in Small Towns." *Brandon Sun* 17 Nov.

Meredith, M.L. 1975. "The Prairie Community System." *Canadian Farm Economics* 10 no. 5: 19–27.

Mulligan, Helen, and Ryder, Wanda. 1985. *Ghost Towns of Manitoba*. Surrey, BC: Heritage House.

Murdie, Robert A. 1965. "Cultural Differences in Consumer Travel." *Economic Geography* 41 (July): 211–33.

Paul, Alexander H. 1977. "Depopulation and Spatial Change in Southern Saskatchewan." In J.E. Spencer, ed., *Saskatchewan Rural Themes*, Regina Geographical Studies No. 1, Department of Geography, University of Regina, 65–85.

Pierce, John T. 1986. "Rural Planning: B.C. Experiences and Beyond." *Operational Geographer* No. 11: 7–8.

Pinfield, L.T., and Etherington, L.D. 1985. "Housing Strategies of Resource Firms in Western Canada." *Canadian Public Policy* 11 no. 1: 93–106.

Porteous, J.D. 1979. "Quality of Life in Resource Towns: Problems and Prospects." In F. Jankunis and B. Sadler, eds., *The Viability and Liveability of Small Urban Centres*, Edmonton: Environmental Council of Alberta, 45–54.

Pressman, N., ed. 1976. *New Communities in Canada: Exploring Planned Environments*. Special issue of *Contact* 8 no. 3.

Qadeer, M., and Chinnery, Kathleen. 1986. "Canadian Towns and Villages: An Economic Profile, 1981." Research and Working Papers 14, Institute of Urban Studies, University of Winnipeg.

Redekop, T. 1983. "Psycho-social Influences of Health Care Professionals in the Single Industry Mining City of Thompson, Manitoba." *Canadian Journal of Community Health* special supplement no. 1: 45–9.

Rees, Ronald. 1974. "Depopulation on the Prairie: Deserted Landscapes of Saskatchewan." *Canadian Geographical Journal* 88 no. 1: 10–19.

– 1988. *New and Naked Land: Making the Prairies Home.* Saskatoon: Western Producer Prairie Books.

Roberts, R., and Fisher, J. 1984. "Canadian Resource Communities: The Residents' Perspective in the 1980s." In *Mining Communities: Hard Lessons for the Future*, Proceedings, 12th Policy Discussion Seminar, Centre for Resource Studies, Kingston, Ont., 151–70.

Robertson, Heather. 1973. *Grass Roots.* Toronto: James Lewis and Samuel.

Rowe, Andy. 1988. "Self-Build: The Informal Sector and Housing Policy in Canada." Paper presented to the Canadian Urban Studies and Housing Conference, University of Winnipeg, Feb. 1988.

Russwurm, L.H., and Bryant, C.R. 1984. "Changing Population Distribution and Rural-Urban Relationships in Canadian Urban Fields, 1941–1976." In M.F. Bunce and M.J. Troughton, eds., *The Pressures of Change in Rural Canada*, Atkinson College, York University, Geographical Monograph No. 14, Downsview, Ont., 113–37.

Smith, Suzanne M. 1970. *An Annotated Bibliography of Small Town Research.* University of Wisconsin, Department of Rural Sociology.

Spelt, Jacob. 1978. "Southern Ontario." In John Warkentin, ed., *Canada: A Geographical Interpretation*, Toronto: Methuen, 334–95.

Stabler, Jack C. 1987. "Non-metropolitan Population Growth and the Evolution of Rural Service Centres in the Canadian Prairie Region." *Regional Studies* 21 no. 1: 43–53.

Todd, D. 1981. "Rural Out-migration in Southern Manitoba: A Simple Path Analysis of Push Factors." *Canadian Geographer* 25 no. 3: 252–66.

– 1982. "Subjective Correlates of Small-town Population Change." *Tijdschrift voor Economische en Sociale Geografie* 73 no. 2: 109–21.

– 1983. "The Small Town Viability Question in a Prairie Context." *Environment and Planning A* 15 no. 7: 903–16.

Turner, Randy. 1988. "Rural Towns Greying Fast." *Winnipeg Free Press* 27 Nov.

Tyman, John L. 1972. *By Section, Township, and Range.* Brandon: Assiniboine Historical Society.

Walker, G. 1988. "Social and Spatial Structures in a Suburbanizing Town: Woodbridge, Ontario." Paper presented at the annual meeting of Canadian Association of Geographers, Halifax, NS.

Walker, J. 1988. *Assessing Labour Market Needs and Facilitating Optimal Use of Existing Resources in Rural Manitoba.* 5 vols. Brandon, Man.: Westarc Group Inc.

Warkentin, J. 1959. "Mennonite Agricultural Settlements of Southern Manitoba." *Geographical Review* 49: 342–68.

Wellar, Barry S. 1978. *The Future of Small- and Medium-Sized Communities in the Prairie Region.* Ottawa: Ministry of State for Urban Affairs.

Welsted, John, Everitt, John, and Stadel, Christoph, eds. 1988. *Brandon: Geographical Perspectives on the Wheat City.* Regina: Canadian Plains Research Centre.

Williams, H.S. "Barn Raisings: An Old Approach Helps Small Towns Today." In F.W. Dykeman, ed., *Rural and Small Town Housing: Issues and Approaches*, Department of Geography, Rural and Small Town Research and Studies Programme, Mount Allison University, 65–72.

NOTES TO CHAPTER FOURTEEN

I wish to thank those with whom I discussed issues and who are cited in the notes and John Gandy, Allan Irving, Jeffrey Patterson, James Struthers, and the editors of this volume for reading drafts of this chapter.

1 Guest (1985: chapter 1). For varied definitions, see O'Brien (1973), A. Rose (1975: 80–6), Wharf (1975: 93), United Way of Canada (1980); and Wills (1989: 19–21).

2 On variations among cities, see Goldberg and Mercer (1986). Among the many works on social policy and fiscal federalism, see Bird (1986), Ismael (1985), Boadway (1986), Milne (1986), and Banting (1987). On municipal finance, see Bird and Slack (1983) and Kitchen (1984).

3 Guest (1985: chapter 2); Irving (1987: 1–15); Artibise (1977); Careless (1984: 51, 59, 71, 73, 100–1, 144–5); Roy (1980: 32, 36); Taylor (1986: 42, 94, 97, 106, 108, 122); Weaver (1982: 29, 31, 54, 69–70, 103–4).

4 Gilbert (1989: graphs 1, 4, 5); Guest (1985: Appendix Table 1); Careless (1984: 187); Foran (1978: 112–13, 116); Roy (1980: 61, 63, 74, 85); Taylor (1986: 156); Weaver (1982: 112); Armstrong and Nelles (1986); Wills (1989: 87–8). Jones (1986); Lemon (1989); Strange (1988: 269–70); Hocken (1914); Riendeau (1984); Kaplan (1982); Magnusson and Sancton (1983).

5 Church, in Toronto (1834–) Council Minutes (1915, 1918, 1920: Appendix C). "Unemployment" as a category was only then becoming firm;

Garraty (1978); Lemon (1985: 19, 21, 23, 25); Piva (1979); Copp (1974); Allen (1971).

6 Guest (1985: Appendix Table 1); Gilbert (1989).

7 Irving (1987: 15–9); Bator (1980); Williams (1984: 1).

8 Struthers (1983: chapter 1); Guest (1985: Appendix Table 1); Oliver (1977: 318–21).

9 Wills (1989: chapters 3 and 4, 208); Taylor (1986: 162); Weaver (1982: 145); Lemon (1985: 23, 25, 33).

10 Irving (1987: 20–5); Struthers (1983: chapters 2–6); Antler (1988); Hunter (1981); Lemon (1985: 60–7); Taylor (1986: 162–4); and Weaver (1982: 135–7).

11 Riendeau (1979); Schulz (1975).

12 MacLennan (1987); Guest (1985: Appendix Table 1, chapter 7); Bacher (1988); Horn (1980); Canada (1940: Book 1, especially 162–77).

13 Wills (1989: chapter 5).

14 Guest (1980: Appendix Table 1).

15 Guest (1985: chapter 8, 135–8); Marsh (1943); Irving (1987: 25–30); Struthers (1987).

16 [Philpott] (1954: 6–7); Lemon (1985: 90–8, 106, 115, 126–8); Weaver (1982: 170); Rose (1958); Frisken and Hauser (1983).

17 Lang (1974); Gilbert (1989); Bennett (1989: 359–70); Guest (1985: Appendix Table 1).

18 Bennett (1980: 370); Boadway (1986); Milne (1986: chapter 5).

19 Guest (1985: chapters 10 and 11); Kitchen (1984: 113).

20 Granatstein (1986: chapter 7, 261–3); Gray (1991); Struthers (1988). On the United States, see for example, Patterson (1986). Gilbert (1988); Milne (1986: chapter 5); Guest (1985: chapter 11).

21 Irving (1987: 28, 30); Sancton (1986).

22 For example, Ley (1974); Taylor (1986: 196–8); Weaver (1982: 182); Lemon (1985: 151–2, 157, 173); Caulfield (1988).

23 Wills (1989: chapter 6); Bookbinder (1971).

24 Drawn from reports in offices of the Social Planning Council (SPC) (1933–   ). Most SPC records are located in City of Toronto Archives, City Hall.

25 Stamp (1982: 217–20, 225–33); Irving (1987: 31–7); Ontario (1975); Ontario Economic Council (1976); Banting (1987: chapter 11).

26 National Council of Welfare (NCW) (1991a; 1991b); Lightman and Irving (1991); SPC (1988); Metropolitan Toronto (1989); Kitchen (1984: 29); Bird and Slack (1983); Bird (1986: 7–8); Canadian Tax Foundation (1988: 3.5).

27 On income distributions, see NCW (1988: especially 105–13); (1990–1: 29). Ross and Shillington (1989: chapters 3 and 4 and on poverty gap, 53, 64). Novick (1989). On poverty lines, see NCW (1989); Ross and Shillington (1989: chapter 2); on provincial grants, Toronto (1988–91).

28 NCW (1988: 55–8, 78); Ross and Shillington (1989: chapter 6); *Economist* (1989).

29 Ross and Shillington (1989: 35, 91); NCW (1988: 39, 19–32, 65).
30 Ross and Shillington (1989: 36).
31 NCW (1988: 33); Ross and Shillington (1989: 34, 36); Savoie (1986); Polèse (1987); SPC (1986); McArthur (1989).
32 Oberlander and Fallick (1988); Ontario (1988a; 1988b); SPC (1989a), and personal communications from David Thornley, Toronto, 1989; Roberts (1989); personal communications from Jim Green, Vancouver, on DERA, 1989; SPC (1989b).
33 On groups across Canada, personal communications from Edward Pennington, Toronto, 1988; Michael Cleague, Vancouver, 1988; Joanne Cook, Halifax, 1988; David Hulchanski, Vancouver, 1989; Sharon Willms, Vancouver, 1989; Judy Orr, Toronto, 1988; Damaris Rose, Montreal, 1989; and Jean Wolfe, Montreal, 1989. Moore Milroy (1991).
34 SPC (1979–80); Lemon (1978); SPC (1984); Laws (1988).
35 Kitchen (1984; 29, 62, 113); Goldberg and Mercer (1986); Lemon (1986); Bird (1986).
36 On long waves, see Goldstein (1988); Batra (1987). In some countries, Keynesian management may have prolonged the latest long wave and perhaps cushioned the downturn, through social welfare payments and military spending.
37 Hirsch (1976); Ellis and Kumar (1983); Leiss (1976).
38 Banting (1987: 213).

REFERENCES FOR
CHAPTER FOURTEEN

Allen, Richard. 1971. *The Social Passion: Religion and Social Reform in Canada 1914–1928* Toronto: University of Toronto Press.

Antler, James A. 1988. "Federal Unemployment Policy under the Conservative Government of R.B. Bennett." MA research paper, Department of Geography, University of Toronto.

Armstrong, Christopher, and Nelles, H.V. 1986. *Monopoly's Moment: The Organization and Regulation of Canadian Utilities, 1830–1930*. Philadelphia: Temple University Press.

Artibise, A.F.J. 1977. *Winnipeg: An Illustrated History* Toronto and Ottawa: Lorimer and Museum of Man.

Bacher, John. 1988. "W.C. Clark and the Politics of Canadian Housing Policy, 1935–52." *Urban History Review* 16: 1–16.

Banting, Keith G. 1987. *The Welfare State and Canadian Federalism*. 2nd edition. Kingston and Montreal: McGill-Queen's University Press.

Bator, Paul A. 1980. "Public Health Reform in Canada and Urban History: A Critical Survey." *Urban History Review* 9: 87–102.

Batra, Ravi. 1987. *The Great Depression of 1990*. New York: Dell.

Bennett, R.J. 1980. *The Geography of Public Finance: Welfare under Fiscal Federation and Local Government Finance*. London and New York: Methuen.

Bird, Richard M. 1986. *Federal Finance in Comparative Perspective*. Toronto: Canadian Tax Foundation.

Bird, Richard M., and Slack, Enid. 1983. *Urban Public Finance in Canada*. Toronto: Butterworths.

Boadway, Robin. 1986. "Federal-Provincial Transfers in Canada: A Critical Review of Existing Arrangements." In Mark Krasnick, ed., *Fiscal Federalism*, Macdonald Commission Research Studies, vol. 65, Toronto: University of Toronto Press.

Bookbinder, Howard. 1971. "The Toronto Social Planning Council and the United Community Fund." In D.I. Davies and Kathleen Herman, eds., *Social Space: Canadian Perspectives*, Toronto: New Press, 196–205.

Canada. Royal Commission on Dominion-Federal Relations. 1940. *Report*. 3 books. Ottawa: Queen's Printer.

Canadian Tax Foundation. 1988. *The National Finance 1987–88*. Toronto: The Foundation.

Careless, J.M.S. 1984. *Toronto to 1918: An Illustrated History*. Toronto and Ottawa: Lorimer, Museum of Man.

Caulfield, Jon. 1988. "'Reform' as a Chaotic Concept." *Urban History Review* 17: 107–11.

Copp, Terry. 1974. *The Anatomy of Poverty: The Condition of the Working Class in Montreal*. Toronto: McClelland and Stewart.

Economist. 1989. "The Working Poor: Poverty's New Stars." 11 Feb.

Ellis, Adrian, and Kumar, Krishan. 1983. *Dilemmas of Liberal Democracies: Studies in Fred Hirsch's 'Social Limits to Growth.'* London: Tavistock.

Foran, Max. 1978. *Calgary: An Illustrated History*. Toronto and Ottawa: Lorimer, Museum of Man.

Frisken, Frances, and Hauser, Dale. 1983. "Local Autonomy vs Fair Share: Intergovernmental Housing Issues in the Toronto Region, 1950–80." Paper presented to Canadian Political Science Association.

Garraty, John A. 1978. *Unemployment in History: Economic Thought and Public Policy*. New York: Harper and Row.

Gilbert, Richard. 1989. "Viewpoint: The Decline of the Canadian Municipality." *Plan Canada* 29, Jan.: 11–13.

Goldberg, Michael A., and Mercer, John, 1986. *The Myth of the North American City: Continentalism Challenged*. Vancouver: University of British Columbia Press.

Goldstein, Jonathan. 1988. *Long Cycles: Prosperity and War in the Modern Age*. New Haven, Conn.: Yale University Press.

Granatstein, J.L. 1986. *Canada, 1957–1967: Years of Uncertainty and Innovation*. Toronto: McClelland and Stewart.

Gray, Gwendolyn. 1991. *Federalism and Health Policy: The Development of Health, Systems in Canada and Australia*. Toronto: University of Toronto Press.

Guest, Dennis. 1985. *The Emergence of Social Security in Canada*. First published 1980. 2nd edition. Vancouver: University of British Columbia Press.

Hirsch, Fred. 1976. *Social Limits to Growth*. Cambridge: Harvard University Press.

Hocken, Horatio C. 1914. "The New Spirit in Municipal Government." In Canadian Club of Ottawa, *Addresses*, Ottawa: Canadian Club, 85–97; part reprinted in Paul Rutherford, ed., *Saving the Canadian City: The First Phase, 1880–1920*, Toronto: University of Toronto Press, 1974, 195–208.

Horn, Michiel. 1980. *The League for Social Reconstruction: Intellectual Origins of the Democratic Left in Canada, 1930–1942*. Toronto: University of Toronto Press.

Hunter, Bernice T. 1981. *That Scatterbrain Booky*. Richmond Hill, Ont.: Scholastic-Tab Publications.

Irving, Allan. 1987. "From No Poor Law to the Social Assistance Review: A History of Social Assistance in Ontario, 1791–1987." Prepared for the Ontario Social Assistance Review, mimeo.

Ismael, Jacqueline S., ed. 1985. *Canadian Social Welfare Policy: Federal and Provincial Dimensions*. Kingston and Montreal: McGill-Queen's University Press.

Jones, Simon. 1986. "A Legitimate Charge on the Municipality? The City of Toronto and the Relief of Unemployment, 1890–1940." MA research paper, Department of Geography, University of Toronto.

Kaplan, Harold. 1982. *Reform, Planning, and City Politics: Montreal, Winnipeg, Toronto*. Toronto: University of Toronto Press.

Kitchen, Harry M. 1984. *Local Government Finance in Canada*. Toronto: Canadian Tax Foundation.

Lang, Vernon. 1974. *The Service State Emerges in Ontario*. Toronto.

Laws, Glenda. 1988. "Privatization and the Local Welfare State: The Case of Toronto's Social Services." Institute of British Geographers, *Transactions*, new series 13: 433–48.

Leiss, William. 1976. *The Limits to Satisfaction: An Essay on the Problems of Needs and Commodities* Toronto: University of Toronto Press.

Lemon, James. 1978. "The Urban Community Movement: Moving toward Public Households." In David Ley and Marwyn S. Samuels, eds., *Humanistic Geography: Prospects and Problems*, Chicago: Maaroufa Press, 319–37.

– 1985. *Toronto to 1918: An Illustated History*. Toronto and Ottawa: Lorimer, Museum of Man.

– 1986. "The American Urban Experience." In D.H. Flaherty and W.R. McKercher, eds., *Southern Exposure: Canadian Perspectives on the United States*, Toronto: McGraw-Hill Ryerson, 181–92.

– 1989. "Plans for Early Twentieth-Century Toronto: Lost in Management." *Urban History Review* 18: 11–31.

Ley, David, ed. 1974. *Community Participation and the Spatial Order of the City*. B.C. Geographical Series 19. Vancouver: Tantalus.

Lightman, Ernie, and Irving, Allan. 1991. "Restructuring Canada's Welfare State." *Journal of Social Policy* 20: 65–86.

McArthur, Jack. 1989. "How Government Spending Supports Poorer Provinces." *Toronto Star* 28 May.

MacLennan, Anne. 1987. "Charity and Change: Montreal English Protestant Charity Faced the Crisis of Depression." *Urban History Review* 16: 1–16.

Magnusson, Warren, and Sancton, Andrew, eds. 1983. *City Politics in Canada.* Toronto: University of Toronto Press.

Marsh, Leonard C. 1943. *Report on Social Security for Canada.* Reprinted 1975. Toronto: University of Toronto Press.

Metropolitan Toronto. 1989. *The Crumbling Partnership: Metropolitan Toronto's Response to Provincial Retrenchment.* Toronto: Municipality of Metropolitan Toronto.

Milne, David. 1986. *Tug of War: Ottawa and the Provinces under Trudeau and Mulroney.* Toronto: Lorimer.

Moore Milroy, Beth. 1991. "People, Urban Space and Advantage." In Trudi Bunting and Pierre Filion, eds., *Canadian Cities in Transition*, Toronto: Oxford University Press, 519–44.

Moscovitch, Allan, and Albert, Jim, eds. 1987. *The "Benevolent State": The Growth of Welfare in Canada.* Toronto: Garamond Press.

National Council of Welfare (NCW). 1988. *Poverty Profile 1988.* Ottawa: Minister of Supply and Services.

– 1989. *1989 Poverty Lines.* Ottawa.

– 1990–91. *Welfare Incomes 1989.* Ottawa.

– 1991a. *Funding Health and Higher Education: Danger Looming.* Ottawa.

– 1991b. *The Canada Assistance Plan: No Time for Cuts.* Ottawa.

Novick, Marvyn. 1989. "Planning to Achieve Social Equity." Presented to CityPlan '91: Forum on the Future of Toronto. May.

Oberlander, H. Peter, and Fallick, Arthur F. 1988. *Homelessness and the Homeless, Responses and Innovations: A Canadian Contribution to IYSH 1987.* Vancouver: University of British Columbia Press.

O'Brien. M.T. 1973. *Planning Local Social Services in Canada.* London, Ont.: Family and Children's Service of London and Middlesex.

Oliver, Peter. 1977. *G. Howard Ferguson: Ontario Tory.* Toronto: University of Toronto Press.

Ontario. 1975. *The Report of the Special Program Review* (Henderson Report). Toronto: Queen's Printer.

– 1988a. Ministry of Housing's Advisory Committee on the International Year of Shelter for the Homeless. *More than a Roof: Action to End Homelessness in Ontario.* Toronto: Queen's Printer.

– 1988b. Social Assistance Review Committee. *Transitions* (Thompson Report). Toronto: Queen's Printer.

Ontario Economic Council. 1976. *Social Security*. Toronto: Council.

Patterson, James T. 1986. *America's Fight against Poverty, 1900–1985*. 2nd edition. Cambridge, Mass.: Harvard University Press.

[Philpott, Florence]. c. 1954. History and Background to the [Toronto] Welfare Council. Manuscript.

Piva, Michael J. 1974. *The Condition of the Working Class in Toronto, 1900–1921*. Ottawa: University of Ottawa Press.

Polèse, Mario. 1987. "Patterns of Regional Economic Development in Canada: Long-Term Trends and Issues." In William J. Coffey and Mario Polèse, eds., *Still Living Together: Recent Trends and Future Directions in Canadian Regional Development*, Montreal: Institute for Research on Public Policy, 13–32.

Riendeau, Roger E. 1979. "A Clash of Interests: Dependency and the Municipal Problem in the Great Depression." *Journal of Canadian Studies* 14: 50–8.

– 1984. "Servicing the Modern City, 1900–30. In Victor L. Russell, ed., *Forging a Consensus: Historical Essays on Toronto*, Toronto: University of Toronto Press, 157–80.

Roberts, Wayne. 1989. "Welfare Business." *Now: Toronto's Weekly News and Entertainment Voice* 8 no. 41 (22–28 June): 8–9, 11.

Rose, Albert. 1958. *Regent Park: A Study of Slum Clearance*. Toronto: University of Toronto Press.

– 1975. "Some Reflections on the History of Social Planning in Ontario." In Ontario, *Sourcebook to Pathways to Social Planning*, Toronto: Ministry of Community and Social Services, 56–91.

Ross, David P., and Shillington, Richard. 1989. *The Canadian Fact Book on Poverty 1989*. Ottawa and Montreal: Canadian Council on Social Development.

Roy, Patricia E. 1980. *Vancouver: An Illustrated History*. Toronto and Ottawa: Lorimer, Museum of Man.

Sancton, Andrew. 1986. *Municipal Government and Social Services: A Case Study on London, Ontario* Local Government Case Studies 2, Political Science Department, University of Western Ontario.

Savoie, Donald J., ed. 1986. *The Canadian Economy: A Regional Perspective*. Toronto: Methuen.

Schulz, Patricia V. 1975. *The East York Workers' Association: A Response to the Great Depression*. Toronto.

Social Planning Council of Metropolitan Toronto (SPC). 1933 – . Various reports. Toronto: Council.

– 1979–80. *Metro's Suburbs in Transition, Part I Evolution and Overview, Part II Planning Agenda for the Eighties*. Written by Marvyn Novick. Toronto: Council.

– 1984. *Caring for Profit: The Commercialization of Human Services in Ontario*. Toronto: Council.

– 1986. "Welfare Benefits: An Inter-Provincial Comparison, 1985." *Social Infopac* 5 no. 1: 1–9.

– 1988. "Perspective on Making Ontario Budgets." *Social Infopac* 7 no. 3: 1–10.
– 1989a. "The SARC Report: Investing in Ontario's Future." *Social Infopac* 8 no. 1: 1–8.
– 1989b. *Target on Training: Meeting Workers' Needs in a Changing Economy*. Toronto: Council.
Stamp, Robert M. 1982. *The Schools of Ontario, 1876–1976*. Toronto: University of Toronto Press.
Strange, Carolyn. 1988. "From Modern Babylon to a City upon a Hill: The Toronto Social Survey Commission of 1915 and the Search for Sexual Order in the City." In Roger Hall, William Westfall, and Laurel Sefton MacDowell, eds., *Patterns of the Past: Interpreting Ontario's History*, Toronto: Dundurn Press, 255–77.
Struthers, James. 1983. *No Fault of Their Own: Unemployment and the Canadian Welfare State, 1914–1941*. Toronto: University of Toronto Press.
– 1987. "Shadows from the Thirties: The Federal Government and Unemployment Assistance, 1941–1956." In Jacqueline S. Ismael, ed., *The Canadian Welfare State: Evolution and Transition*, Edmonton: University of Alberta Press, 3–32.
– 1988. "War and Social Policy in Australia and Canada: Review Essay." *Australian-Canadian Studies* 6: 119–24.
Taylor, John H. 1986. *Ottawa: An Illustrated History*. Toronto and Ottawa: Lorimer, Museum of Civilization.
Toronto. 1834–   . *Council Minutes* Toronto: City of Toronto.
– 1988-91. *Realty Tax Information for the 1988 [1989, 1990, 1991] Final Realty Tax Bill*. Toronto: City of Toronto.
United Way of Canada. 1980. *Social Planning and the United Way: Roles and Relationships*. Ottawa: United Way.
Weaver, John C. 1982. *Hamilton: An Illustrated History*. Toronto and Ottawa: Lorimer, Museum of Man.
Wharf, Brian. 1975. "Social Planning Functions and Social Planning Organizations in Ontario." In *Sourcebook to Pathways to Social Planning*, Toronto: Ministry of Community and Social Services, 93–121.
Williams, Clifford J. 1984. *Decades of Services: A History of the Ontario Ministry of Community and Social Services, 1930–1980*. Toronto: Ministry.
Wills, Jacquelyn Gale. 1989. "Efficiency, Feminism and Cooperative Democracy: Origins of the Toronto Social Planning Council, 1918–1957." PhD thesis, Faculty of Social Work, University of Toronto.

NOTES TO CHAPTER FIFTEEN

1 For an excellent review see Mackie (1981).
2 There is, of course, considerable interest in the re-emergence of production within the home, captured in the popular imagination by the vision of the "electronic

cottage," the home wired to the workplace via a computer and telecommunications. We return to this and related developments in the last section of the chapter.

3 See Goheen (1970) on the increasing separation between home and work in Toronto during the later nineteenth century. Detailed maps of the journey to work for employees of particular establishments in 1860 and 1890 document increasing separation between home and work.

4 Bradbury (1984).   5 Medjuck (1980).   6 Harris (1989).   7 Miron (1989).   8 Miron (1989: 87).   9 Sennett (1970a).   10 Mackenzie (1980); Rutherford (1974).   11 Cited in Social Planning Council of Metropolitan Toronto (1979: 44–8).   12 Social Planning Council (1979: 56–7).   13 Social Planning Council (1979: 57).   14 Gans (1972); Willmott and Young (1973).   15 Rainwater (1966).   16 Sennett (1970).   17 Willmott and Young (1973).   18 Pratt (1986b).   19 Ward (1975: 148).   20 James S. Duncan and Nancy Duncan (1976).   21 Seeley, Sim and Loosley (1963: 384–94).   22 Gans (1967).   23 Holdsworth (1983).   24 Pratt (1981).   25 Michelson (1977).   26 Klodawski, Spector, and Rose (1985); Rose and LeBourdais (1986).   27 Michelson (1977).   28 Mackenzie and Rose (1983); Wekerle, Peterson, and Morley (1980).   29 Hayden (1981).   30 Wekerle (1988).   31 Wekerle (1988); Novac (1988).   32 Mills (1989).   33 Mackenzie (1987).   34 Dyck (1989).   35 Forrest and Murie (1987).   36 Fried (1963); Clairmont and Magill (1987).   37 Clairmont and Magill (1987).   38 Jones (1988: D2).   39 Cited in Holdsworth (1977: 192).   40 Michelson (1977).   41 Kirk and Kirk (1981).   42 Thernstrom (1974).   43 Harris (1992).   44 Adams (1984); Agnew (1984); Perin (1977).   45 Harris (1988: 118).   46 Harris (1988: 85–6); Cox (1982).   47 Moore and Skaburskis (1989).   48 Saunders (1978).   49 Ball (1983).   50 Bourne (1981: 50–4).   51 Schellenberg (1987).   52 Harris (1986a).   53 Harris (1986a).   54 Harris and Choko (1988).   55 Burgess (1927); Choko and Harris (1990).   56 Hoyt (1933); Steele (1979).   57 Harris and Choko (1988); Choko and Harris (1990); Langlois (1961).   58 Harris and Hamnett (1987).   59 Harris (1989; 1990).   60 Harris (1986c).   61 Miron (1988).   62 Calculated from Miron (1988: 199).   63 Rose and LeBourdais (1986).   64 Fallick (1988). See Dear and Wolch, chapter 16, in this volume.   65 Kemeny (1981).   66 Hulchanski (1986).   67 Hulchanski (1986).   68 Belec, Holmes, and Rutherford (1987: 217).   69 Dowler (1983).   70 Belec, Holmes, and Rutherford (1987); Hayden (1984); Mackenzie and Rose (1983).   71 Quoted from Holdsworth (1977: 193).   72 Progressive Conservative Party of Canada (1979).   73 Pratt (1986b).   74 Edel, Sclar, and Luria (1984); Rose (1980).   75 Dennis and Fish (1972); Rose (1980).   76 Ruttle (1983).   77 This has been documented more fully in the United States. See Checkoway (1980); Weiss (1987).   78 Relph

(1981: 99).   79 Magnusson and Sancton (1983: 3–57); Higgins (1986: 232–56).   80 Magnusson and Sancton (1983); Gutstein (1983).   81 Patterson (1989).   82 Ley (1985: 144–52).   83 Ley and Mercer (1980).   84 Harris (1988: 85–6).   85 Harris (1988: 86).   86 City of Vancouver Planning Department (1987).   87 Wekerle (1988).   88 Goldberg (1983).   89 Pratt (1986a); Michelson (1985).   90 McAfee (1987).   91 Klodowski, Spector, and Rose (1985); Rose and LeBourdais (1986).   92 Boris and Daniels (1989); Johnson and Johnson (1982); Walker (1989).   93 Cited in Walker (1989).   94 Hanson and Pratt (1988).   95 Hirsch (1989).

REFERENCES FOR
CHAPTER FIFTEEN

Adams, John. 1984. "The Meaning of Housing in America." *Annals of the Association of American Geographers* 74: 515–27.

Agnew, John. 1984. "Homeownership and the Capitalist Social Order." In Michael Dear and Allen Scott, eds., *Urbanization and Urban Planning in Capitalist Society*, New York: Methuen, 457–80.

Bacher, John. 1986. "Canadian Housing 'Policy' in Perspective." *Urban History Review* 15 no. 1: 3–18.

Badcock, B.A. 1989. "Homeownership and the Accumulation of Real Wealth." *Environment and Planning D: Society and Space* 7: 69–91.

Ball, Michael. 1983. *Housing Policy and Economic Power: The Political Economy of Owner-Occupation*. New York: Methuen.

Belec, John, Holmes, John, and Rutherford, Tod. 1987. "The Rise of Fordism and the Transformation of Consumption Norms: Mass Consumption and Housing in Canada, 1930–1945." In Richard Harris and Geraldine Pratt, eds., *Housing Tenure and Social Class*, Gävle: National Swedish Institute for Building Research, 187–237.

Boris, Eileen, and Daniels, Cynthia, eds. 1989. *Homework*. Urbana and Chicago: University of Illinois Press.

Bourne, Larry S. 1981. *The Geography of Housing*. London: Arnold.

Bradbury, Bettina. 1984. "Pigs, Cows, and Boarders: Non-Wage Forms of Survival among Montreal Families." *Labor/Le Travail* 14: 9–46.

Burgess, Ernest W. 1927. "The Determination of Gradient in the Growth of the City." *Publications of the American Sociological Society* 21: 178–83.

Checkoway, Barry. 1980. "Large Builders, Federal Housing Programmes, and Post-War Suburbanization." *International Journal of Urban and Regional Research* 4: 21–45.

Choko, Marc, and Dansereau, Francine. 1987. *Restauration résidentielle et copropriété au Centre-Ville de Montréal*, Études et documents 53, INRS-Urbanisation, Montreal.

Choko, Marc, and Harris, Richard. 1990. "The Local Culture of Property: A

Comparative History of Housing Tenure in Montreal and Toronto." *Annals, Association of American Geographers* 80 no. 1: 73–95.

City of Vancouver Planning Department. 1987. "Reports to Council: Artists' Live/Work Studios." City of Vancouver Planning Department, Vancouver, July 2.

Clairmont, Donald H., and Magill, Dennis William. 1987. *Africville: The Life and Death of a Canadian Black Community.* 2nd edition. Toronto: Canadian Scholar's Press.

Cox, Kevin. 1982. "Housing Tenure and Neighborhood Activism." *Urban Affairs Quarterly* 18 no. 1: 107–29.

Dennis, Michael, and Fish, Susan. 1972. *Programs in Search of a Policy.* Toronto: Hakkert.

Divay, G., and Richard, L. 1981. *L'aide gouvernementale au logement et sa distribution sociale: bilan sommaire pour les années soixante-dix au Québec.* Études et documents 26, INRS-Urbanisation, Montreal.

Doucet, Michael, and Weaver, John. 1985. "Material Culture and the North American House: The Era of the Common Man, 1870–1920." *Journal of American History* 72 no. 3: 560–87.

Dowler, Robert. 1983. *Housing-Related Tax Expenditures: An Overview and Evaluation.* Major Report No. 22, Centre for Urban and Community Studies, University of Toronto.

Duncan, James S., and Duncan, Nancy. 1976. "Housing as Presentation of Self and the Structure of Social Networks." In Gary T. Moore and Reginald G. Golledge, eds., *Environmental Knowing,* Stroudsberg, Penn.: Dowden, Hutchinson and Ross, 247–53.

Duncan, Simon. 1986. "House Building, Profits and Social Efficiency in Sweden and Britain." *Housing Studies* 1 no. 1: 11–33.

Dyck, Isabel. 1989. "Integrating Home and Wage Workplace: Women's Daily Lives in a Canadian Suburb." *Canadian Geographer* 33 no. 4: 329–41.

Edel, Matthew, Sclar, Elliott D., and Luria, D. 1984. *Shaky Palaces: Homeownership and Social Mobility in Boston's Suburbanization.* New York: Columbia University Press.

Fallick, Arthur. 1988. "Homelessness and the Homeless in Canada: A Geographic Perspective." PhD thesis, Department of Geography, University of British Columbia.

Forrest, R., and Murie, A. 1987. "The Affluent Homeowner, Labour-Market Position and the Shaping of Housing Histories." In Nigel Thrift and Peter Williams, eds., *Class and Space,* London: Routledge and Kegan Paul, 330–59.

Fried, M. 1963. "Grieving for a Lost Home." In L.J. Duhl, ed., *The Urban Condition,* New York: Basic Books, 151–71.

Gans, Herbert. 1967. *The Levittowners: Ways of Life and Politics in a New Suburban Community.* New York: Pantheon.

Goheen, Peter. 1970. *Victorian Toronto, 1850 to 1900.* University of Chicago, Department of Geography Research Paper No. 127.

Goldberg, Michael. 1983. *The Housing Problem: A Real Crisis?* Vancouver: University of British Columbia Press.

Gutstein, Donald. 1983. "Vancouver." In Magnusson and Sancton (1983) 189–221.

Hanson, Susan, and Pratt, Geraldine. 1988. "On the Links between Home and Work in Urban Geography: Review, Critique, Agenda." *Economic Geography* 64: 299–321.

Hardy, Dennis, and Ward, Colin. 1984. *Arcadia for All: The Legacy of a Makeshift Landscape.* London and New York: Mansell.

Harris, Richard. 1986a. "Homeownership and Class in Modern Canada." *International Journal of Urban and Regional Research* 19 no. 1: 167–86.

– 1986b. "Working Class Homeownership and Housing Affordability across Canada in 1931." *Histoire sociale/Social History* 19: 121–38.

– 1986c. "Boom and Bust: The Effects of House Price Inflation on Homeownership Patterns in Montreal, Toronto, and Vancouver." *Canadian Geographer* 30 no. 4: 302–15.

– 1988. *Democracy in Kingston: A Social Movement in Urban Politics.* Montreal and Kingston: McGill-Queen's University Press.

– 1989. *The Family Home in Working-Class Life.* Research Paper No. 171, Centre for Urban and Community Studies, University of Toronto.

– 1990. "Self-Building and the Social Geography of Toronto, 1901–1913: A Challenge for Urban Theory." *Transactions, Institute of British Geographers* 15: 387–402.

– 1992. "The End Justified the Means: Boarding and Rooming in a City of Homes, 1890–1951." *Journal of Social History* (forthcoming).

Harris, Richard, and Choko, Marc. 1988. *The Evolution of Housing Tenure in Montreal and Toronto since the Mid-Nineteenth Century.* Research Paper No. 166, Centre for Urban and Community Studies, University of Toronto.

Harris, Richard, and Hamnett, Chris. 1987. "The Myth of the Promised Land: The Social Diffusion of Homeownership in Britain and North America." *Annals of the Association of American Geographers* 77 no. 2: 173–90.

Hayden, Dolores. 1981. *The Grand Domestic Revolution: A History of Feminist Designs for American Homes, Neighborhoods, and Cities.* Cambridge, Mass.: M.I.T. Press.

– 1984. *Redesigning the American Dream: The Future of Housing, Work and Family Life.* New York: Norton.

Higgins, Donald J.H. 1986. *Local and Urban Politics in Canada.* Toronto: Gage.

Hirsch, Kathleen. 1989. *Songs from the Alley.* New York: Ticknor and Fields.

Holdsworth, Deryck. 1977. "House and Home in Vancouver: Images of West Coast Urbanism 1886–1929." In Gilbert Stelter and Alan F.J. Artibise, eds., *The Canadian City: Essays in Urban History,* Toronto: McClelland and Stewart, 186–211.

– 1983. "Appropriating the Past? Heritage Designation and Inner City Revitalisation." Paper presented at the Annual Meetings of the Canadian Association of Geographers, Winnipeg, Man.

Hoyt, Homer. 1933. *One Hundred Years of Land Values in Chicago*. Chicago: University of Chicago Press.

Hulchanski, J. David. 1986. "The 1935 Dominion Housing Act: Setting the Stage for a Permanent Federal Presence in Canada's Housing Sector." *Urban History Review* 15 no. 1: 19–39.

– 1989. "New Forms of Owning and Renting." In J. Miron, ed., *Housing Progress in Canada since 1945*, Report to CMHC, Ottawa.

Johnson, Laura C., with Johnson, Robert E. 1982. *The Seam Allowance: Industrial Home Sewing in Canada*. Toronto: Women's Press.

Jones, D. 1988. "Africville: Nova Scotia's Blacks Remember." *Globe and Mail* 2 July, D2.

Karn, Valerie, Kemeny, John, and Williams, Peter. 1985. *Homeownership in the Inner City*. Aldershot, England: Gower.

Kemeny, Jim. 1981. *The Myth of Home Ownership*. London: Routledge and Kegan Paul.

Kirk, C.T., and Kirk, W.K. 1981. "The Impact of the City on Homeownership: A Comparison of Immigrants and Native Whites at the Turn of the Century." *Journal of Urban History* 7: 471–87.

Klodawski, Fran, Spector, Aron, and Rose, Damaris. 1985. *Canadian Housing Policies and Single-Parent Families: How Mothers Lose*. Ottawa: Canada Mortgage and Housing Corp.

Langlois, C. 1961. "Problems of Urban Growth in Greater Montreal." *Canadian Geographer* 5 no. 3: 1–11.

Ley, David. 1985. *Gentrification in Canadian Inner Cities: Patterns, Analysis, Impacts and Policy*. Ottawa: CMHC.

Ley, David, and Mercer, J. 1980. "Locational Conflict and the Politics of Consumption." *Economic Geography* 56 no. 2: 89–109.

Linteau, Paul-André. 1985. *The Promoters' City: Building the Industrial Town of Maisonneuve, 1881–1918*. Toronto: Lorimer.

McAfee, A. 1987. "Secondary Suites: The Issues." *Quarterly Review*, Vancouver: City of Vancouver Planning Department, 16–18.

Mackenzie, Suzanne. 1980. *Women and the Reproduction of Labour Power in the Industrial City: A Case Study*. Working Paper No. 23, Urban and Regional Studies, University of Sussex, Brighton, England.

– 1987. "Neglected Spaces in Peripheral Places: Homeworkers and the Creation of a New Economic Centre." *Cahiers de géographie du Québec* 31: 247–60.

Mackenzie, Suzanne, and Rose, Damaris. 1983. "Industrial Change, the Domestic Economy and Home Life." In J. Anderson, S. Duncan, and R. Hudson eds., *Redundant Spaces in Cities and Regions*, New York: Academic Press, 155–200.

Mackie, Kathleen A. 1981. "An Exploration of the Idea of Home in Human Geography." MA thesis, University of Toronto.

Magnusson, Warren, and Sancton, Andrew, eds. 1983. *City Politics in Canada*. Toronto: University of Toronto Press.

Medjuck, Sheva. 1980. "The Importance of Boarding for the Structure of the Household in the Nineteenth Century: Moncton, New Brunswick and Hamilton, Canada West." *Histoire sociale/Social History* 13: 207–13.

Michelson, William. 1977. *Environmental Choice, Human Behavior and Residential Satisfaction.* New York: Oxford University Press.

– 1985. *From Sun to Sun: Daily Obligations and Community Structure in the Lives of Employed Mothers and Their Families.* Ottawa: Towman and Allenheld.

Mills, Caroline. 1989. "Interpreting Gentrification: Post Industrial, Post Patriarchal, Post Modern?" PhD thesis, Department of Geography, University of British Columbia.

Miron, John. 1988. *Housing in Postwar Canada: Demographic Change, Household Formation, and Housing Demand.* Kingston and Montreal: McGill-Queen's University Press.

– ed. 1989. *Housing Progess in Canada since 1945.* Report to CMHC, Ottawa.

Moore, Eric, and Skaburskis, André. 1989. "Measuring Transitions in the Housing Stock." In J. Miron, ed., *Housing Progress in Canada since 1945*, Report to CMHC, Ottawa.

Novac, Sylvia. 1988. "Building Feminist Communities." Paper presented in special session on Women's Experience of Community at the Canadian Urban and Housing Studies Conference in Winnipeg, Man.

Pahl, R.E. 1984. *Divisions of Labour.* Oxford: Blackwell.

Patterson, Jeffrey. 1989. "Housing and Community Development Policies." In John Miron, ed., *Housing Progress in Canada since 1945*, Report to CMHC, Ottawa.

Perin, Constance. 1977. *Everything in Its Place.* Princeton: Princeton University Press.

Pratt, Geraldine. 1981. "The House as an Expression of Social Worlds." In James S. Duncan, ed., *Housing and Identity: Cross-cultural Perspectives*, London: Croom Helm, 135–80.

– 1986a. "Against Reductionism: The Relations of Consumption as a Mode of Social Structuration." *International Journal of Urban and Regional Research* 10: 377–400.

– 1986b. "Housing Tenure and Social Cleavages in Urban Canada." *Annals of the Association of American Geographers* 76: 366–80.

Progressive Conservative Party of Canada. 1979. *Homeowners and Buyers Deserve a Break ... Economic Recovery through Stimulus.* N.p.: n.p.

Rainwater, Lee. 1966. "Fear and the House-as-Haven in the Lower Class." *Journal of the American Institute of Planners* 32: 23–31.

Relph, Edward. 1981. *Rational Landscapes.* London: Croom Helm.

Rose, Albert. 1980. *Canadian Housing Policies 1935–1980.* Toronto: Butterworth.

Rose, Damaris. 1980. "Toward a Reevaluation of the Political Significance of Homeownership in Britain." Paper presented to the Conference of Socialist Economists, London.

Rose, Damaris, and LeBourdais, Celine. 1986. "The Changing Conditions of Single-Parenthood in Montreal's Inner City and Suburban Neighborhoods." *Urban Resources* 3 no. 2: 45–52.

Rowe, Andrew. 1989. "Self-Help Housing Provision: Production, Consumption, Accumulation and Policy in Atlantic Canada." *Housing Studies* 4: 75–91.

Rutherford, Paul. 1974. *Saving the Canadian City: The First Phase, 1880–1920*. Toronto: University of Toronto Press.

Ruttle, Carole. 1983. "Turning back the Clock: British Columbia Housing Policy, 1975–1983." Paper presented at the Annual Meetings of the CAG, Winnipeg, Man.

Rybczynski, Witold. 1986. *Home. A Short History of an Idea*. New York: Viking.

Saunders, Peter. 1978. "Domestic Property and Social Class." *International Journal of Urban and Regional Research* 2: 233–51.

Saywell, John. 1975. *Housing Canadians: Essays on the History of Residential Construction in Canada*. Discussion Paper No. 24, Economic Council of Canada, Ottawa.

Schellenberg, Kathryn. 1987. "The Persistence of the Homeownership Norm and the Implications of Mortgage Debt." *Journal of Urban Affairs* 9 no. 4: 355–66.

Seeley, J.R., Sims, R.A., and Loosley, E.W. 1963. *Crestwood Heights: A Study of the Culture of Suburban Life*. New York: John Wiley.

Sennett, Richard. 1970. *Families against the City: Middle Class Homes of Industrial Chicago 1872–1890*. Cambridge, Mass.: Harvard University Press.

Social Planning Council of Metropolitan Toronto. 1979. *Metro's Suburbs in Transition*. Toronto: Council.

Steele, Marion. 1979. *The Demand for Housing in Canada*. Ottawa: Statistics Canada.

Streich, Patricia. 1989. "The Affordablity of Housing in Postwar Canada." In John Miron, ed., *Housing Progress in Canada since 1945*, Report to CMHC, Ottawa.

Thernstrom, Stephan. 1974. *Poverty and Progress. Social Mobility in the Nineteenth Century City*. Cambridge, Mass.: Harvard University Press.

Walker, J. 1989. "Production of Exchange Values in the Home." *Environment and Planning A* 21: 685–88.

Walker, Richard. 1981. "A Theory of Suburbanization: Capitalism and the Construction of Urban Space in the United States." In Michael J. Dear and Allen Scott, eds., *Urbanization and Urban Planning in Capitalist Society*, London and New York: Methuen, 383–429.

Ward, David. 1975. "Victorian Cities: How Modern?" *Journal of Historical Geography* 1 no. 2: 135–51.

Weiss, Marc. 1987. *The Rise of the Community Builders: The American Real Estate Industry and Urban Land Planning*. New York: Columbia University Press.

Wekerle, Gerda. 1988. *Women's Housing Projects in Eight Canadian Cities*. Ottawa: Canada Mortgage and Housing Corp.

Wekerle, Gerda, Peterson, Rebecca, and Morley, David, eds., 1980. *New Space for Women*. Boulder, Col.: Westview.

Wilmott, P., and Young, M. 1973. *The Symmetrical Family*. New York: Pantheon.

NOTES TO CHAPTER SIXTEEN

The assistance of Jeff Heilman is gratefully acknowledged. This research was supported by a grant from the US National Science Foundation. Thanks also to Larry Bourne and David Ley for valuable comments on an earlier draft.

1 See, for example, Splane (1965), Moscovitch (1983), and Dear and Wolch (1987).

2 An excellent account of the post-1945 period in Ontario is to be found in Lang (1974).

3 The Ontario Ministry of Housing (1987) estimates that approximately forty-five deaths each year in Ontario are caused by homelessness. The death of Drina Joubert in Toronto in 1986 brought widespread national attention to the hardships suffered by the homeless; see *Globe and Mail* (1986b); also Scott (1987a) and Winnipeg *Free Press* (1986b).

4 City of Toronto Alternative Housing Subcommittee (1985: 2).   5 Oberlander and Fallick (1988: 11); see also Oberlander and Fallick (1987).   6 Oberlander and Fallick (1988: 12); see also McLaughlin (1987: 24).   7 Ontario Ministry of Housing (1987: 2).   8 Canadian Council on Social Development (CCSD) (1987).   9 CCSD (1987: 3–5).   10 CCSD (1987: Table 3, p. 5).   11 See Fallick (1987).   12 Ontario Ministry of Housing (1987: 2).   13 Metropolitan Toronto Assisted Housing Study (MTAHS) (1983).   14 *Maclean's* (1987: 36).   15 CCSD (1987: 8). See also the special issue of *Women & Environments* (1987).   16 *Maclean's* (1986: 28).   17 Winnipeg *Free Press* (1986b).   18 *Maclean's* (1987: 41).   19 CCSD (1987: Table 5, p. 3).   20 Brundrige (1987).

21 See, for instance, the Canadian Association of Housing and Renewal Officals (1988), Fallick (1987), and Ontario Ministry of Housing (1988) on the special needs of Native Canadians.

22 An equivalent account, dealing with US experience, is to be found in Wolch, Dear, and Akita (1988).

23 CCSD (1987: 9); McLaughlin (1987).

24 Deinstitutionalization included many other groups, such as the mentally retarded, the dependent elderly, and the physically disabled. See Dear and Wolch (1987: chapter 4).

25 *Globe and Mail* (1986c).   26 *Globe and Mail* (1986c).   27 Winnipeg *Free Press* (1986b).   28 CCSD (1987: 10).   29 *Globe and Mail* (1986f).   30 Toronto *Star* (1987).   31 Brundrige (1987). See also Fallick (1987); Canadian Association of Housing and Renewal Officials (1988); and Ontario Ministry of Housing (1988).   32 Daly (1988: 10).   33 *Macleans* (1987: 40).   34 Hamilton *Spectator* (1986).   35 CCSD (1987: 3).   36 *Macleans*

(1986: 29).   37 Cf. Dear and Taylor (1982).   38 Scott (1987b).
39 *Globe and Mail* (1986f). The case of group home location is different; see Dear and Laws (1986).   40 *Globe and Mail* (1986a). See also
Dear and Laws (1986); Dear and Wolch (1987: chapter 4); and Joseph and Hall
(1985).   41 McLaughlin (1987: 25).   42 Daly (1988: 10). For a general
discussion of Toronto's housing crisis, see *City Planning* (1985).   43 *Macleans*
(1986: 28).

44 MTAHS (1983: iii–iv); see also Dear and Wolch (1987: chapter 8).

45 See, for example, Dear and Wolch (1987: chapter 7); *Phoenix Rising*
(1987); and Daly (1988: 13–18).

46 *Women & Environments* (1987); Toronto *Star* (1986); Watson and Austerberry (1986).

47 Cf. Wolch, Dear, and Akita (1988) and Dear and Wolch (1987: chapters
8–9), for a fuller treatment. For a general consideration of homelessness policy
in Canada, see Daly (1989); also Fallick (1987), Canadian Association of
Housing and Renewal Officials (1988), and Ontario Ministry of Housing
(1988).

48 In Los Angeles, homeless people have been attacked and even shot; shelters
for the homeless have been deliberately set on fire by local opponents. On the
potential for large-scale social unrest see Wolch (1987).

49 Cf. Dear and Wolch (1987: chapter 7).

50 *Globe and Mail* (1986d, 1986e).

51 On the social dynamics of the contemporary city, see Wolch and Dear (1989);
for the specific case of housing, see Bourne (1981).

52 This account draws on Dear and Wolch (1987).

53 There is a growing geographical literature on "service-dependent" populations and their needs. See, for example, Austin, Smith, and Wolpert (1970),
White (1979), Massey (1980), Wolpert (1980), Smith (1983), and Smith
and Giggs (1989).

54 For a detailed presentation on planning in the face of community opposition,
see Dear and Wolch (1987: chapter 9).

55 Cf. Taylor et al. (1984).

56 The large majority of the urban homeless are indigenous. In one Toronto survey,
80 per cent of hotels residents came from Metropolitan Toronto; MTAHS
(1987: iii).

57 See Ward (1989).

REFERENCES FOR
CHAPTER SIXTEEN

Austin, M., Smith, T., and Wolpert, J. 1970. "The Implementation of Controversial
    Facility-Complex Programs." *Geographical Analysis* 3: 15–29.
Bourne, L.S. 1981. *The Geography of Housing*. New York: V.H. Winston.
Brundrige, R. 1987. "Homelessness." *City Magazine* 9 no. 2: 10–18.

Canadian Association of Housing and Renewal Officials and the International Council on Social Welfare-Canada. 1988. *New Partnership: Building for the Future*. Ottawa: Association.

Canadian Council on Social Development (CCSD). 1987. "Homelessness in Canada: The Report of the National Inquiry." *Social Development Overview* 5 no. 1: 1–16.

City of Toronto Alternative Housing Committee. 1985. *Off the Streets: A Case for Long-Term Housing*. Toronto: City of Toronto.

City Planning. 1985. 3 no. 2.

Daly, G. 1988. *A Comparative Assessment of Programs Dealing with the Homeless Population in the United States, Canada and Britain*. Ottawa: Canada Mortgage and Housing Corp.

– 1989. "No Place Like Home: Dealing with Homelessness in Canada." *Plan Canada* 29 no. 3: 17–19.

Dear, M. 1987. "Social Welfare in the City." In M. Dear, J. Drake, and L. Reeds, eds., *Steel City: Hamilton and Region*, Toronto: University of Toronto Press, 190–201.

Dear, M., and Laws, G. 1986. "Anatomy of a Decision: Recent Land Use Zoning Appeals and Their Effect on Group Home Locations in Ontario." *Canadian Journal of Community Mental Health* 5 no. 1: 5–17.

Dear, M. and Taylor, S.M. 1982. *Not on Our Street: Community Attitudes to Mental Health Care*. London: Pion.

Dear, Michael, and Wolch, Jennifer. 1987. *Landscapes of Despair: From Deinstitutionalization to Homelessness*. Princeton, NJ: Princeton University Press.

Fallick, A., ed. 1987. *A Place to Call Home*. School of Community and Regional Planning, University of British Columbia, Vancouver.

Globe and Mail. 1986a. "Bitter Experiences Ease Way for Increase in Group Homes." 12 May: A-13.

– 1986b. "Frozen Body of Bag Lady Is Found in Outside Stairwell." 29 Jan.: A-17.

– 1986c. "Lack of Help Takes Heavy Toll on Ex-psychiatric Patients." 31 Jan.: A-13.

– 1986d. "Mentally Ill Languish behind Bars." 25 March: A-17.

– 1986e. "Plight of Mentally Ill in Prisons Must be Faced." 28 March: A-4.

– 1986f. "Singles' Housing Needs Leadership." 16 July: A-8.

Hamilton *Spectator*. 1986. "Group Fears Hotels Will Evict Residents for Expo 86 Guests." 19 Feb.: D-9.

Joseph, A., and Hall, G. 1985. "The Locational Concentration of Group Home Networks in Toronto, Canada." *Professional Geographer* 37: 143–55.

Lang, V. 1974. *The Service State Emerges in Ontario 1945–73*. Toronto: Ontario Economic Council.

McLaughlin, M. 1987. "Homelessness in Canada." *Perception* 10 no. 2: 24–7.

Maclean's. 1986. "The Growing Ranks of the Homeless." 13 Jan.: 28–9.

– 1987. "The Search for a Future." 16 Feb.: 34–43.

Massey, D.S. 1980. "Residential Segregation and Spatial Distribution of a
  Non-Labor Force Population: The Needy, Elderly and Disabled." *Economic
  Geography* 56: 190–200.

Metropolitan Toronto Assisted Housing Study (MTAHS). 1987. *No Place to Go:
  A Study of Homelessness in Metropolitan Toronto.* Toronto: Policy & Planning Division.

Moscovitch, A. 1983. *The Welfare State in Canada: A Selected Bibliography
  1840–1978.* Waterloo, Ont.: Wilfrid Laurier University Press.

Oberlander, H.P., and Fallick, A.L. 1987. *Shelter or Homes: A Contribution
  to the Search for Solutions to Homelessness in Canada.* Centre for Human Settlements, University of British Columbia, Vancouver.

– 1988. *Homelessness & the Homeless: Responses & Innovations.* Centre for
  Human Settlements, University of British Columbia, Vancouver.

Ontario Ministry of Housing. 1987. *Some Facts about Homelessness.* Toronto:
  IYSH Ontario Secretariat.

– 1988. *More than Just a Roof: Action to End Homelessness in Ontario.* Toronto:
  Ministry.

Perception. 1987. 10 no. 4.

Phoenix Rising. 1987. 6 no. 4.

Scott, R. 1987a. "Living – and Dying – on the Street." *Phoenix Rising* 10 no. 4:
  6–8.

– 1987b. "Zoning Our Housing out." *Phoenix Rising* 10 no. 4: 27.

Smith, C.J. 1983. "Innovation in Mental Health Policy: The Political Economy
  of the Community Mental Health Movement, 1965–80." *Society & Space*
  1: 447–68.

Smith, C.J., and Giggs, J. 1989. *Location and Stigma.* Boston: Unwin Hyman.

Splane, R.B. 1965. *Social Welfare in Ontario 1791–1893.* Toronto: University of
  Toronto Press.

Taylor, S.M., Hall, G.B., Hughes, R.C., and Dear, M.J. 1984. "Predicting Community Reaction to Mental Health Facilities." *Journal of the American Planning Association* 50: 36–47.

Toronto *Star.* 1986. "Female Tenants Seen as Trouble." 20 Feb.: A-9.

– 1987. "Refugees Boosting Ranks of Homeless." 25 Jan.: A-8.

Ward, J. 1989. *Organizing for the Homeless.* Ottawa: Canadian Council on Social
  Development.

Watson, S., and Austerberry, H. 1986. *Housing and Homelessness.* London:
  Routledge Kegan Paul.

White, A. 1979. "Accessibility and Public Facility Location." *Economic Geography*
  55: 18–35.

Winnipeg *Free Press.* 1986a. "Debate on Aid to Homeless Stirred by Deaths of
  Punks." 6 June: 12.

– 1986b. "Patients Stay Lost in Shuffle." 8 June: 1.

Wolch, J. 1987. "Landscapes of Despair: The Homelessness Crisis in U.S. cities."
  Society & Space 5: 119–22.

Wolch, J. and Dear, M., eds. 1989. *The Power of Geography: How Territory Shapes Social Life*. Boston: Unwin Hyman.

Wolch, J., Dear, M., and Akita, A. 1988. "Explaining Homelessness." *Journal of the American Planning Association* 54 no. 4: 443–53.

Wolpert, J. 1980. "The Dignity of Risk." *Transactions of the Institute of British Geographers* 5: 391–410.

Women & Environments 1987. 10 no. 1.

NOTES TO CHAPTER SEVENTEEN

1 Health and Welfare Canada (1986); Taylor (1990).   2 Health and Welfare Canada (1986).   3 Health and Welfare Canada (1988).   4 Lalonde (1974).   5 Health and Welfare Canada (1986: 3).   6 Greenberg (1987); Jones and Moon (1987); Learmonth (1987); Meade, Florin, and Gesler (1988).   7 IBG/AAG (1985; 1986; 1988).   8 Townsend and Davidson (1982).   9 Better Health Commission (1986).   10 Wigle and Mao (1980); Wilkins (1980); Millar (1983); and Saveland (1983).   11 Wigle and Mao (1980).   12 Millar (1983).   13 Ames (1972); Copp (1974); Henripin (1969); Loslier (1976); and Semiatycki (1976).   14 Wilkins (1980; 1983).   15 Liaw, Wort, and Hayes (1989).   16 City of Toronto (1985).   17 Wilkins and Adams (1983).   18 Wilkins (1983).   19 Townsend and Davidson (1982).   20 Wilkins (1980) reports that the ratio of the relative mortality rate for males between the lowest and highest occupational classes is remarkably similar in Canada (1.9) and the United Kingdom (1.8).   21 Daiches (1981).   22 Townsend and Davidson (1982: 112–34).   23 Frank, Gibson, and McPherson (1988).   24 Shannon et al. (1988).   25 Cecilioni (1976).   26 Kerigan, Goldsmith, and Pengelly (1986); Pengelly et al. (1984).   27 Pengelly et al. (1984).   28 Spitzer et al. (1986); Scott (1986).   29 Hertzman et al. (1987).   30 Spitzer et al. (1986).   31 Scott (1986).   32 Hayes and Taylor (1986).   33 Vyner (1988).   34 For example, Dear and Taylor (1982); Tefft, Segal, and Trute (1987).   35 Dear and Wolch (1987).   36 Kearns (1987).   37 Kearns (1987: 219).   38 Kearns, Taylor, and Dear (1987).   39 Laws and Dear (1988).   40 Taylor, Elliott, and Kearns (1989).   41 Eyles and Smith (1988).   42 Lord, Schnarr, and Hutchinson (1987).   43 Kearns (1987: 165–6).   44 Kearns (1987: 167).   45 Joseph and Phillips (1984).   46 Rosenberg (1988).   47 Thouez (1987).   48 Clarke, Marrett, and Kreiger (1987).   49 Taylor et al. (1989).

REFERENCES FOR
CHAPTER SEVENTEEN

Ames, Herbert, B. 1972. *The City below the Hill: A Sociological Study of a Portion of the City of Montreal, Canada*. Toronto: University of Toronto Press.

Better Health Commission. 1986. *Looking Forward to Better Health*. Canberra: Australian Government Publications.

Cecilioni, V. 1976. "Occupational Health Hazards and Community Health Problems." *Canadian Journal of Public Health* 67 Supplement 2 53–60.

City of Toronto. 1985. *Health Status Report 1983*. Toronto: Department of Public Health.

Clarke, E.A., Marrett, L.D., and N. Kreiger. 1987. *Twenty Years of Cancer Incidence 1964–1983: The Ontario Cancer Registry*. Toronto: Ontario Cancer Treatment and Research Foundation.

Copp, Terry. 1974. *The Anatomy of Poverty: The Condition of the Working Class in Montreal, 1897–1929*. Toronto: McClelland and Stewart.

Daiches, Sol. 1981. *People in Distress: A Geographical Perspective on Psychological Well-being*. University of Chicago, Department of Geography Research Paper No. 197.

Dear, Michael J., and Taylor, S. Martin. 1982. *Not on Our Street: Community Attitudes to Mental Health Care*. London: Pion.

Dear, Michael J., and Wolch, Jennifer. 1987. *Landscapes of Despair*. Princeton, NJ: Princeton University Press.

Eyles, John, and Smith, David. eds.. 1988. *Qualitative Methods in Human Geography*. London: Polity Press.

Frank, John W., Gibson, B., and McPherson, M. 1988. "Information Needs in Epidemiology: Detecting the Health Effects of Environmental Chemical Exposure." In C.D. Fowle, A.P. Grima, and R.E. Munn, eds., *Information Needs for Risk Management*, Toronto: Institute of Environmental Studies, University of Toronto, 129–44.

Greenberg, Michael, 1987. *Public Health and the Environment*. New York: Guilford Press.

Hayes, Michael J., and Taylor, S. Martin. 1986. "Do Health Studies of Persons Exposed to Toxic Waste Sites Do More Harm Than Good?" In *Proceedings of the Second International Symposium on Medical Geography*, New Brunswick, NJ: Rutgers University, vol. 1, 248–66.

Health and Welfare Canada. 1986. *Achieving Health for All: A Framework For Health Promotion*. Ottawa: Supply and Services Canada.

– 1988. *Mental Health for Canadians: Striking a Balance*. Ottawa: Supply and Services Canada.

Henripin, Jacques. 1969. "L'inégalité sociale devant la mort: la mortinatalité et la mortalité infantile à Montréal." *Recherches sociographiques* 2 no. 1: 3–34.

Hertzman, Clyde, Hayes, Mike, Singer, Joel, and Highland, Joseph. 1987. "Upper Ottawa Street Landfill Site Health Study." *Environmental Health Perspectives* 75: 173–95.

Institute of British Geographers/Association of American Geographers (IBG/AAG. 1985. *Proceedings of the First International Symposium in Medical Geography*. Department of Geography, University of Nottingham, Nottingham, England, 15–19 July.

— 1986. *Proceedings of the Second International Symposium in Medical Geography.* Department of Geography, Rutgers University, New Brunswick, NJ, 14–18 July.

— 1988. *Proceedings of the Third International Symposium in Medical Geography.* Department of Geography, Queen's University, Kingston, Ont., 7–12 Aug.

Jones, Kelvyn, and Moon, Graham. 1987. *Health, Disease and Society: An Introduction to Medical Geography.* London: Routledge and Kegan Paul.

Joseph, Alun, and Phillips, David. 1984. *Accessibility and Utilization: Geographical Perspectives on Health Care Delivery.* New York: Harper and Row.

Kearns, Robin A. 1987. "In the Shadow of Illness: A Social Geography of the Chronically Mentally Disabled in Hamilton, Ontario." Doctoral thesis, McMaster University, Hamilton, Ont.

Kearns, Robin A., Taylor, S. Martin, and Dear, Michael. 1987. "Coping and Satisfaction among the Chronically Mentally Disabled." *Canadian Journal of Community Mental Health* 6: 13–24.

Kerigan, Anthony T., Goldsmith, Charles H., and Pengelly, L. David. 1986. "A Three Year Cohort Study of the Role of Environmental Factors in the Respiratory Health of Children in Hamilton, Ontario." *American Review of Respiratory Disease* 133: 987–93.

Lalonde, Marc. 1974. *A New Perspective on the Health of Canadians.* Ottawa: Health and Welfare Canada.

Laws, Glenda, and Dear, Michael. 1988. "Coping in the Community: A Review of the Factors Influencing the Lives of Ex-psychiatric Patients." In C. Smith and J. Giggs, eds., *Location and Stigma: Emerging Trends in the Study of Mental Health and Mental Illness*, London: George Allen and Unwin, 83–102.

Learmonth, Andrew. 1987. *Disease Ecology: An Introduction to Ecological Medical Geography.* London: Basil Blackwell.

Liaw, Kao-Lee, Wort, Shelley, and Hayes, Michael. 1989. "Intraurban Mortality Variation and Income Disparity: A Case Study of Hamilton-Wentworth Region." *Canadian Geographer* 33 no. 2: 131–45.

Lord, J., Schnarr, A., and Hutchinson, P. 1987. "The Voice of the People: Qualitative Research and the Needs of Consumers." *Canadian Journal of Community Mental Health* 6: 25–36.

Loslier, Luc. 1976. *La mortalité dans les aires sociales de la région métropolitaine de Montréal.* Québec: Ministère des affaires sociales.

Meade, Melinda, Florin, John, and Gesler, Wilbert. 1988. *Medical Geography.* New York: Guilford Press.

Millar, W.J. 1983. "Sex Differential in Mortality by Income Level in Urban Canada." *Canadian Journal of Public Health* 74: 329–34.

Pengelly, L. David, Kerigan, Antony T., Goldsmith, Charles H., and Inman, Elizabeth E. 1984. "The Hamilton Study: Distribution of Factors Confounding the Relationships between Air Quality and Respiratory Health." *Journal of the Air Pollution Control Association* 34 no. 10: 1,039–43.

Rosenberg, Mark. 1988. "Linking the Geographical, the Medical and the Political in Analyzing Health Care Delivery Systems." *Social Science and Medicine* 26: 179–86.

Saveland, Walter. 1983. "Mortality of Canadians: The Geographic and Socioeconomic Distribution of Mortality. Part II. Areal Variations in Canadian Mortality: Metropolitan Neighbourhoods." In B. Cooper, D. McCalla, and F. Mustard, eds., *Proceedings of the Population Health Special Lecture Series*, Hamilton: McMaster University, Faculty of Health Sciences, 28–36.

Scott, Fran. 1986. "The Junction Triangle Health Study." MSc thesis, Faculty of Health Science, McMaster University, Hamilton, Ont.

Semiatycki, Jack. 1976. "The Distribution of Disease." *McGill Medical Journal* 44 no. 2/3: 9–19.

Shannon, Harry, Hertzman, Clyde, Julian, Jim, Hayes, Michael, Henry, Nancy, Charters, J., Cunningham, I., Gibson, E.S., and Sackett, D.L. 1988. "Lung Cancer and Air Pollution in an Industrial City: A Geographical Analysis." *Canadian Journal of Public Health* 79 no. 4: 255–9.

Spitzer, Walter, Suissa, Samy, Eastridge, Lily, Shenker, Stephanie, Germanson, Terry, Murdie, Robert, and Macpherson, Alexander. 1986. "The Toronto Junction Triangle Study: A Response to a Community Health Emergency." *Canadian Journal of Public Health* 77: 257–62.

Taylor, S. Martin. 1990. "Geographical Perspectives on National Health Challenges." *Canadian Geographer* 34: 334–8.

Taylor, S. Martin, Elliott, Susan, and Kearns, Robin. 1989. "The Housing Experience of Chronically Mentally Disabled Clients in Hamilton, Ontario." *Canadian Geographer* 33 no. 2: 146–55.

Taylor, S. Martin, Frank, John, Haight, Murray, Streiner, David, Walter, Stephen, White, Norman, Willms, Dennis, Birnie, Susan, and Elliott, Susan. 1989. *The Psychosocial Impacts of Exposure to Environmental Contaminants in Ontario: A Feasibility Study*. Research Report to the Ontario Ministry of the Environment, Toronto.

Tefft, Bruce, Segal, Alex, and Trute, Barry. 1987. "Neighbourhood Response to Community Mental Health Facilities for the Chronically Mentally Disabled." *Canadian Journal of Community Mental Health* 6 no. 2: 37–50.

Thouez, Jean-Paul. 1987. *Organisation spatiale des systèmes de santé*. Montréal: Presses de l'Université de Montréal.

Townsend, Peter, and Davidson, Nick. 1982. *Inequalities in Health*. Harmondsworth; England: Penguin Books.

Vyner, Henry. 1988. *Invisible Trauma: The Psychosocial Effects of the Invisible Environmental Contaminants*. Lexington: D.C. Heath Lexington Books.

Wigle, D.T., and Mao, Y. 1980. *Mortality by Income Level in Urban Canada*. Ottawa: Health Protection Branch, Department of National Health and Welfare.

Wilkins, Russell. 1980. *Health Status in Canada, 1926–76: Rising Life Expectancies, Diminishing Regional Differences, Persistent Social Disparities*. Institute for Research on Public Policy, Occasional Paper No. 13, Montreal.

– 1983. "The Burden of Ill Health in Canada: Socio-economic Inequalities in the Healthfulness of Life." In B. Cooper, D. McCalla, and F. Mustard, eds., *Proceedings of the Population Health Special Lecture Series*, Hamilton: Faculty of Health Sciences, McMaster University, 37–55.

Wilkins, Russell, and Adams, Owen. 1983. "Health Expectancy in Canada, Late 1970s: Demographic, Regional and Social Dimensions." *American Journal of Public Health* 73 no. 9: 1,073–80.

NOTES TO CHAPTER EIGHTEEN

1 See Statistics Canada (1986); Health and Welfare Canada (1986).

2 As is evident below, such care takes a variety of forms, from licensed centre care with regular fees to care by neighbours or relatives in their own homes which may be paid for in cash and/or the exchange of services and goods. We exclude regular or occasional care in the child's own home by babysitters or nannies.

3 Brown et al (1974); Hodgson and Doyle (1978); Holmes, Williams, and Brown (1972).

4 See Health and Welfare Canada (1986) for a detailed discussion.

5 The seminal critique is Teitz (1968).

6 Dear and Taylor (1982); Dear (1978).

7 Hodgson and Oppong (1989); Smith (1977).

8 Castells (1977).

9 Dear (1978).

10 The home-and-neighbourhood sphere was breached to some extent by critiques of the model of "optimizing economic man" who animated Pareto-optimal models. This figure came out of this unexamined realm first as the behaviourists' "satisficer" and later incorporated some of the attributes suggested by humanists and historical materialists and feminists: intentionality, moral values, and a social context. Most recently, this man has even become human and incorporated the activities and creativity of women as well. On these various models of human nature, see Pred (1967); Ley and Samuels (1978); Sayer (1979); Klodawsky and Mackenzie (1987); Mackenzie (1989a).

11 On human services literature, see, for example, Johnson and Dineen (1981); Finch and Groves (1983); On feminist methodology, see, for example, Miles and Finn (1982); Bowles and Klein (1983); Harding (1987); Adamson, Briskin, and McPhail (1988); Eichler (1988).

12 One of the most extensive Canadian studies is Dyck (1989). On Britain, see Tivers, (1985); Mackenzie (1989b).

13 Aries (1962); Parr (1982).

14 On Canada see Katz (1975) and Luxton (1980). On the United States see Hayden (1981; 1984). On Britain see Mackenzie (1989b).

15 For more on this process in Canadian suburbs, see Mackenzie (1988).

16 Schulz (1978). Also see Stapleford (1976); Krashinsky (1977); Rose (1990).

17 Schulz (1978).

18 Armstrong (1984); Eichler (1983).

19 On the Canadian women's movement, see Finn and Miles (1982); Adamson, Briskin, and McPhail (1988). Canadian literature on women and environments includes Gilbert and Rose (1987); Andrew and Moore Milroy (1988). Reviews of Anglo–North American work include Bowlby et al. (1989) and Mackenzie (1989c); Bondi (1990).

20 See, for example, Michelson (1985); Dyck (1989).

21 See, for example, Abella (1984); Anderson (1987); Canadian Advisory Council on the Status of Women (1987).

22 See Townson (1985); Ontario Ministry of Community and Social Services (1987).

23 Revenue Canada (1990: 270).

24 Data compiled from Community Information Centre of Metropolitan Toronto (1971; 1990). Only centres offering full-time care are included, and the number of spaces includes only full-time spaces.

25 Female labour-force participation rate (60.7 per cent) from Statistics Canada Cat. 71-001, Dec. 1990; number of children under 6 years (150,380), from 1986 census of Canada.

26 This sample from forty-three daycare centres was obtained in early 1985. It includes only children attending full time, not school-age children or centres. Data from Truelove (1989b).

27 The Daycare Resource and Research Unit, University of Toronto, reports twenty-eight work-related childcare centres in Metropolitan Toronto in September 1991.

28 Social Planning Council of Metropolitan Toronto (1984: 1). See also Friendly (1989).

29 Johnson (1977); Lero (1981; 1985). *The Status of Day Care in Canada* (Hearth and Welfare Canada 1990:4) states that about 50 per cent of those not using formal care would do so if given the opportunity. The advantages and disadvantages of all forms of daycare are discussed in Kaiser and Rasminsky (1991). Other Canadian studies of childcare are Kivikink and Schell (1987); Lero and Kyle (1989); Kanaroglou and Rhodes (1990); and Rose (1990).

30 This material was collected as part of a larger project on home work and community development funded by the Social Science and Humanities Research Council of Canada and carried out in 1984–85. None of the caregivers was formally licensed, although 20 (11 in Kingston and 9 in Nelson) were inspected and assisted by government-sponsored resource centres.

31 On other areas of Canada, see Johnson and Dineen (1981); Gentleman (1983).

32 None of the caregivers cited legal requirements as a primary reason for these renovations. Renovations were carried out "voluntarily," in order to provide a better work space.

33 On similar assessments by mothers in Coquitlam, British Columbia, see Dyck (1989).

34 Home-caregivers generally earn relatively little, although a large number of the women, especially in the Trail-Nelson area, supplemented household resources through bartering childcare services for other goods and services. See Mackenzie (1987).

35 This issue is discussed in the context of urban community movements generally in Lemon (1978).

36 Canadian Day Care Advocacy Association (1987); Ontario Coalition for Better Day Care (1987).

REFERENCES FOR CHAPTER EIGHTEEN

Abella, Rosalie. 1984. *Report of the Royal Commission on Equality in Employment.* Ottawa: Minister of Supply and Services.

Adamson, Nancy, Briskin, Linda, and McPhail, Margaret, eds. 1988. *Feminist Organizing for Change: The Contemporary Women's Movement in Canada.* Toronto: Oxford University Press.

Anderson, Doris, ed. 1987. *Women and Public Policy.* Reprints from Policy Options Politiques. Ottawa: Institute for Research on Public Policy/L'Institut de recherches politiques.

Andrew, Caroline, and Moore Milroy, Beth, eds. 1988. *Life Spaces: Gender, Household, Employment.* Vancouver: University of British Columbia Press.

Aries, Philippe. 1962. *Centuries of Childhood: A Social History of Family Life.* New York: Vintage.

Armstrong, Pat. 1984. *Labour Pains: Women's Work in Crisis.* Toronto: Women's Press.

Bondi, Liz. 1990. "Progress in Geography and Gender: Feminism and Difference." *Progress in Human Geography* 14 no. 3: 438–45.

Bowlby, Sophie, Lewis, Jane, McDowell, Linda, and Foord, Linda. 1989. "The Geography of Gender." In Richard Peet and Nigel Thrift, eds., *New Models in Geography,* New York: Allen and Unwin. 157–75.

Bowles, Gloria, and Klein, Renate, eds. 1983. *Theories of Women's Studies.* London: Routledge.

Brown, Lawrence, Williams, Forrest, Youngmann, Carl, Holmes, John, and Walby, Karen. 1974. "The Location of Urban Population Service Facilities: A Strategy and Its Application." *Social Science Quarterly* 54 no. 4: 784–99.

Canadian Advisory Council on the Status of Women. 1987. *National Symposium on Women and the Family.* Proceedings of the Symposium. Ottawa: Council.

Canadian Day Care Advocacy Association. 1987. *New Federal Policy.* Ottawa: Association.

Castells, Manuel. 1977. *The Urban Question: A Marxist Approach.* London: Edward Arnold.

Community Information Centre of Metropolitan Toronto. 1971–90. *Directory of Child Care Services in Metropolitan Toronto*. Annual publication. Toronto.

Dear, Michael. 1978. "Planning for Mental Health Care: A Reconsideration of Public Facility Location Theory." *International Regional Science Review*. 3 no. 2: 93–111.

Dear, Michael, and Taylor, Martin. 1982. *Not on Our Street*. London: Pion.

Dyck, Isabel. 1989. "Integrating Home and Wage Workplace: Women's Daily Lives in a Canadian Suburb." *Canadian Geographer* 33 no. 4: 329–41.

Eichler, Margrit. 1983. *Families in Canada Today: Recent Changes and Their Policy Consequences*. Toronto: Gage.

– 1988. *Nonsexist Research Methods: A Practical Guide*. Boston: Allen and Unwin.

Finch, Janet, and Groves, Dulcie, eds. 1983. *A Labour of Love: Women, Work and Caring*. London: Routledge and Kegan Paul.

Finn, Geraldine, and Miles, Angela. 1982. *Feminism in Canada: From Pressure to Politics*. Montreal: Black Rose Books.

Friendly, Martha. 1989. *Assessing the Community Need for Child Care*. Toronto: Childcare Resource and Research Unit, Centre for Urban and Community Studies, University of Toronto.

Gentleman, Gail. 1983. *Child Care Family Resource Centres: A Background Paper*. Toronto: Ontario Ministry of Community and Social Services.

Gilbert, Anne, and Rose, Damaris, eds. 1987. "Espaces et femmes" (numéro spécial). *Cahiers de géographie du Québec* 31 no. 83.

Harding, Sandra, ed. 1987. *Feminism and Methodology: Social Science Issues*. Bloomington: Indiana University Press.

Hayden, Dolores. 1981. *The Grand Domestic Revolution: A History of Feminist Designs for American Homes, Neighborhoods and Cities*. Cambridge, Mass.: M.I.T. Press.

– 1984. *Redesigning the American Dream*. New York: Norton.

Health and Welfare Canada. 1973–90. *Status of Day Care in Canada*. Annual. Ottawa: National Day Care Information Centre, Health and Welfare Canada.

Hodgson, John, and Doyle, Pat. 1978. "The Location of Public Services Considering the Mode of Travel." *Socio-Economic Planning Science* 12: 49–54.

Hodgson, John, and Oppong, Joseph. 1989. "Some Efficiency and Equity Effects of Boundaries in Location-Allocation Models." *Geographical Analysis* 21 no. 2: 167–78.

Holmes, John, Williams, Forrest B., and Brown, Larry. 1972. "Facility Location under Maximum Travel Restriction: An Example Using Day Care Facilities." *Geographical Analysis* 4: 258–66.

Johnson, Laura. 1977. *Who Cares?* Toronto: Social Planning Council of Metropolitan Toronto.

Johnson, Laura, and Dineen, Janice. 1981. *The Kin Trade: The Day Care Crisis in Canada*. Toronto: McGraw-Hill Ryerson.

Kaiser, Barbara, and Rasminsky, Judy S. 1991. *The Daycare Handbook: A Parents' Guide to Finding and Keeping Quality Daycare in Canada*. Toronto: Little, Brown.

Kanaroglou, Pavlos S., and Rhodes, Steven A. 1990. "The Demand and Supply of Child Care: The Case of the City of Waterloo, Ontario." *Canadian Geographer* 34 no. 3: 209–24.

Katz, Michael. 1975. *The People of Hamilton, Canada West: Family and Class in a Mid–Nineteenth Century City*. Cambridge, Mass.: Harvard University Press.

Kivikink, Ronald, and Schell, Bernadette. 1987. "Demographic, Satisfaction, and Commitment Profiles of Day Care Users, Nursery Schools Users, and Babysitter Users in a Medium-sized Canadian City." *Child and Youth Care Quarterly* 16: 116–30.

Klodawsky, Fran, and Mackenzie, Suzanne. 1987. "Gender Sensitive Theory and the Housing Needs of Mother-Led Families: Some Concepts and Some Buildings." *Feminist Perspectives* 8. Ottawa: Canadian Research Institute for the Advancement of Women.

Krashinsky, Michael. 1977. *Day Care and Public Policy in Ontario*. Toronto: Ontario Economic Council.

– 1981. *User Charges in the Social Services: An Economic Theory of Need and Inability*. Toronto: Ontario Economic Council.

Lemon, James. 1978. "The Urban Community Movement: Moving toward Public Households." In David Ley and Marwyn Samuels, eds., *Humanistic Geography: Prospects and Problems*, Chicago: Maaroufa. 319–37.

Lero, Donna. 1981. "Factors Influencing Parents' Preferences for and Use of Alternative Child Care Arrangements for Pre–School Age Children." Final report of a project funded by Health and Welfare Canada, Ottawa.

– 1985. *Parents' Needs, Preferences and Concerns about Child Care: Case Studies of 336 Canadian Families*. Final report of a study commissioned by the Task Force on Child Care. Ottawa: Task Force.

Lero, Donna, and Kyle, Irene. 1989. *Families and Children in Ontario: Supporting the Parenting Role*. Toronto: Child, Youth and Family Policy Research Centre.

Ley, David, and Samuels, Marwyn, eds. 1978. *Humanistic Geography: Problems and Prospects*. Chicago: Maaroufa.

Luxton, Meg. 1980. *More Than a Labour of Love: Three Generations of Women's Work in the Home*. Toronto: Women's Press.

Mackenzie, Suzanne. 1987. "Neglected Spaces in Peripheral Places: Homeworkers and the Creation of a New Economic Centre." *Cahiers de géographie du Québec* 31 no. 83: 247–60.

– 1988. "Building Women, Building Cities: Toward Gender Sensitive Theory in the Environmental Disciplines." In Caroline Andrews and Beth Moore Milroy, eds., *Life Spaces: Gender, Household, Employment*, Vancouver: University of British Columbia Press, 13–30.

– 1989a. "Restructuring the Relations of Work and Life: Women as Environmental Actors, Feminism as Geographic Analysis." In Audrey Kobayashi and Suzanne Mackenzie, eds., *Remaking Human Geography*, London: Unwin and Hyman, 40–61.

– 1989b. *Visible Histories: Women and Environments in a Post-War British City*. Montreal: McGill-Queen's University Press.

– 1989c. "Women in the City." In Richard Peet and Nigel Thrift, eds., *New Models in Geography*, New York: Allen and Unwin, 109–26.

Michelson, William. 1985. *From Sun to Sun*. Totowa, NJ: Rowman and Allenheld.

Miles, Angela, and Finn, Geraldine, eds. 1982. *Feminism in Canada: From Pressure to Politics*. Montreal: Black Rose Books.

Ontario Coalition for Better Day Care. 1987. "Development of Non-Profit Child Care in Ontario." Brief to the Select Committee on Health, Ontario Coalition for Better Day Care, Toronto.

Ontario Ministry of Community and Social Services. 1987. *New Directions for Child Care*. Toronto: Ministry, Child Care Branch.

Parr, Joy, ed. 1982. *Childhood and Family in Canadian History*. Toronto: McClelland and Stewart.

Pred, Alan. 1967. *Behaviour and Location: Foundations for a Geographic and Dynamic Location Theory Parts I and II*. Lund, Sweden: C.W.K. Gleerup.

Revenue Canada. 1990. *Taxation Statistics, 1988 Tax Year*. Ottawa: Minister of Supply and Services.

Rose, Damaris. 1990. "'Collective Consumption' Revisited: Analysing Modes of Provision and Access to Childcare Services in Montreal, Quebec." *Political Geography Quarterly* 9: 353–80.

Sayer, Andrew. 1979. "Espistemology and Conception of People and Nature in Geography." *Geoforum* 10 no. 1: 19–43.

Schulz, Pat. 1978. "Day Care in Canada: 1850–1962." In K.G. Ross, ed., *Good Day Care*, Toronto: Women's Press, 137–58.

Smith, David. 1977. *Human Geography: A Welfare Approach*. London: Edward Arnold.

Social Planning Council of Metropolitan Toronto. 1984. "Measuring Day Care Need: The Number Game." *Social Infopac* 3 no. 3 (June).

Stapleford, E.M. 1976. *History of the Day Nurseries Branch*. Toronto: Ontario Ministry of Community and Social Services.

Statistics Canada. 1986. *Family Characteristics and Labour Force Activity*. Cat. 71-533. Ottawa: Minister of Supply and Services.

Teitz, Michael. 1968. "Toward a Theory of Urban Public Facility Location." *Papers and Proceedings of the Regional Science Association*. 21: 35–44.

Tivers, Jacqueline. 1985. *Women Attached: The Daily Lives of Women with Young Children*. London: Croom Helm.

Townson, Monica. 1985. *Financing Child Care through the Canada Assistance Plan*. Ottawa: Task Force on Child Care.

Truelove, Marie. 1984. "Constraints for Subsidized Daycare Users." *Women and Environments* 6 no. 3: 12–13.

– 1989a. "Day Care Centres in Metropolitan Toronto: Policies and Availability." *Ohio Geographers* 17: 26–42.

– 1989b. "Journey to Day-care Centres in Metropolitan Toronto." *Ohio Geographers* 17: 65–83.

NOTES TO CHAPTER NINETEEN

1 The historical development of urban planning practice is described by Perks and Jamieson (1991: 487–513). For another useful overview, though with more emphasis on the recent past, see Kiernan (1990: 58–77).

2 Hodge (1991: 2). Hodge (1991) and Cullingworth (1987) provide detailed descriptions of Canadian planning institutions and procedures.

3 Roweis (1983: 148–55). "Law" is used here in the sense of judicial procedures for settling disputes.

4 Higgins (1986: xi–xii). In his discussion of urban politics, Higgins draws repeatedly on planning examples.

5 Kiernan (1982: 15–16).

6 Ley (1983: 280).

7 Lash (1976: 10).

8 Robinson (1979: 204–16). See also Richardson (1981: 573–84) on the "retreat from provincial planning" in Ontario.

9 For assessments of Toronto's attempt at regional growth management see Bordessa and Cameron (1982: 127–45); Richardson (1981: 568–86); and several contributions to a special issue of *Plan Canada* 24 (Dec. 1984).

10 Cameron (1979: 246–57).

11 Frisken (1990).

12 For a detailed examination of transportation planning issues in the Toronto region see Pill (1979).

13 Robinson and Webster (1985: 29–30) and B. Richardson (1972: 108–28). For case studies illustrating the difficulties of growth management in the Niagara and Edmonton regions see Gayler (1982: 321–38) and Smith (1982: 207–22).

14 Ottawa's situation is discussed by several contributors in Wesche and Kugler-Gagnon (1978).

15 Sewell (1977: 20–7) and Social Planning Council of Metropolitan Toronto (1979: 35–58).

16 For technical information on neighbourhood planning and design see CMHC (1981); for an empirical evaluation of neighbourhood planning principles see Vischer (1987: 130–40).

17 Carver (1962: 50–108).

18 Sewell (1977: 28–38).

19 Smith and Harasym (1975: 176–85).

20 Cervero (1986).

21 Aspects of the decentralization of office development in Vancouver are considered by Hutton and Davis (1985) and by Ley (1985).

22 See Greenberg and Maguire (1988: 120–3); Gorrie (1991); and Relph (1991).

23 For an illustration of this planning logic see Dakin (1969: 5–24). The implications for Montreal, Toronto, and Winnipeg are examined by Kaplan (1982).

24 The example of the St James Town district in Toronto is described by Whitzman (1989: 32–3).

25 Spragge (1983: 37–43).

26 For examples see Goldrick (1978); McGibbon (1986); and Harris (1988).

27 The Spadina Expressway case is discussed by Leo (1977: 31–42) and by Pill (1979: 38–43); Milton-Park by Helman (1987); Strathcona by Wong (1978); and Portage and Main by Walker (1979).

28 Axworthy (1979). See also Roussopoulos (1982), with reference to Montreal and Toronto.

29 McGibbon and Smith (1991).

30 Dyson (1984) and Johnson and Fairbairn (1990: 79–87).

31 See, for example, Frisken (1988: 95–101).

32 For an overview of the urban renewal program see Smith (1990: 191–203).

33 Rose (1958: 35–102).

34 Pickett (1968: 233–41).

35 Success is always in the eye of the beholder; for critical views of the Civic Square project see Freeman (1982: 103–9) and *City Magazine* 4 no. 3 (1980) 110–21.

36 The issue of national housing policy in general, and urban renewal in particular, came to a head in 1968 with the appointment of the Task Force on Housing and Urban Development; see Axworthy (1971).

37 Robertson (1973: 51–91) presents a compelling case study of the effects of displacement under a public redevelopment scheme in Victoria.

38 The urban renewal controversy of the late 1960s spawned a large literature, most of it impressionistic and anecdotal. For a selection see Lorimer and Baird (1968); Lorimer and Phillips (1971); Fraser (1972); B. Richardson (1972: 173–200); and Sewell (1972).

39 NIP's aims and achievements are reviewed by Lyon and Newman (1986).

40 Filion (1989: 91–106) contrasts the situations in Toronto, which lived up to the ideal of participatory planning, and Montreal, where a more autocratic, "top-down" style of planning prevailed.

41 For case studies of neighbourhood planning in Ottawa, Toronto, and Vancouver, respectively, see Andrew and Moore Milroy (1986); Jacobs (1971: 292–305); Lewinberg (1986); and Horsman and Raynor (1978).

42 The essays in Oberlander and Fallick (1987) offer a generally appreciative assessment of MSUA's activities. For an account of its first venture see Swain (1979).

43 Contrary assessments of the Core Area Initiative are provided by Kiernan (1987) and by Gerecke and Barton (1990).

44 Smith and Wang (1988). For an overview of the pedestrianization of the central areas of Canadian cities, especially in the context of the livable winter cities movement, see Pressman (1988).

45 Schuyler and Ircha (1987) and Kiernan (1990: 75–7).

46 For an example see Sewell (1977: 39–48) and Social Planning Council of Metropolitan Toronto (1979: 55–248).

REFERENCES FOR CHAPTER NINETEEN

Andrew, Caroline, and Moore Milroy. Beth 1986. "Making Policies and Plans for Neighbourhoods: Ottawa's Experience." *Plan Canada* 26: 34–9.

Axworthy, Lloyd. 1971. "The Housing Task Force: A Case Study." In G. Bruce Doern and Peter Aucoin, eds., *The Structures of Policy-Making in Canada*, Toronto: Macmillan, 130–53.

– 1979. "The Politics of Urban Populism: A Decade of Reform." In William T. Perks and Ira M. Robinson, eds., *Urban and Regional Planning in a Federal State: The Canadian Experience*, 1982. New York: McGraw-Hill, 282–92.

Bordessa, Ronald, and Cameron, James. 1982. "Growth Management Conflicts in the Toronto-Centred Region." in Kevin R. Cox and R.J. Johnston, eds., *Conflict, Politics and the Urban Scene*, London: Longman 1982, 127–45.

Cameron, Kenneth D. 1979. "Planning in a Metropolitan Framework: The Toronto Experience." In William T. Perks and Ira M. Robinson, eds., *Urban and Regional Planning in a Federal State: The Canadian Experience*, New York: McGraw-Hill, 246–57.

Canada Mortgage and Housing Corp. (CMHC). 1981. *Residential Site Development: Advisory Document.* Ottawa: CMHC.

Carver, Humphrey. 1962. *Cities in the Suburbs.* Toronto: University of Toronto Press.

Cervero, Robert. 1986. "Urban Transit in Canada: Integration and Innovation at Its Best." *Transportation Quarterly* 40: 293–316.

Cullingworth, J. Barry. 1987. *Urban and Regional Planning in Canada.* New Brunswick, NJ: Transaction Books.

Dakin, John. 1969. "Metropolitan Toronto Planning." *Town Planning Review* 40: 3–24.

Dyson, Brian. 1984. "LRT: A Streetcar Named Speculation." *City Magazine* 7 No. 1: 30–4.

Filion, Pierre. 1989. "The Neighbourhood Improvement Program in Montreal and Toronto: Two Approaches to Publicly Sponsored Upgrading." In Trudi E. Bunting and Pierre Filion, eds. *The Changing Canadian Inner City*, Geography

Publication Series No. 31, Department of Geography, University of Waterloo, 87–106.

Fraser, Graham. 1972. *Fighting Back: Urban Renewal in Trefann Court*. Toronto: Hakkert.

Freeman, Bill. 1982. "Hamilton: How the City Was Built." In Dimitrios Roussopoulos, ed., *The City and Radical Social Change*, Montreal: Black Rose Books, 92–109.

Frisken, Frances. 1988. *City Policy-Making in Theory and Practice: The Case of Toronto's Downtown Plan*. Local Government Case Studies No. 3, Department of Political Science, University of Western Ontario, London.

– 1990. *Planning and Servicing the Greater Toronto Area: The Interplay of Provincial and Municipal Interests*. Urban Studies Working Paper No. 12, Urban Studies Program, York University, North York, Ont.

Gayler, H.J. 1982. "Conservation and Development in Urban Growth: The Preservation of Agricultural Land in the Rural-Urban Fringe of Ontario." *Town Planning Review* 53: 321–41.

Gerecke, Kent, and Reid, Barton. 1990. "False Prophets and Golden Idols in Canadian City Planning." *City Magazine* 12 no. 1: 16–22.

Goldrick, Michael. 1978. "The Anatomy of Urban Reform in Toronto." *City Magazine* 3 nos. 4 and 5: 29–39.

Gorrie, Peter. 1991. "North York's Instant Downtown." *Canadian Geographic* 111 no. 2: 66–73.

Greenberg, Ken, and Maguire, Robert. 1988. "The Architecture of Emerging Public Spaces: Suburban Intensification in Toronto, Canada." In Madis Pihlak, ed., *The City of the 21st Century*, Tempe: Department of Planning, Arizona State University, 119–24.

Harris, Richard. 1988. *Democracy in Kingston: A Social Movement in Urban Politics*. Kingston and Montreal: McGill-Queen's University Press.

Helman, Claire. 1987. *The Milton-Park Affair: Canada's Largest Citizen-Developer Confrontation*. Montreal: Véhicule Press.

Higgins, Donald J.H. 1986. *Local and Urban Politics in Canada*. Toronto: Gage.

Hodge, Gerald. 1991. *Planning Canadian Communities: An Introduction to the Principles, Practice and Participants*. 2nd edition. Toronto: Nelson Canada.

Horsman, A., and Raynor, P. 1978. "Citizen Participation in Local Area Planning: Two Vancouver Cases." In L.J. Evenden, ed., *Vancouver: Western Metropolis*, Western Geographical Series Vol. 16, Department of Geography, University of Victoria, 239–53.

Hutton, Thomas A., and Davis, H. Craig. 1985. "The Role of Office Location in Regional Town Centre Planning and Metropolitan Multinucleation: The Case of Vancouver." *Canadian Journal of Regional Science* 8: 17–34.

Jacobs, Dorene E. 1971. "The Annex Ratepayers' Association: Citizens' Efforts to Exercise Social Choice in Their Urban Environment." In James A. Draper, ed. *Citizen Participation: Canada. A Book of Readings*, Toronto: New Press 1971, 288–306.

Johnson, Denis B., and Fairbairn, Kenneth J. 1990. "LRT Systems in Edmonton and Calgary." In P.J. Smith and Edgar L. Jackson, eds. *A World of Real Places: Essays in Honour of William C. Wonders*, Edmonton: Department of Geography, University of Alberta, 77–94.

Kaplan, Harold. 1972. *Reform, Planning and City Politics: Montreal, Winnipeg, Toronto*. Toronto: University of Toronto Press.

Kiernan, Matthew J. 1982. "Ideology and the Precarious Future of the Canadian Planning Profession." *Plan Canada* 22: 14–24.

– 1987. "Intergovernmental Innovation: Winnipeg's Core Area Initiative." *Plan Canada* 27: 23–31.

– 1990. "Land-Use Planning." In Richard A. Loreto and Trevor Price, eds., *Urban Policy Issues: Canadian Perspectives*, Toronto: McClelland and Stewart, 58–85.

Lash, Harry, 1976. *Planning in a Human Way: Personal Reflections on the Regional Planning Experience in Greater Vancouver*. Urban Prospects Series. Ottawa: Ministry of State for Urban Affairs.

Leo, Christopher. 1977. *The Politics of Urban Development: Canadian Urban Expressway Disputes*. Monographs on Canadian Public Administration No. 3, Institute of Public Administration of Canada, Toronto.

Lewinberg, Frank. 1986. "Neighbourhood Planning: The Reform Years in Toronto." *Plan Canada* 26: 40–5.

Ley, David. 1983. *A Social Geography of the City*. New York: Harper and Row.

– 1985. "Downtown or the Suburbs? A Comparative Study of Two Vancouver Head Offices." *Canadian Geographer* 29 no. 1: 30–43.

Lorimer, James, and Baird, G. 1968. "Ask the People Who Live There What to Do with Your Urban Renewal Area, and They'll Tell You." *Architecture Canada* 45 no. 5: 46–7.

Lorimer, James, and Phillips, Myfanwy. 1971. *Working People: Life in a Downtown Neighbourhood*. Toronto: James Lewis & Samuel.

Lyon, Deborah, and Newman, Lynda H. 1986. *The Neighbourhood Improvement Program, 1973–1983: A National Review of an Intergovernmental Initiative*. Research and Working Papers No. 15, Institute of Urban Studies, University of Winnipeg.

McGibbon, Michael. 1986. "Citizen Protest in the Urban Planning Process in Edmonton." In *Current Research by Western Canadian Geographers: The University of Victoria Papers 1985*, B.C. Geographical Series No. 42, Tantalus Research Limited, Vancouver, 103–13.

McGibbon, Michael, and Smith, P.J. 1991. *Effects of Neighbourhood Planning on Housing Quality in Edmonton*. Ottawa: CMHC.

Oberlander, H. Peter, and Fallick, Arthur L., eds. 1987. *The Ministry of State for Urban Affairs: A Courageous Experiment in Public Administration.* Vancouver: Centre for Human Settlements, University of British Columbia.

Perks, William T. and Jamieson, Walter. 1991. "Planning and Development in Canadian Cities." In Trudi Bunting and Pierre Filion, eds., *Canadian Cities in Transition*, Toronto: Oxford University Press, 487–518.

Pickett, Stanley H. 1968. "An Appraisal of the Urban Renewal Programme in Canada." *University of Toronto Law Journal* 18: 233–47.

Pill, Juri. 1979. *Planning and Politics: The Metro Toronto Transportation Plan Review.* Cambridge: Mass.: M.I.T. Press.

Pressman, Norman. 1988. "Winter Policies, Plans and Designs: The Canadian Experience." In Jorma Mänty and Norman Pressman, eds., *Cities Designed for Winter*, Helsinki: Building Book Ltd, 35–64.

Relph, Edward. 1991. "Suburban Downtowns of the Greater Toronto Area." *Canadian Geographer* 35 no. 4: 421–5.

Richardson, Boyce. 1972. *The Future of Canadian Cities.* Toronto: New Press.

Richardson, N.H. 1981. "Insubstantial Pageant: The Rise and Fall of Provincial Planning in Ontario." *Canadian Public Administration* 24: 563–86.

Robertson, R.W. 1973. "Anatomy of a Renewal Scheme." In C.N. Forward, ed., *Residential and Neighbourhood Studies in Victoria*, Western Geographical Series Vol. 5, Department of Geography, University of Victoria, 40–100.

Robinson, Ira M. 1979. "Trends in Provincial Land Planning, Control, and Management." In William T. Perks and Ira M. Robinson, eds.; *Urban and Regional Planning in a Federal State: The Canadian Experience*, New York: McGraw-Hill, 204–27.

Robinson, Ira M., and Webster, Douglas R. 1985. "Regional Planning in Canada: History, Practice, Issues and Prospects." *Journal of the American Planning Association* 51: 23–33.

Rose, Albert. 1958. *Regent Park: A Study in Slum Clearance.* Toronto: University of Toronto Press.

Roussopoulos, Dimitrios, ed. 1982. *The City and Radical Social Change.* Montreal: Black Rose Books.

Roweis, S.T. 1983. "Urban Planning as Professional Mediation of Territorial Politics." *Environment and Planning D: Society and Space* 1: 139–62.

Schuyler, George, and Ircha, Michael. 1987. "Market Square: Downtown Economic Revival." *Plan Canada* 27: 16–22.

Sewell, John. 1972. *Up against City Hall.* Toronto: James Lewis & Samuel.
– 1977. "The Suburbs." *City Magazine* 2 no. 6: 19–55.

Smith, P.J. 1982. "Municipal Conflicts over Territory and the Effectiveness of the Regional Planning System in the Edmonton Metropolitan Area." In H. Becker, ed., *Kulturgeographische Prozessforschung in Kanada*, Bamberger Geographische Schriften 4, Bamberg, 207–23.

– 1990. "Theory and Practice of Urban Renewal Planning in Canada." In P.J. Smith and Edgar L. Jackson, eds., *A World of Real Places: Essays in Honour of William C. Wonders*, Edmonton: Department of Geography, University of Alberta, 191–206.

Smith, P.J., and Harasym, D.G. 1975. "Planning for Retail Services in New Residential Areas since 1945." In B.M. Barr, ed., *Calgary: Metropolitan Structure and Influence*, Western Geographical Series Vol. 11, Victoria: Department of Geography, University of Victoria 157–92.

Smith, P.J., and Wang, Shuguang. 1988. "Planning for Pedestrian Facilities in a Winter City: The Central Area of Edmonton, Alberta." In Madis Pihlak, ed., *The City of the 21st Century*, Tempe: Department of Planning, Arizona State University, 135–41.

Social Planning Council of Metropolitan Toronto. 1979. *Metro's Suburbs in Transition: Evolution and Overview*. Toronto: Council.

Spragge, Godfrey. 1983. "Exploring a Planning Methodology: Policies for Whitepainted Neighbourhoods." *Plan Canada* 23: 36–50.

Swain, Harry. 1979. "The Halifax-Dartmouth Waterfront Project." In William T. Perks and Ira M. Robinson, eds., *Urban and Regional Planning in a Federal State: The Canadian Experience*, New York: McGraw-Hill, 271–81.

Vischer, Jacqueline C. 1987. "The Changing Canadian Suburb." *Plan Canada* 27: 130–40.

Walker, David C. 1979. *The Great Winnipeg Dream: The Redevelopment of Portage and Main*. Oakville, Ont.: Mosaic Press.

Wesche, Rolf, and Kugler-Gagnon, Marianne, eds. 1978. *Ottawa-Hull: Spatial Perspectives and Planning*. Ottawa: University of Ottawa Press.

Whitzman, Carolyn. 1989. "Against the Solution: St James Town." *City Magazine* 11 no. 1: 32–6.

Wong, S.T. 1978. "Urban Redevelopment and Rehabilitation in the Strathcona Area: A Case Study of an East Vancouver Community." In L.J. Evenden, ed., *Vancouver: Western Metropolis*, Western Geographical Series Vol. 16, Victoria: Department of Geography, University of Victoria, 255–69.

Index